AF389210

Computer Vision and Machine Learning for Intelligent Sensing Systems

Computer Vision and Machine Learning for Intelligent Sensing Systems

Editor

Jing Tian

MDPI • Basel • Beijing • Wuhan • Barcelona • Belgrade • Manchester • Tokyo • Cluj • Tianjin

Editor
Jing Tian
National University of Singapore,
Singapore

Editorial Office
MDPI
St. Alban-Anlage 66
4052 Basel, Switzerland

This is a reprint of articles from the Special Issue published online in the open access journal *Sensors* (ISSN 1424-8220) (available at: https://www.mdpi.com/journal/sensors/special_issues/machinelearning_intelligent).

For citation purposes, cite each article independently as indicated on the article page online and as indicated below:

LastName, A.A.; LastName, B.B.; LastName, C.C. Article Title. *Journal Name* **Year**, *Volume Number*, Page Range.

ISBN 978-3-0365-7868-2 (Hbk)
ISBN 978-3-0365-7869-9 (PDF)

Contents

About the Editor . . . . **vii**

Jing Tian
Computer Vision and Machine Learning for Intelligent Sensing Systems
Reprinted from: *Sensors* **2023**, *23*, 4214, doi:10.3390/s23094214 . . . . **1**

Alexey Kokhanovskiy, Nikita Shabalov, Alexandr Dostovalov and Alexey Wolf
Highly Dense FBG Temperature Sensor Assisted with Deep Learning Algorithms
Reprinted from: *Sensors* **2021**, *21*, 6188, doi:10.3390/s21186188 . . . . **3**

Shintaro Shiba, Yoshimitsu Aoki and Guillermo Gallego
Event Collapse in Contrast Maximization Frameworks
Reprinted from: *Sensors* **2022**, *22*, 5190, doi:10.3390/s22145190 . . . . **13**

Fangni Chen, Anding Wang, Yu Zhang, Zhengwei Ni and Jingyu Hua
Energy Efficient SWIPT Based Mobile Edge Computing Framework for WSN-Assisted IoT
Reprinted from: *Sensors* **2021**, *21*, 4798, doi:10.3390/s21144798 . . . . **33**

Zhaofeng Niu, Yuichiro Fujimoto, Masayuki Kanbara, Taishi Sawabe, Hirokazu Kato
DFusion: Denoised TSDF Fusion of Multiple Depth Maps with Sensor Pose Noises
Reprinted from: *Sensors* **2022**, *22*, 1631, doi:10.3390/s22041631 . . . . **47**

Manzoor Ahmed Hashmani, Mehak Maqbool Memon, Kamran Raza, Syed Hasan Adil, Syed Sajjad Rizvi and Muhammad Umair
Content-Aware SLIC Super-Pixels for Semi-Dark Images (SLIC++)
Reprinted from: *Sensors* **2022**, *22*, 906, doi:10.3390/s22030906 . . . . **61**

Van-Hung Le and Rafal Scherer
Human Segmentation and Tracking Survey on Masks for MADS Dataset
Reprinted from: *Sensors* **2021**, *21*, 8397, doi:10.3390/s21248397 . . . . **87**

Van Nhiem Tran, Shen-Hsuan Liu, Yung-Hui Li and Jia-Ching Wang
Heuristic Attention Representation Learning for Self-Supervised Pretraining
Reprinted from: *Sensors* **2022**, *22*, 5169, doi:10.3390/s22145169 . . . . **109**

Effat Jalaeian Zaferani, Mohammad Teshnehlab, Amirreza Khodadadian, Clemens Heitzinger, Mansour Vali, Nima Noii and Thomas Wick
Hyper-Parameter Optimization of Stacked Asymmetric Auto-Encoders for Automatic Personality Traits Perception
Reprinted from: *Sensors* **2022**, *22*, 6206, doi:10.3390/s22166206 . . . . **131**

Kai Hu, Yiwu Ding, Junlan Jin, Min Xia and Huaming Huang
Multiple Attention Mechanism Graph Convolution HAR Model Based on Coordination Theory
Reprinted from: *Sensors* **2022**, *22*, 5259, doi:10.3390/s22145259 . . . . **153**

Jaekwang Oh, Youngkeun Lee, Jisang Yoo and Soonchul Kwon
Improved Feature-Based Gaze Estimation Using Self-Attention Module and Synthetic Eye Images
Reprinted from: *Sensors* **2022**, *22*, 4026, doi:10.3390/s22114026 . . . . **171**

Jingyu Song and Joonwoong Lee
Online Self-Calibration of 3D Measurement Sensors Using a Voxel-Based Network
Reprinted from: *Sensors* **2022**, *22*, 6447, doi:10.3390/s22176447 . . . . **189**

Marco A. Moreno-Armendáriz, Hiram Calvo, Carlos A. Duchanoy, Arturo Lara-Cázares, Enrique Ramos-Diaz and Víctor L. Morales-Flores
Deep-Learning-Based Adaptive Advertising with Augmented Reality
Reprinted from: *Sensors* **2022**, *22*, 63, doi:10.3390/s22010063 . . . . **213**

About the Editor

Jing Tian

Tian Jing is a faculty member at the National University of Singapore in the areas of artificial intelligence, computer vision, and machine learning. He received his B.Eng. and M.Eng. degrees (Electronic and Information Engineering) from the School of Electronic and Information Engineering, South China University of Technology, Guangzhou, China, and Ph.D. degree (Electrical and Electronic Engineering) from the School of Electrical and Electronic Engineering, Nanyang Technological University, Singapore. Prior to joining NUS, he spent a number of years in research work with a focus in the areas of computer vision and artificial intelligence at the Institute for Infocomm Research (A*STAR), Temasek Laboratories, NTU, Singapore, respectively. He also served as a technical leader in numerous research projects with both major local government agencies and SME, as well as international collaborators. He serves as an active reviewer for prestigious international journals and conferences (IEEE T-IP, IEEE T-CSVT, ICIP, CVPR, etc.). He has published 100+ publications and 2000+ citations (Google Scholar).

Editorial

Computer Vision and Machine Learning for Intelligent Sensing Systems

Jing Tian

Institute of Systems Science, National University of Singapore, Singapore 119615, Singapore; tianjing@nus.edu.sg

Intelligent sensing systems have been fueled to make sense of visual sensory data to handle complex and difficult real-world sense-making challenges due to the rapid growth of computer vision and machine learning technologies. We can now interpret visual sensory data more effectively thanks to recent developments in machine learning algorithms. This means that in related research, significant attention is now being paid to problems in this field, such as visual surveillance, smart cities, etc.

The Special Issue offers a selection of high-quality research articles that tackle the major difficulties in computer vision and machine learning for intelligent sensing systems from both a theoretical and practical standpoint. It includes intelligent sensing techniques [1–5], twelve foundational investigations into sense-making methods [6–10], and particular uses of intelligent sensing systems in autonomous driving [11] and virtual reality [12].

Intelligent sensing techniques

- Kokhanovskiy et al. [1] demonstrated the application of deep neural networks to process the reflectance spectrum from a fiberoptic temperature sensor.
- Shiba et al. [2] proposed collapse metrics by using the first principles of space–time deformation based on differential geometry and physics for contrast maximization, which provided state-of-the-art results on several event-based computer vision tasks.
- Chen et al. [3] presented a system that integrates mobile edge computing technology and simultaneous wireless information and power transfer technology to improve the service supply capability of WSN-assisted IoT applications.
- Niu et al. [4] proposed a new fusion network to minimize the influences from the two most common sensor noises, i.e., depth noises and pose noises.
- Hashmani et al. [5] presented a novel SLIC extension based on a hybrid distance measure to retain content-aware information for semi-dark images.

Intelligent sense-making techniques

- Le and Scherer [6] performed a comprehensive survey of many studies, methods, datasets, and results for human segmentation and tracking in video.
- Tran et al. [7] proposed a heuristic attention representation learning framework relying on the joint embedding architecture, in which the two neural networks are trained to produce similar embedding results for different augmented views of the same image.
- Zaferani et al. [8] proposed a method for the automatic hyper-parameter tuning of a stacked asymmetric auto-encoder to extract personality perception from speech.
- Hu et al. [9] developed two attention modules that work together to extract the coordination characteristics in the process of motion and strengthen the attention of the model to the more important joints in the process of moving.
- Oh et al. [10] suggested estimating gaze by detecting eye region landmarks through a single eye image. It learns representations of images at various resolutions, and the self-attention module is used to obtain a refined feature map with better spatial information.

Citation: Tian, J. Computer Vision and Machine Learning for Intelligent Sensing Systems. *Sensors* **2023**, *23*, 4214. https://doi.org/10.3390/s23094214

Received: 12 April 2023
Accepted: 17 April 2023
Published: 23 April 2023

Applications of intelligent sensing systems

- Song and Lee [11] studied autonomous driving and proposed a novel algorithm for online self-calibration between sensors using voxels and three-dimensional convolution kernels.
- Moreno-Armendáriz et al. [12] described a system composed of deep neural networks that analyze characteristics of customers based on their face, as well as the ambient temperature, to generate a personalized signal to potential buyers who pass in front of a beverage establishment.

In conclusion, through the wide range of research presented in this Special Issue, we would like to boost fundamental and practical research on applying computer vision and machine learning for intelligent sensing systems.

Conflicts of Interest: The author declares no conflict of interest.

References

1. Kokhanovskiy, A.; Shabalov, N.; Dostovalov, A.; Wolf, A. Highly Dense FBG Temperature Sensor Assisted with Deep Learning Algorithms. *Sensors* **2021**, *21*, 6188. [CrossRef] [PubMed]
2. Shiba, S.; Aoki, Y.; Gallego, G. Event Collapse in Contrast Maximization Frameworks. *Sensors* **2022**, *22*, 5190. [CrossRef] [PubMed]
3. Chen, F.; Wang, A.; Zhang, Y.; Ni, Z.; Hua, J. Energy Efficient SWIPT Based Mobile Edge Computing Framework for WSN-Assisted IoT. *Sensors* **2021**, *21*, 4798. [CrossRef] [PubMed]
4. Niu, Z.; Fujimoto, Y.; Kanbara, M.; Sawabe, T.; Kato, H. DFusion: Denoised TSDF Fusion of Multiple Depth Maps with Sensor Pose Noises. *Sensors* **2022**, *22*, 1631. [CrossRef] [PubMed]
5. Hashmani, M.A.; Memon, M.M.; Raza, K.; Adil, S.H.; Rizvi, S.S.; Umair, M. Content-Aware SLIC Super-Pixels for Semi-Dark Images (SLIC++). *Sensors* **2022**, *22*, 906. [CrossRef] [PubMed]
6. Le, V.-H.; Scherer, R. Human Segmentation and Tracking Survey on Masks for MADS Dataset. *Sensors* **2021**, *21*, 8397. [CrossRef] [PubMed]
7. Tran, V.N.; Liu, S.-H.; Li, Y.-H.; Wang, J.-C. Heuristic Attention Representation Learning for Self-Supervised Pretraining. *Sensors* **2022**, *22*, 5169. [CrossRef] [PubMed]
8. Zaferani, E.J.; Teshnehlab, M.; Khodadadian, A.; Heitzinger, C.; Vali, M.; Noii, N.; Wick, T. Hyper-Parameter Optimization of Stacked Asymmetric Auto-Encoders for Automatic Personality Traits Perception. *Sensors* **2022**, *22*, 6206. [CrossRef] [PubMed]
9. Hu, K.; Ding, Y.; Jin, J.; Xia, M.; Huang, H. Multiple Attention Mechanism Graph Convolution HAR Model Based on Coordination Theory. *Sensors* **2022**, *22*, 5259. [CrossRef] [PubMed]
10. Oh, J.; Lee, Y.; Yoo, J.; Kwon, S. Improved Feature-Based Gaze Estimation Using Self-Attention Module and Synthetic Eye Images. *Sensors* **2022**, *22*, 4026. [CrossRef] [PubMed]
11. Song, J.; Lee, J. Online Self-Calibration of 3D Measurement Sensors Using a Voxel-Based Network. *Sensors* **2022**, *22*, 6447. [CrossRef] [PubMed]
12. Moreno-Armendáriz, M.A.; Calvo, H.; Duchanoy, C.A.; Lara-Cázares, A.; Ramos-Diaz, E.; Morales-Flores, V.L. Deep-Learning-Based Adaptive Advertising with Augmented Reality. *Sensors* **2021**, *22*, 63. [CrossRef] [PubMed]

Communication

Highly Dense FBG Temperature Sensor Assisted with Deep Learning Algorithms

Alexey Kokhanovskiy [1,*], Nikita Shabalov [1,2], Alexandr Dostovalov [1,2] and Alexey Wolf [1,2]

[1] Novosibirsk State University, 1 Pirogova Street, Novosibirsk 630090, Russia; gefest776@gmail.com (N.S.); dostovalov@iae.nsk.su (A.D.); alexey.a.wolf@gmail.com (A.W.)

[2] Institute of Automation and Electrometry of the SB RAS, 1 Academic Koptyug Avenue, Novosibirsk 630090, Russia

* Correspondence: alexey.kokhanovskiy@gmail.com

Abstract: In this paper, we demonstrate the application of deep neural networks (DNNs) for processing the reflectance spectrum from a fiberoptic temperature sensor composed of densely inscribed fiber bragg gratings (FBG). Such sensors are commonly avoided in practice since close arrangement of short FBGs results in distortion of the spectrum caused by mutual interference between gratings. In our work the temperature sensor contained 50 FBGs with the length of 0.95 mm, edge-to-edge distance of 0.05 mm and arranged in the 1500–1600 nm spectral range. Instead of solving the direct peak detection problem for distorted signal, we applied DNNs to predict temperature distribution from entire reflectance spectrum registered by the sensor. We propose an experimental calibration setup where the dense FBG sensor is located close to an array of sparse FBG sensors. The goal of DNNs is to predict the positions of the reflectance peaks of the reference sparse FBG sensors from the reflectance spectrum of the dense FBG sensor. We show that a convolution neural network is able to predict the positions of FBG reflectance peaks of sparse sensors with mean absolute error of 7.8 pm that is slightly higher than the hardware reused interrogator equal to 5 pm. We believe that dense FBG sensors assisted with DNNs have a high potential to increase spatial resolution and also extend the length of a fiber optical sensors.

Keywords: fiber bragg grating; optical fiber sensor; distributed temperature sensor; deep learning algorithms; fully connected neural network; convolutional neural network

Citation: Kokhanovskiy, A.; Shabalov, N.; Dostovalov, A.; Wolf, A. Highly Dense FBG Temperature Sensor Assisted with Deep Learning Algorithms. *Sensors* **2021**, *21*, 6188. https://doi.org/10.3390/s21186188

Academic Editor: Jing Tian

Received: 16 July 2021
Accepted: 13 September 2021
Published: 15 September 2021

Publisher's Note: MDPI stays neutral with regard to jurisdictional claims in published maps and institutional affiliations.

1. Introduction

Fiber bragg grating sensors are widely used for distributed temperature and strain sensing [1–3]. The performance of FBG-based sensors significantly depends on the accuracy of the peak detection algorithms that provide the possibility of converting a registered signal into temperature/strain values. Commonly, an array of FBGs is sparsely inscribed on a stretch of fiber in spatial and spectral domains to avoid mutual interference between neighboring FBGs. In this case, the peaks positions and their shapes are relatively simple to distinguish and plenty of algorithms may be applied for processing [4–6].

However, there is always a trade-off between peak detection accuracy and the number of FBG sensors in the sensing channel. The limited capacity of a sensing channel results in the limited length and spatial resolution of an FBG sensor. Increasing the number of sensing channels limits the sensor interrogation rate and increases the cost of a final device.

At the same time, FBG sensors with high spatial resolution are attractive for various applications including damage processes monitoring [7], healthcare [8], and heat localization [9]. The spatial resolution of a fiber optical sensor may be raised by implementation of chirped FBGs [10] or the implementation more advanced interrogation technique [11].

Software solutions are mainly focused on the coupling of the reflectance spectra of several FBG sensors into one sensing channel and processing the resulting signal for peaks discrimination [12,13]. Today, the scientific community is paying significant attention to machine-learning (ML) algorithms, which have already shown good performance at various fundamental levels and practical applications [14,15]. Particularly, data-driven algorithms are capable of operating with large-scale high-dimensional data and finding hidden intrinsic features and dependencies. There have already been successive attempts to apply ML algorithms for interpretation the of overlapped reflectance spectra from sparse FBG sensors including: extreme learning machines [12], least squares support vector regression [16], convolutional neural networks [17], particle swarm optimization algorithms, long short-term memory algorithms [18] and others [19,20].

However, in most of the presented works, the performances of the algorithms were demonstrated on the model spectra of FBGs, where various additional spectrum distortions associated, for example, with mutual interference between neighboring FBGs, are not considered. Additionally, FBG arrays with a small number of gratings (up to 4 FBGs) were used in experiments, while a real network of sensors can contain more than 50 individual sensing points, for which the presented algorithms will have a significant root mean square error and a significantly longer signal processing time (~s), which makes it difficult to use these algorithms for real-time measurement applications. The possibility of interrogating a sensors network containing 60 FBGs was recently demonstrated [20], but in this case spectral bandwidth was divided into 30 independent regions without crosstalk containing only two paired FBGs with spectral overlap. Each FBG has a bandwidth of $\approx$0.25 nm to ensure the absence of crosstalk between adjacent regions (1–3 nm), so a length of FBG was ~5 mm, which limits the spatial resolution of measurements. In the case of short FBGs (<1 mm) used for high spatial resolution measurements, the spectral width is much higher (~1.5 nm) and for this reason the spectra will significantly affect each other and therefore these algorithms will give a large error.

Here, we investigate alternative method to process experimental reflection spectra of a highly dense FBG temperature sensor. Fifty closely inscribed FBGs allowed us to increase the length and spatial resolution of the temperature sensor interrogated by using a single optical channel of interrogator. To fabricate the sensor, we used the femtosecond point-by-point inscription technique, allowing high-precision FBG positioning and wavelength resonance specification. For adequate interpretation of a complex reflection signal we do not solve the peak detection problem, but apply deep learning algorithms in order to match the whole reflectance spectrum of the dense FBG sensor with temperature distribution. We also propose an experimental setup based on optical interrogator and Peltier cells for training procedures of deep learning algorithms. By applying the developed algorithms, we show that the capacity of the optical channel of the interrogator can be increased without significant loss of accuracy in the FBG peak detection.

2. Experimental Setup

In the following experiments we used two different types of FBG-based temperature sensors inscribed in Fibercore SM1500(9/125)P polyimide-coated fiber by using femtosecond IR laser pulses [21]. The first one was a highly dense sensor composed of 50 uniform FBGs equidistantly arranged along 50 mm fiber segment. Each of the gratings had a length of 0.95 mm and was separated from the neighboring by 0.05 mm, as shown in Figure 1. The resonant wavelengths of the FBGs in the array were chosen to uniformly fill the spectral range of the used 8-channels HBM FS22-SI interrogator (1500–1600 nm). The interrogator operated in optical spectrum analyzer mode providing 1 Spectrum per second refresh rate, 20,001 points per spectrum, and 5 pm resolution. The second type of FBG array was an array containing only intermediate elements of a highly dense array, as shown in Figure 1. Due to the decrease in the FBG density, the reflection spectra the array possess a less noisy shape, making it possible to use such a sensor as a reference when processing the data of a highly dense sensor.

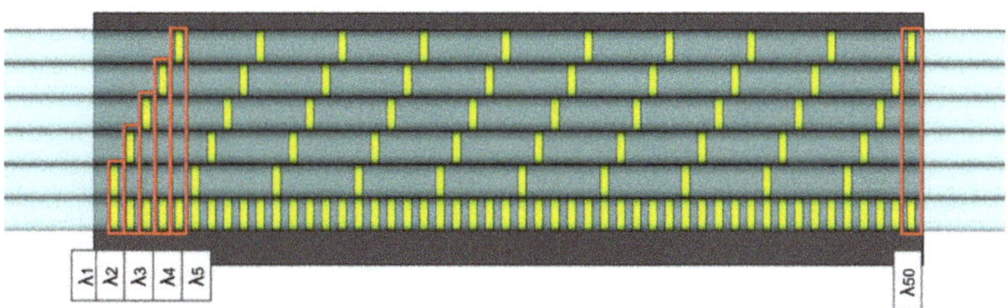

Figure 1. Arrangement of FBGs in highly dense and sparse temperature sensors.

Figure 2a shows a fragment of a reflection spectrum of the highly dense FBG temperature sensor and one of the sparse FBG temperature sensors measured by the optical interrogator. The whole reflectance spectrums are depicted in Figure 2b. As can be seen from the spectra, the close spectral arrangement of the FBGs in a highly dense array leads to mutual interference between adjacent gratings, which consequently increases the noise level of the resonance peaks compared to a sparse FBG array.

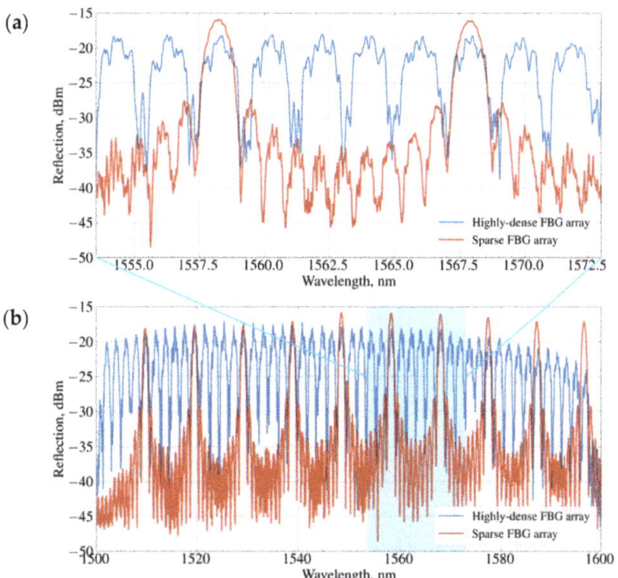

Figure 2. Reflection spectra of a highly dense and one of the sparse FBG sensors: zoomed (**a**) and full (**b**) spectral ranges.

We used five calibrated sparse temperature sensors and additional channels of the interrogator to calibrate the dense FBG sensor. The general scheme of the experimental setup is depicted in Figure 3.

FBG sensors were glued closely to each other (as shown at Figure 1) on an aluminum plate with attached Peltier cells on the back surface. By controlling the current and polarity of the electrical power supplies, as well as the positions of Peltier cells, we applied different temperature gradients to the plate, some of which are presented on Figure 4. The temperature of the plate varied in the range from 10 to 80 °C with spatial gradients in the range from −0.38 °C/mm to 0.44 °C/mm. The diversity of the temperature gradients was exploited during the training procedure, thus improving the ability of the deep neural network to generalize incoming data and improve the sensitivity of the dense FBG sensor.

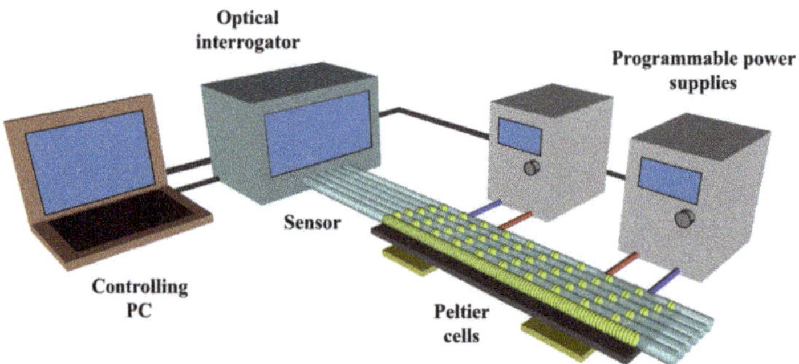

Figure 3. Principle scheme of the experimental setup.

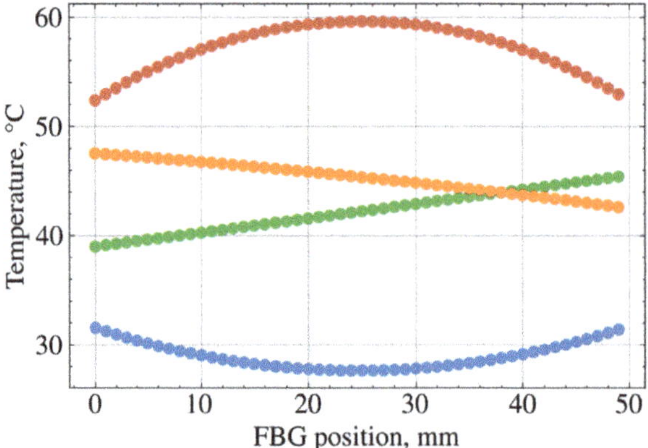

Figure 4. Examples of exploited temperature gradients during training procedure.

3. Architectures of Deep Neural Network

The task for deep neural networks was to predict the reflectance peak positions of 50 FBGs contained in 5 sparse sensors by the optical reflectance spectrum of the highly dense sensor. We investigated the performances of full-connected and convolutional neural networks, which were built using the TensorFlow software package [22]. A fully connected neural network (FCNN) was selected as the most common and simple architecture. Our FCNN consisted of input, hidden and output layers, as shown in Figure 5. The size of the input layer was 20,001 neurons corresponding to the size of the signal array from the interrogator, the output layer had 50 neurons corresponding to the number of the FBGs of the dense sensor. The size of the hidden layer was optimized in order to reduce computational time and maintain the precision of the algorithm. During the grid search procedure of hyperparameters of FCNN we used Mean Squared Error as a loss function and Adamax optimizer for neuron weights optimization. FCNN with 500 neurons at the hidden layer with sigmoid activation function showed the best performance (Section 4).

We also chose a convolutional neural network (CNN) for the task, because of its potential capability to reveal hidden features related to interference between reflectance patterns of nearby FBGs (Figure 6). The input layer was a two-dimensional array of the reflectance pattern, after which there was a layer with one convolutional filter of size 2×2 and Relu activation function. The 2-dimensional convoluted image was then flattened and transferred to two fully connected layers. The size of the last output layer was 50 neurons.

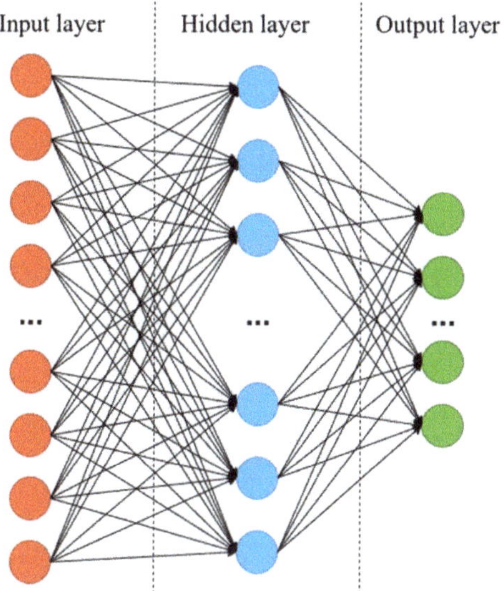

Figure 5. Feed forward neural network with three layers. The final architecture had 20,001 neurons at the input layer, 500 neurons at the hidden layer, 50 neurons at the output layer.

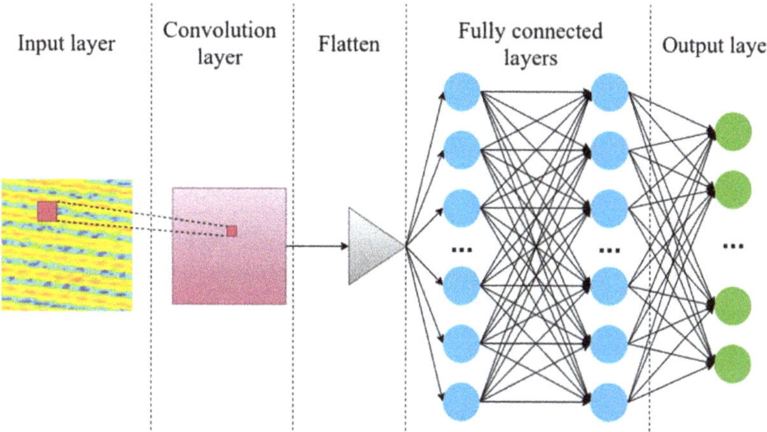

Figure 6. Convolutional neural network consisted of six layers.

Unlike FCNN, where we have used a 1-dimensional array as an input, we transformed the input signal array into a two-dimensional image. This was undertaken in order to reduce the size of a convolutional filter. Our first attempt was to slice 1-dimensional into 50 windows with centered reflectance peaks of FBGs (Figure 7a). However, this approach leads to greater challenges during signal processing due to the aperiodic arrangement of the peaks. Instead, we sliced the input signal into 59 parts, since 59 is a prime factor for 20,001 (Figure 7b). Despite the image losing its physical meaning, such an approach is more robust and straightforward.

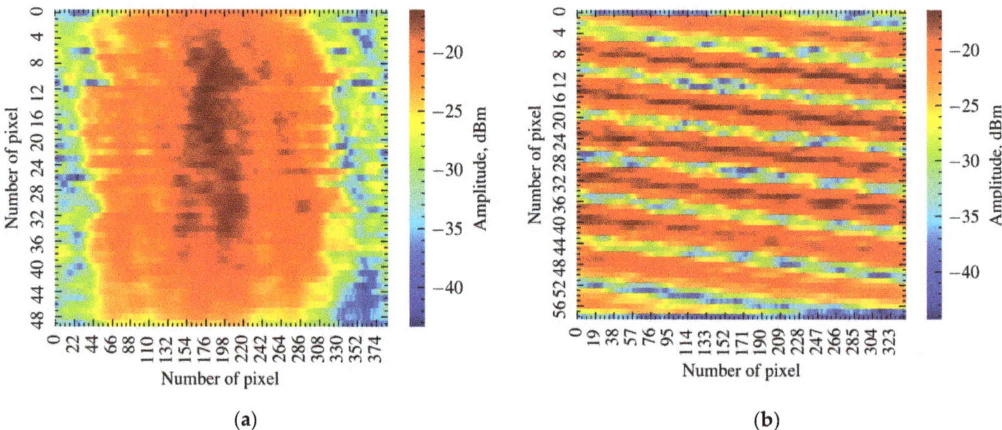

Figure 7. (**a**) 2D reflectance spectrum of FBG sensor with centered reflectance peaks; (**b**) 2D reflectance pattern of FBG obtained by slicing the reflectance spectrum into 59 pieces.

The input data was scaled for both neural networks in following order: first, each sample of a reflectance spectra was shifted by mean value of the intensity, then it was normalized by standard deviation of the intensity. Sample size of training, validation and testing datasets were 2000, 500 and 1000, respectively.

4. Results

The learning curve of the FCNN on a training dataset and prediction error on a validation dataset are depicted at Figure 8a. We plotted the curves on a logarithmic scale for the convenience of their analysis. During training epochs loss function (mean absolute error) drops down and saturates after 100 epochs. The evolution of the prediction error (orange line) shows that the neural network was not overfitted. Figure 8b shows mean absolute error rates for different activation functions and number of nodes at the hidden layer. Figure 8c demonstrates the predicted positions of the reflectance peaks for all 50 FBGs of the sparse sensors against measured ones in scaled units. Figure 8c is the same curve for a single FBG in nm units. The curves are in close proximity to the straight line corresponding to the ideal case when predicted values are equal to measured values. We found out that FCNN is able to predict the positions of the reflectance peak of the sparse FBG sensors with a mean absolute error equal to 10.9 pm. The root mean square error (RMSE) was 18 pm and the coefficient of determination (R2) was 0.9988.

In the same way we analyzed the performance of CNN (Figure 9). It can be seen that neural networks have similar performance; however, CNN shows the lowest mean absolute error, equal to 7.4 pm, RMSE equal to 14 pm and R2 equal to 0.9993.

The RMSE metric is more sensitive to large errors comparing to MAE. It is clearly seen from Figures 8c and 9c that the mismatch between real, predicted and measured values is not uniformly distributed along different FBGs. For some FBGs, the mean absolute error does not exceed 5 pm; however, for one FBG the mean absolute error reaches 14 pm. We attribute this to the violation of the uniformity of the temperature field across fiber sensors during the mechanical translation of the Peltier cells. Indeed, CNN performed worse at convex temperature distribution when only one Peltier cell was used. The RMSE was equal to 14.48 pm and R2 was equal to 0.9967 for convex temperature distribution comparing to 8.48 pm and 0.9993 for raised gradient temperature distribution. The issue may be solved by adding more Peltier cells with lower size or building more complicated heating/cooling systems, for instance, using laser heating in combination with a spatial light modulator.

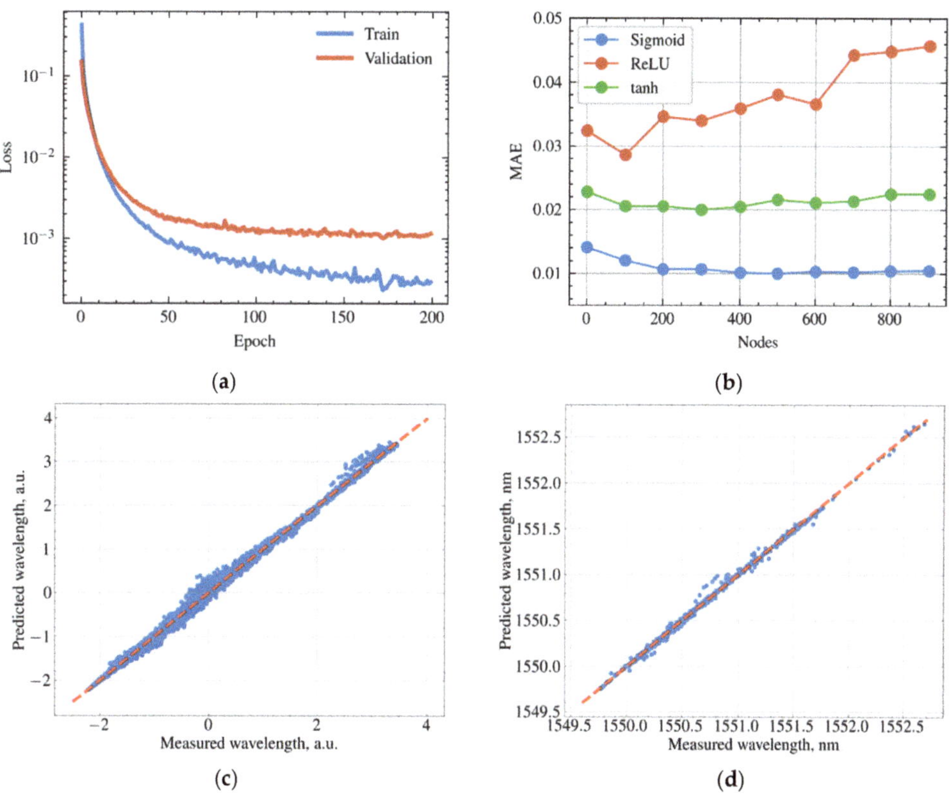

Figure 8. Performance of the FCNN. (**a**) Learning curve of the model on training dataset and evolution of the loss function of validation dataset. (**b**) Mean absolute error of FCNN for different activation functions and different number of nodes of the hidden layer. (**c**) Predicted reflectance peak positions of the FBGs of the sparse sensors against measured values at normalized scales. (**d**) Predicted positions of the reflectance peak for single FBG of sparse sensor against measured value.

Computational complexities of FCNN and CNN may be estimated as follows:

$$C_{FCNN} = N_{input} \cdot N_{hidden} + N_{hiden} * N_{output}, \tag{1}$$

$$C_{CNN} = N_{input} \cdot D^2 + N_{input} * N_{fcl1} + N_{fcl1} \cdot N_{fcnn2} + N_{fcnn2} * N_{output}, \tag{2}$$

where N_{input}, N_{output}, N_{hidden}—numbers of neurons of the input, output layers and hidden layers in FCNN, D—dimension of the convolution filter, N_{fcl1} and N_{fcl2}—numbers of neurons of the fully connected layers in CNN architecture. Calculation of FCNN output takes around 10 million operations, while calculation of CNN output takes around 16 million operations. Better performance of CNN may be related to increased complexity of the architecture. At any case the computational time of the neural networks output is negligible comparing to hardware acquisition time of reflectance spectrum. Computation time of the CNN output from a single sample of the registered reflectance spectrum takes in average 37 milliseconds running on modest graphical processor unit NVIDEA GeForce GTX 950M.

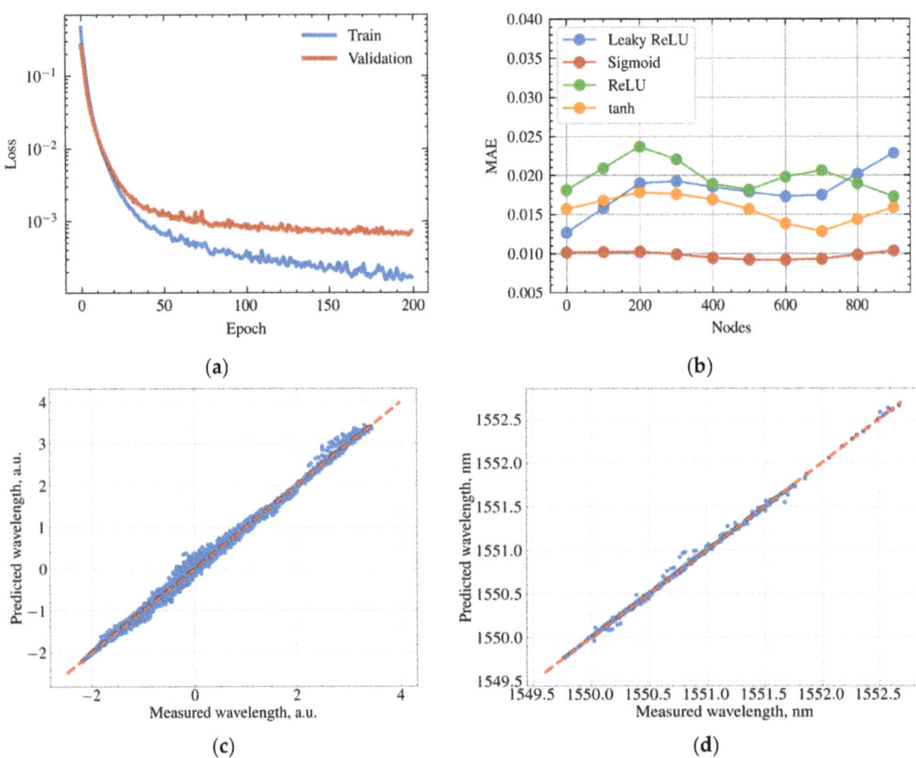

Figure 9. Performance of the CNN (**a**) Learning curve of the model on training dataset and evolution of the loss function of validation dataset. (**b**) Mean absolute error of CNN for different activation functions and different number of nodes of the fully connected layers. (**c**) Predicted reflectance peak positions of the FBGs of the sparse sensors against measured values at normalized scales. (**d**) Predicted positions of the reflectance peak for single FBG of sparse sensor against measured value.

5. Conclusions

Thus, the calibration method of a highly dense FBG temperature sensor is proposed in the paper. It provides a possibility for increasing the spatial resolution of a fiberoptic sensor, avoiding the complications of FBG manufacturing or of an interrogation setup. The method is an alternative to the more common approach, wherein several sparse FBGs sensors are coupled into one optical channel. It was shown that deep learning algorithms are capable of mapping the complex reflectance spectrum of the dense sensor with 50 peaks to position of reflectance peaks of the sparse calibrated FBG temperature sensors. The relatively simple architecture of convolutional neural network allowed us to increase the spatial resolution of the dense FBG sensor by five times while maintaining a high temperature resolution close to hardware resolution. Future improvements of the method may be associated with complication of the architecture of the neural network and increasing the uniformity of the temperature distribution across fiber sensors.

Author Contributions: Conceptualization, A.K. and A.W.; methodology, A.K. and A.W.; software, N.S.; resources, A.D., A.W., A.K.; writing—original draft preparation, A.K., N.S., A.W.; writing—review and editing, A.K., A.W., visualization, N.S.; supervision A.K. and A.W. All authors have read and agreed to the published version of the manuscript.

Funding: A. Dostovalov and A. Wolf acknowledge the support of the Ministry of Science and Higher Education of the Russian Federation (14.Y26.31.0017). The work of A. Kokhanovskiy was supported by the Russian Science Foundation (Grant No. 17-72-30006-П).

Institutional Review Board Statement: Not applicable.

Informed Consent Statement: Not applicable.

Data Availability Statement: The data presented in this study is available on request to the corresponding author.

Conflicts of Interest: The authors declare no conflict of interest.

References

1. Zaynetdinov, M.; See, E.M.; Geist, B.; Ciovati, G.; Robinson, H.D.; Kochergin, V. A fiber bragg grating temperature sensor for 2-400 K. *IEEE Sens. J.* **2015**, *15*, 1908–1912. [CrossRef]
2. Rao, Y.J.; Webb, D.J.; Jackson, D.A.; Zhang, L.; Bennion, I. In-fiber bragg-grating temperature sensor system for medical applications. *J. Light. Technol.* **1997**, *15*, 779–784. [CrossRef]
3. Zhang, B.; Kahrizi, M. High-temperature resistance Fiber bragg grating temperature sensor fabrication. *IEEE Sens. J.* **2007**, *7*, 586–591. [CrossRef]
4. Zhang, W.; Zhang, M.; Wang, X.; Zhao, Y.; Jin, B.; Dai, W. The analysis of FBG central wavelength variation with crack propagation based on a self-adaptive multi-peak detection algorithm. *Sensors* **2019**, *19*, 1056. [CrossRef] [PubMed]
5. Tosi, D. Review and analysis of peak tracking techniques for fiber bragg grating sensors. *Sensors* **2017**, *17*, 2368. [CrossRef]
6. Mohammed, N.A.; Ali, T.A.; Aly, M.H. Evaluation and performance enhancement for accurate FBG temperature sensor measurement with different apodization profiles in single and quasi-distributed DWDM systems. *Opt. Lasers Eng.* **2014**, *55*, 22–34. [CrossRef]
7. Yashiro, S.; Okabe, T.; Toyama, N.; Takeda, N. Monitoring damage in holed CFRP laminates using embedded chirped FBG sensors. *Int. J. Solids Struct.* **2007**, *44*, 603–613. [CrossRef]
8. Saccomandi, P.; Varalda, A.; Gassino, R.; Tosi, D.; Massaroni, C.; Caponero, M.A.; Pop, R.; Korganbayev, S.; Perrone, G.; Diana, M.; et al. Linearly chirped fiber bragg grating response to thermal gradient: From bench tests to the real-time assessment during in vivo laser ablations of biological tissue. *J. Biomed. Opt.* **2017**, *22*, 097002. [CrossRef] [PubMed]
9. Nand, A.; Kitcher, D.J.; Wade, S.A.; Nguyen, T.B.; Baxter, G.W.; Jones, R.; Collins, S.F. Determination of the position of a localized heat source within a chirped fibre bragg grating using a Fourier transform technique. *Meas. Sci. Technol.* **2006**, *17*, 1436–1445. [CrossRef]
10. Tosi, D. Review of chirped fiber bragg grating (CFBG) fiber-optic sensors and their applications. *Sensors* **2018**, *18*, 2147. [CrossRef]
11. Sancho, J.; Chin, S.; Barrera, D.; Sales, S.; Thévenaz, L. Time-frequency analysis of long fiber bragg gratings with low reflectivity. *Opt. Express* **2013**, *21*, 7171. [CrossRef]
12. Qi, Y.; Li, C.; Jiang, P.; Jia, C.; Liu, Y.; Zhang, Q. Research on demodulation of FBGs sensor network based on PSO-SA algorithm. *Optik* **2018**, *164*, 647–653. [CrossRef]
13. Jiang, H.; Chen, J.; Liu, T. Wavelength detection in spectrally overlapped fbg sensor network using extreme learning machine. *IEEE Photonics Technol. Lett.* **2014**, *26*, 2031–2034. [CrossRef]
14. Genty, G.; Salmela, L.; Dudley, J.M.; Daniel, B.; Kokhanovskiy, A.; Sergei, K.; Turitsyn, S.K. Machine learning and applications in ultrafast photonics. *Nat. Photonics* **2020**, *15*, 91–101. [CrossRef]
15. Zhang, W.; Li, H.; Li, Y.; Liu, H.; Chen, Y.; Ding, X. *Application of Deep Learning Algorithms in Geotechnical Engineering: A Short Critical Review*; Springer: Dordrecht, The Netherlands, 2021; ISBN 0123456789.
16. Chen, J.; Jiang, H.; Liu, T.; Fu, X. Wavelength detection in FBG sensor networks using least squares support vector regression. *J. Opt.* **2014**, *16*, 045402. [CrossRef]
17. Melnikov, A.D.; Tsentalovich, Y.P.; Yanshole, V.V. Deep Learning for the Precise Peak Detection in High-Resolution LC-MS Data. *Anal. Chem.* **2020**, *92*, 588–592. [CrossRef]
18. Manie, Y.C.; Li, J.W.; Peng, P.C.; Shiu, R.K.; Chen, Y.Y.; Hsu, Y.T. Using a machine learning algorithm integrated with data de-noising techniques to optimize the multipoint sensor network. *Sensors* **2020**, *20*, 1070. [CrossRef]
19. Djurhuus, M.S.E.; Werzinger, S.; Schmauss, B.; Clausen, A.T.; Zibar, D. Machine learning assisted fiber bragg grating-based temperature sensing. *IEEE Photonics Technol. Lett.* **2019**, *31*, 939–942. [CrossRef]
20. Jiang, H.; Zeng, Q.; Chen, J.; Qiu, X.; Liu, X.; Chen, Z.; Miao, X. Wavelength detection of model-sharing fiber bragg grating sensor networks using long short-term memory neural network. *Opt. Express* **2019**, *27*, 20583. [CrossRef]
21. Dostovalov, A.V.; Wolf, A.A.; Parygin, A.V.; Zyubin, V.E.; Babin, S.A. Femtosecond point-by-point inscription of bragg gratings by drawing a coated fiber through ferrule. *Opt. Express* **2016**, *24*, 16232. [CrossRef]
22. Abadi, M.; Agarwal, A.; Barham, P.; Brevdo, E.; Chen, Z.; Citro, C.; Corrado, G.S.; Davis, A.; Dean, J.; Devin, M.; et al. TensorFlow: Large-Scale Machine Learning on Heterogeneous Systems. *arXiv* **2016**, arXiv:1603.04467.

Article

Event Collapse in Contrast Maximization Frameworks

Shintaro Shiba [1,2,*], Yoshimitsu Aoki [1] and Guillermo Gallego [2,3]

[1] Department of Electronics and Electrical Engineering, Faculty of Science and Technology, Keio University, 3-14-1, Kohoku-ku, Yokohama 223-8522, Kanagawa, Japan; aoki@elec.keio.ac.jp
[2] Department of Electrical Engineering and Computer Science, Technische Universität Berlin, 10587 Berlin, Germany; guillermo.gallego@tu-berlin.de
[3] Einstein Center Digital Future and Science of Intelligence Excellence Cluster, 10117 Berlin, Germany
* Correspondence: sshiba@keio.jp

Abstract: Contrast maximization (CMax) is a framework that provides state-of-the-art results on several event-based computer vision tasks, such as ego-motion or optical flow estimation. However, it may suffer from a problem called event collapse, which is an undesired solution where events are warped into too few pixels. As prior works have largely ignored the issue or proposed workarounds, it is imperative to analyze this phenomenon in detail. Our work demonstrates event collapse in its simplest form and proposes collapse metrics by using first principles of space–time deformation based on differential geometry and physics. We experimentally show on publicly available datasets that the proposed metrics mitigate event collapse and do not harm well-posed warps. To the best of our knowledge, regularizers based on the proposed metrics are the only effective solution against event collapse in the experimental settings considered, compared with other methods. We hope that this work inspires further research to tackle more complex warp models.

Keywords: computer vision; intelligent sensors; robotics; event-based camera; contrast maximization; optical flow; motion estimation

Citation: Shiba, S.; Aoki, Y.; Gallego, G. Event Collapse in Contrast Maximization Frameworks. *Sensors* **2022**, *22*, 5190. https://doi.org/10.3390/s22145190

Academic Editor: Jing Tian

Received: 9 June 2022
Accepted: 7 July 2022
Published: 11 July 2022

Publisher's Note: MDPI stays neutral with regard to jurisdictional claims in published maps and institutional affiliations.

1. Introduction

Event cameras [1–3] offer potential advantages over standard cameras to tackle difficult scenarios (high speed, high dynamic range, low power). However, new algorithms are needed to deal with the unconventional type of data they produce (per-pixel asynchronous brightness changes, called events) and unlock their advantages [4]. Contrast maximization (CMax) is an event processing framework that provides state-of-the-art results on several tasks, such as rotational motion estimation [5,6], feature flow estimation and tracking [7–11], ego-motion estimation [12–14], 3D reconstruction [12,15], optical flow estimation [16–19], motion segmentation [20–24], guided filtering [25], and image reconstruction [26].

The main idea of CMax and similar event alignment frameworks [27,28] is to find the motion and/or scene parameters that align corresponding events (i.e., events that are triggered by the same scene edge), thus achieving motion compensation. The framework simultaneously estimates the motion parameters and the correspondences between events (data association). However, in some cases CMax optimization converges to an undesired solution where events accumulate into too few pixels, a phenomenon called event collapse (Figure 1). Because CMax is at the heart of many state-of-the-art event-based motion estimation methods, it is important to understand the above limitation and propose ways to overcome it. Prior works have largely ignored the issue or proposed workarounds without analyzing the phenomenon in detail. A more thorough discussion of the phenomenon is overdue, which is the goal of this work.

Contrary to the expectation that event collapse occurs when the event transformation becomes sufficiently complex [16,27], we show that it may occur even in the simplest case of one degree-of-freedom (DOF) motion. Drawing inspiration from differential geometry and electrostatics, we propose principled metrics to quantify event collapse and discourage it by incorporating penalty terms in the event alignment objective function. Although event collapse depends on many factors, our strategy aims at modifying the objective's landscape to improve the well-posedness of the problem and be able to use well-known, standard optimization algorithms.

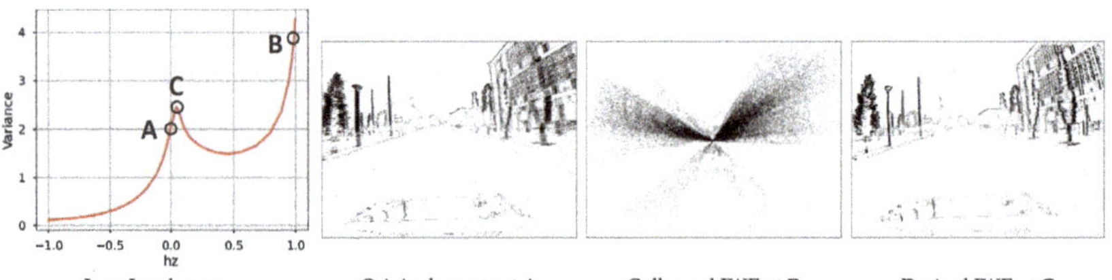

Figure 1. *Event Collapse.* **Left**: Landscape of the image variance loss as a function of the warp parameter h_z. **Right**: The IWEs at the different h_z marked in the landscape. (**A**) Original events (identity warp), accumulated over a small Δt (polarity is not used). (**B**) Image of warped events (IWE) showing event collapse due to maximization of the objective function. (**C**) Desired IWE solution using our proposed regularizer: sharper than (**A**) while avoiding event collapse (**C**).

In summary, our contributions are:

1. A study of the event collapse phenomenon in regard to event warping and objective functions (Sections 3.3 and 4).
2. Two principled metrics of event collapse (one based on flow divergence and one based on area-element deformations) and their use as regularizers to mitigate the above-mentioned phenomenon (Sections 3.4 to 3.6).
3. Experiments on publicly available datasets that demonstrate, in comparison with other strategies, the effectiveness of the proposed regularizers (Section 4).

To the best of our knowledge, this is the first work that focuses on the paramount phenomenon of event collapse, which may arise in state-of-the-art event-alignment methods. Our experiments show that the proposed metrics mitigate event collapse while they do not harm well-posed warps.

2. Related Work

2.1. Contrast Maximization

Our study is based on the CMax framework for event alignment (Figure 2, bottom branch). The CMax framework is an iterative method with two main steps per iteration: transforming events and computing an objective function from such events. Assuming constant illumination, events are triggered by moving edges, and the goal is to find the transformation/warping parameters θ (e.g., motion and scene) that achieve motion compensation (i.e., alignment of events triggered at different times and pixels), hence revealing the edge structure that caused the events. Standard optimization algorithms (gradient ascent, sampling, etc.) can be used to maximize the event-alignment objective. Upon convergence, the method provides the best transformation parameters and the transformed events, i.e., motion-compensated image of warped events (IWE).

The first step of the CMax framework transforms events according to a motion or deformation model defined by the task at hand. For instance, camera rotational motion estimation [5,29] often assumes constant angular velocity ($\theta \equiv \omega$) during short time spans, hence events are transformed following 3-DOF motion curves defined on the image plane by candidate values of ω. Feature tracking may assume constant image velocity $\theta \equiv \mathbf{v}$ (2-DOF) [7,30], hence events are transformed following straight lines.

In the second step of CMax, several event-alignment objectives have been proposed to measure the goodness of fit between the events and the model [10,13], establishing connections between visual contrast, sharpness, and depth-from-focus. Finally, the choice of iterative optimization algorithm also plays a big role in finding the desired motion-compensation parameters. First-order methods, such as non-linear conjugate gradient (CG), are a popular choice, trading off accuracy and speed [12,21,22]. Exhaustive search, sampling, or branch-and-bound strategies may be affordable for low-dimensional (DOF) search spaces [14,29]. As will be presented (Section 3), our proposal consists of modifying the second step by means of a regularizer (Figure 2, top branch).

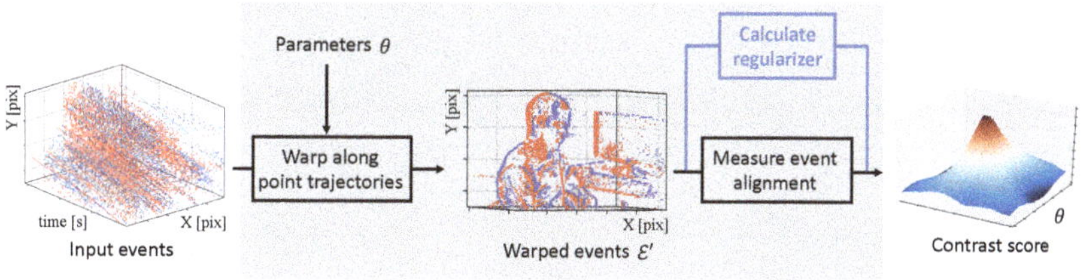

Figure 2. Proposed modification of the contrast maximization (CMax) framework in [12,13] to also account for the degree of regularity (collapsing behavior) of the warp. Events are colored in red/blue according to their polarity. Reprinted/adapted with permission from Ref. [13], 2019, Gallego et al.

2.2. Event Collapse

In which estimation problems does event collapse appear? At first look, it may appear that event collapse occurs when the number of DOFs in the warp becomes large enough, i.e., for complex motions. Event collapse has been reported in homographic motions (8 DOFs) [27,31] and in dense optical flow estimation [16], where an artificial neural network (ANN) predicts a flow field with $2N_p$ DOFs (N_p pixels), whereas it does not occur in feature flow (2 DOFs) or rotational motion flow (3 DOFs). However, a more careful analysis reveals that this is not the entire story because event collapse may occur even in the case of 1 DOF, as we show.

How did previous works tackle event collapse? Previous works have tackled the issue in several ways, such as: (i) initializing the parameters sufficiently close to the desired solution (in the basin of attraction of the local optimum) [12]; (ii) reformulating the problem, changing the parameter space to reduce the number of DOFs and increase the well-posedness of the problem [14,31]; (iii) providing additional data, such as depth [27], thus changing the problem from motion estimation given only events to motion estimation given events and additional sensor data; (iv) whitening the warped events before computing the objective [27]; and (v) redesigning the objective function and possibly adding a strong classical regularizer (e.g., Charbonnier loss) [10,16]. Many of the above mitigation strategies are task-specific because it may not always be possible to consider additional data or reparametrize the estimation problem. Our goal is to approach the issue without the need for additional data or changing the parameter space, and to show how previous objective functions and newly regularized ones handle event collapse.

3. Method

Let us present our approach to measure and mitigate event collapse. First, we revise how event cameras work (Section 3.1) and the CMax framework (Section 3.2), which was informally introduced in Section 2.1. Then, Section 3.3 builds our intuition on event collapse by analyzing a simple example. Section 3.4 presents our proposed metrics for event collapse, based on 1-DOF and 2-DOF warps. Section 3.5 specifies them for higher DOFs, and Section 3.6 presents the regularized objective function.

3.1. How Event Cameras Work

Event cameras, such as the Dynamic Vision Sensor (DVS) [2,3,32], are bio-inspired sensors that capture pixel-wise intensity changes, called events, instead of intensity images.

An event $e_k \doteq (\mathbf{x}_k, t_k, p_k)$ is triggered as soon as the logarithmic intensity L at a pixel exceeds a contrast sensitivity $C > 0$,

$$L(\mathbf{x}_k, t_k) - L(\mathbf{x}_k, t_k - \Delta t_k) = p_k\, C, \tag{1}$$

where $\mathbf{x}_k \doteq (x_k, y_k)^\top$, t_k (with µs resolution) and polarity $p_k \in \{+1, -1\}$ are the spatio-temporal coordinates and sign of the intensity change, respectively, and $t_k - \Delta t_k$ is the time of the previous event at the same pixel $\mathbf{x}_k$. Hence, each pixel has its own sampling rate, which depends on the visual input.

3.2. Mathematical Description of the CMax Framework

The CMax framework [12] transforms events in a set $\mathcal{E} = \{e_k\}_{k=1}^{N_e}$ geometrically

$$e_k \doteq (\mathbf{x}_k, t_k, p_k) \quad \overset{W}{\mapsto} \quad e_k' \doteq (\mathbf{x}_k', t_{\text{ref}}, p_k), \tag{2}$$

according to a motion model $\mathbf{W}$, producing a set of warped events $\mathcal{E}' = \{e_k'\}_{k=1}^{N_e}$. The warp $\mathbf{x}_k' = \mathbf{W}(\mathbf{x}_k, t_k; \boldsymbol{\theta})$ transports each event along the point trajectory that passes through it (Figure 2, left), until t_{ref} is reached. The point trajectories are parametrized by $\boldsymbol{\theta}$, which contains the motion and/or scene unknowns. Then, an objective function [10,13] measures the alignment of the warped events $\mathcal{E}'$. Many objective functions are given in terms of the count of events along the point trajectories, which is called the image of warped events (IWE):

$$I(\mathbf{x}; \boldsymbol{\theta}) \doteq \sum_{k=1}^{N_e} b_k\, \delta(\mathbf{x} - \mathbf{x}_k'(\boldsymbol{\theta})). \tag{3}$$

Each IWE pixel $\mathbf{x}$ sums the values of the warped events $\mathbf{x}_k'$ that fall within it: $b_k = p_k$ if polarity is used or $b_k = 1$ if polarity is not used. The Dirac delta δ is in practice replaced by a smooth approximation [33], such as a Gaussian, $\delta(\mathbf{x} - \boldsymbol{\mu}) \approx \mathcal{N}(\mathbf{x}; \boldsymbol{\mu}, \epsilon^2)$ with $\epsilon = 1$ pixel. A popular objective function $G(\boldsymbol{\theta})$ is the visual contrast of the IWE (3), given by the variance

$$G(\boldsymbol{\theta}) \equiv \text{Var}\big(I(\mathbf{x}; \boldsymbol{\theta})\big) \doteq \frac{1}{|\Omega|} \int_\Omega \big(I(\mathbf{x}; \boldsymbol{\theta}) - \mu_I\big)^2 d\mathbf{x}, \tag{4}$$

with mean $\mu_I \doteq \frac{1}{|\Omega|} \int_\Omega I(\mathbf{x}; \boldsymbol{\theta}) d\mathbf{x}$ and image domain Ω. Hence, the alignment of the transformed events $\mathcal{E}'$ (i.e., the candidate "corresponding events", triggered by the same scene edge) is measured by the strength of the edges of the IWE. Finally, an optimization algorithm iterates the above steps until the best parameters are found:

$$\boldsymbol{\theta}^* = \arg\max_{\boldsymbol{\theta}} G(\boldsymbol{\theta}). \tag{5}$$

3.3. Simplest Example of Event Collapse: 1 DOF

To analyze event collapse in the simplest case, let us consider an approximation to a translational motion of the camera along its optical axis Z (1-DOF warp). In theory, translational motions also require the knowledge of the scene depth. Here, inspired by the 4-DOF in-plane warp in [20] that approximates a 6-DOF camera motion, we consider a simplified warp that does not require knowledge of the scene depth. In terms of data, let us consider events from one of the driving sequences of the standard MVSEC dataset [34] (Figure 1).

For further simplicity, let us normalize the timestamps of $\mathcal{E}$ to the unit interval $t \in [t_1, t_{N_e}] \mapsto \tilde{t} \in [0, 1]$, and assume a coordinate frame at the center of the image plane, then the warp $\mathbf{W}$ is given by

$$\mathbf{x}_k' = (1 - \tilde{t}_k h_z)\, \mathbf{x}_k, \tag{6}$$

where $\boldsymbol{\theta} \equiv h_z$. Hence, events are transformed along the radial direction from the image center, acting as a virtual focus of expansion (FOE) (cf. the true FOE is given by the data).

Letting the scaling factor in (6) be $s_k \doteq 1 - \tilde{t}_k h_z$, we observe the following: (i) s_k cannot be negative since it would imply that at least one event has flipped the side on which it lies with respect to the image center; (ii) if $s_k > 1$ the warped event gets away from the image center ("expansion" or "zoom-in"); and (iii) if $s_k \in [0, 1)$ the warped event gets closer to the image center ("contraction" or "zoom-out"). The equivalent conditions in terms of h_z are: (i) $h_z < 1$, (ii) $h_z < 0$ is an expansion, and (iii) $0 < h_z < 1$ is a contraction.

Intuitively, event collapse occurs if the contraction is large ($0 < s_k \ll 1$) (see Figures 1C and 3a). This phenomenon is not specific of the image variance; other objective functions lead to the same result. As we see, the objective function has a local maximum at the desired motion parameters (Figure 1B). The optimization over the entire parameter space converges to a global optimum that explains the event collapse.

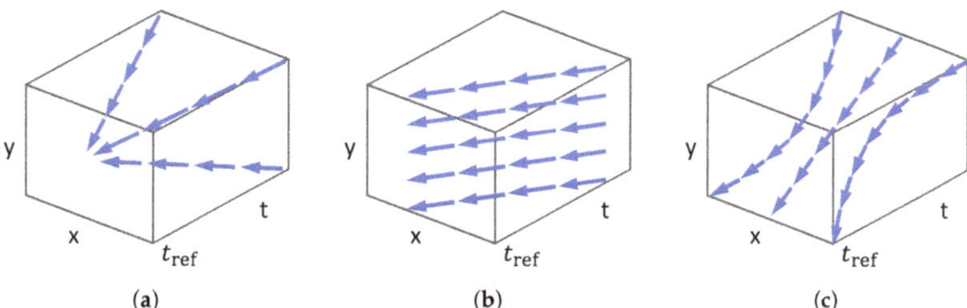

Figure 3. *Point trajectories* (streamlines) defined on $x - y - t$ image space by various warps. (**a**) Zoom in/out warp from image center (1 DOF). (**b**) Constant image velocity warp (2 DOF). (**c**) Rotational warp around X axis (3 DOF).

Discussion

The above example shows that event collapse is enabled (or disabled) by the type of warp. If the warp does not enable event collapse (contraction or accumulation of flow vectors cannot happen due to the geometric properties of the warp), as in the case of feature flow (2 DOF) [7,30] (Figure 3b) or rotational motion flow (3 DOF) [5,29] (Figure 3c), then the optimization problem is well posed and multiple objective functions can be designed to achieve event alignment [10,13]. However, the disadvantage is that the type of warps that satisfy this condition may not be rich enough to describe complex scene motions.

On the other hand, if the warp allows for event collapse, more complex scenarios can be described by such a broader class of motion hypotheses, but the optimization framework designed for non-event-collapsing scenarios (where the local maximum is assumed to be the global maximum) may not hold anymore. Optimizing the objective function may lead to an undesired solution with a larger value than the desired one. This depends on multiple elements: the landscape of the objective function (which depends on the data, the warp parametrization, and the shape of the objective function), and the initialization and search strategy of the optimization algorithm used to explore such a landscape. The challenge in this situation is to overcome the issue of multiple local maxima and make the problem better posed. Our approach consists of characterizing event collapse via novel metrics and including them in the objective function as weak constraints (penalties) to yield a better landscape.

3.4. Proposed Regularizers

3.4.1. Divergence of the Event Transformation Flow

Inspired by physics, we may think of the flow vectors given by the event transformation $\mathcal{E} \mapsto \mathcal{E}'$ as an electrostatic field, whose sources and sinks correspond to the location of electric charges (Figure 4). Sources and sinks are mathematically described by the divergence operator $\nabla \cdot$. Therefore, the divergence of the flow field is a natural choice to characterize event collapse.

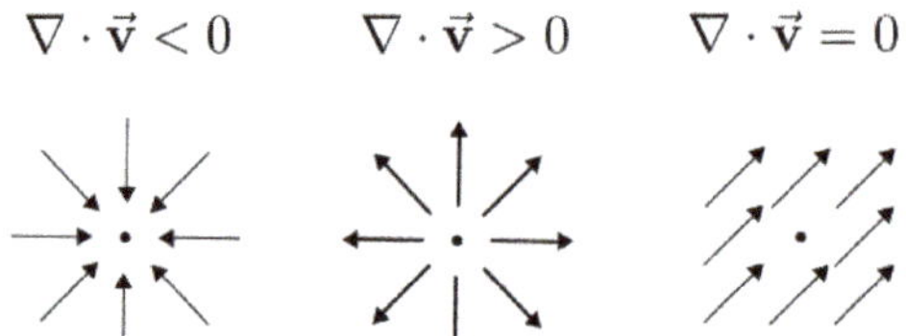

Figure 4. *Divergence of different vector fields,* $\nabla \cdot \mathbf{v} = \partial_x \mathbf{v}_x + \partial_y \mathbf{v}_y$. From left to right: contraction ("sink", leading to event collapse), expansion ("source"), and incompressible fields. Image adapted from khanacademy.org (accessed on 6 July 2022).

The warp $\mathbf{W}$ is defined over the space-time coordinates of the events, hence its time derivative defines a flow field over space-time:

$$\mathbf{f} \doteq \frac{\partial \mathbf{W}(\mathbf{x}, t; \boldsymbol{\theta})}{\partial t}. \tag{7}$$

For the warp in (6), we obtain $\mathbf{f} = -h_z \mathbf{x}$, which gives $\nabla \cdot \mathbf{f} = -h_z \nabla \cdot \mathbf{x} = -2h_z$. Hence, (6) defines a constant divergence flow, and imposing a penalty on the degree of concentration of the flow field accounts to directly penalizing the value of the parameter h_z.

Computing the divergence at each event gives the set

$$\mathcal{D}(\mathcal{E}, \boldsymbol{\theta}) \doteq \{\nabla \cdot \mathbf{f}_k\}_{k=1}^{N_e}, \tag{8}$$

from which we can compute statistical scores (mean, median, min, etc.):

$$R_D(\mathcal{E}, \boldsymbol{\theta}) \doteq \frac{1}{N_e} \sum_{k=1}^{N_e} \nabla \cdot \mathbf{f}_k. \qquad \text{(mean)} \tag{9}$$

To have a 2D visual representation ("feature map") of collapse, we build an image (like the IWE) by taking some statistic of the values $\nabla \cdot \mathbf{f}_k$ that warp to each pixel, such as the "average divergence per pixel":

$$\mathrm{DIWE}(\mathbf{x}; \mathcal{E}, \boldsymbol{\theta}) \doteq \frac{1}{N_e(\mathbf{x})} \sum_k (\nabla \cdot \mathbf{f}_k)\, \delta(\mathbf{x} - \mathbf{x}'_k), \tag{10}$$

where $N_e(\mathbf{x}) \doteq \sum_k \delta(\mathbf{x} - \mathbf{x}'_k)$ is the number of warped events at pixel $\mathbf{x}$ (the IWE). Then we aggregate further into a score, such as the mean:

$$R_{\mathrm{DIWE}}(\mathcal{E}, \boldsymbol{\theta}) \doteq \frac{1}{|\Omega|} \int_\Omega \mathrm{DIWE}(\mathbf{x}; \mathcal{E}, \boldsymbol{\theta}) d\mathbf{x}. \tag{11}$$

In practice we focus on the collapsing part by computing a trimmed mean: the mean of the DIWE pixels smaller than a margin α (-0.2 in the experiments). Such a margin does not penalize small, admissible deformations.

3.4.2. Area-Based Deformation of the Event Transformation

In addition to vector calculus, we may also use tools from differential geometry to characterize event collapse. Building on [12], the point trajectories define the streamlines of the transformation flow, and we may measure how they concentrate or disperse based on how the area element deforms along them. That is, we consider a small area element $dA = dx\,dy$ attached to each point along the trajectory and measure how much it deforms when transported to the reference time: $dA' = |\det(\mathtt{J})|\, dA$, with the Jacobian

$$\mathtt{J}(\mathbf{x}, t; \boldsymbol{\theta}) \doteq \frac{\partial \mathbf{W}(\mathbf{x}, t; \boldsymbol{\theta})}{\partial \mathbf{x}} \tag{12}$$

(see Figure 5). The determinant of the Jacobian is the amplification factor: $|\det(J)| > 1$ if the area expands, and $|\det(J)| < 1$ if the area shrinks.

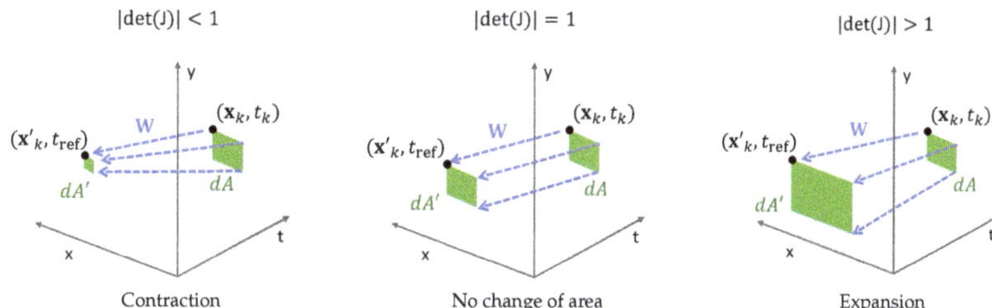

Figure 5. *Area deformation of various warps.* An area of dA pix^2 at $(\mathbf{x}_k, t_k)$ and is warped to t_{ref}, giving an area $dA' = |\det(J_k)| dA$ pix^2 at $(\mathbf{x}'_k, t_{\mathrm{ref}})$, where $J_k \equiv J(e_k) \equiv J(\mathbf{x}_k, t_k; \boldsymbol{\theta})$ (see (12)). From left to right, increasing area amplification factor $|\det(J)| \in [0, \infty)$.

For the warp in (6), we have the Jacobian $J = (1 - \tilde{t}h_z)\mathrm{Id}$, and so $\det(J) = (1 - \tilde{t}h_z)^2$. Interestingly, the area deformation around event e_k, $J(e_k) \equiv J(\mathbf{x}_k, t_k; \boldsymbol{\theta})$, is directly related to the scaling factor s_k: $\det(J(e_k)) = s_k^2$.

Computing the amplification factors at each event gives the set

$$\mathcal{A}(\mathcal{E}, \boldsymbol{\theta}) \doteq \{|\det(J(e_k))|\}_{k=1}^{N_e}, \tag{13}$$

from which we can compute statistical scores. For example,

$$R_A(\mathcal{E}, \boldsymbol{\theta}) \doteq \frac{1}{N_e} \sum_{k=1}^{N_e} |\det(J(e_k))| \qquad \text{(mean)} \tag{14}$$

gives an average score: $R_A > 1$ for expansion, and $R_A < 1$ for contraction.

We build a deformation map (or image of warped areas (IWA)) by taking some statistic of the values $|\det(J(e_k))|$ that warp to each pixel, such as the "average amplification per pixel":

$$\mathrm{IWA}(\mathbf{x}) \doteq 1 + \frac{1}{N_e(\mathbf{x})} \sum_{k=1}^{N_e} (|\det(J(e_k))| - 1)\, \delta(\mathbf{x} - \mathbf{x}'_k). \tag{15}$$

This assumes that if no events warp to a pixel $\mathbf{x}_p$, then $N_e(\mathbf{x}_p) = 0$, and there is no deformation ($\mathrm{IWA}(\mathbf{x}_p) = 1$). Then, we summarize the deformation map into a score, such as the mean:

$$R_{\mathrm{IWA}}(\mathcal{E}, \boldsymbol{\theta}) \doteq \frac{1}{|\Omega|} \int_\Omega \mathrm{IWA}(\mathbf{x}; \mathcal{E}, \boldsymbol{\theta}) d\mathbf{x}. \tag{16}$$

To concentrate on the collapsing part, we compute a trimmed mean: the mean of the IWA pixels smaller than a margin α (0.8 in the experiments). The margin approves small, admissible deformations.

3.5. Higher DOF Warp Models

3.5.1. Feature Flow

Event-based feature tracking is often described by the warp $\mathbf{W}(\mathbf{x}, t; \boldsymbol{\theta}) = \mathbf{x} + (t - t_{\mathrm{ref}})\boldsymbol{\theta}$, which assumes constant image velocity $\boldsymbol{\theta}$ (2 DOFs) over short time intervals. As expected, the flow for this warp coincides with the image velocity, $\mathbf{f} = \boldsymbol{\theta}$, which is independent of the space-time coordinates $(\mathbf{x}, t)$. Hence, the flow is incompressible ($\nabla \cdot \mathbf{f} = 0$): the streamlines given by the feature flow do not concentrate or disperse; they are parallel. Regarding the area deformation, the Jacobian $J = \partial(\mathbf{x} + (t - t_{\mathrm{ref}})\boldsymbol{\theta})/\partial\mathbf{x} = \mathrm{Id}$ is the identity matrix. Hence $|\det(J)| = 1$, that is, translations on the image plane do not change the area of the pixels around a point.

In-plane translation warps, such as the above 2-DOF warp, are well-posed and serve as reference to design the regularizers that measure event collapse. It is sensible for well-designed regularizers to penalize warps whose characteristics deviate from those of the reference warp: zero divergence and unit area amplification factor.

3.5.2. Rotational Motion

As the previous sections show, the proposed metrics designed for the zoom in/out warp produce the expected characterization of the 2-DOF feature flow (zero divergence and unit area amplification), which is a well-posed warp. Hence, if they were added as penalties into the objective function they would not modify the energy landscape. We now consider their influence on rotational motions, which are also well-posed warps. In particular, we consider the problem of estimating the angular velocity of a predominantly rotating event camera by means of CMax, which is a popular research topic [5,14,27–29]. By using calibrated and homogeneous coordinates, the warp is given by

$$\mathbf{x}^{h\prime} \sim \mathsf{R}(t\boldsymbol{\omega})\,\mathbf{x}^h, \tag{17}$$

where $\boldsymbol{\theta} \equiv \boldsymbol{\omega} = (\omega_1, \omega_2, \omega_3)^\top$ is the angular velocity, $t \in [0, \Delta t]$, and R is parametrized by using exponential coordinates (Rodrigues rotation formula [35,36]).

Divergence: It is well known that the flow is $\mathbf{f} = B(\mathbf{x})\,\boldsymbol{\omega}$, where $B(\mathbf{x})$ is the rotational part of the feature sensitivity matrix [37]. Hence

$$\nabla \cdot \mathbf{f} = 3(x\omega_2 - y\omega_1). \tag{18}$$

Area element: Letting $\mathbf{r}_3^\top$ be the third row of R, and using (32)–(34) in [38],

$$\det(\mathsf{J}) = (\mathbf{r}_3^\top \mathbf{x}^h)^{-3}. \tag{19}$$

Rotations around the Z axis clearly present no deformation, regardless of the amount of rotation, and this is captured by the proposed metrics because: (i) the divergence is zero, thus the flow is incompressible, and (ii) $\det(\mathsf{J}) = 1$ since $\mathbf{r}_3 = (0, 0, 1)^\top$ and $\mathbf{x}^h = (x, y, 1)^\top$. For other, arbitrary rotations, there are deformations, but these are mild if the rotation angle $\Delta t\|\boldsymbol{\omega}\|$ is small.

3.5.3. Planar Motion

Planar motion is the term used to describe the motion of a ground robot that can translate and rotate freely on a flat ground. If such a robot is equipped with a camera pointing upwards or downwards, the resulting motion induced on the image plane, parallel to the ground plane, is an isometry (Euclidean transformation). This motion model is a subset of the parametric ones in [12], and it has been used for CMax in [14,27]. For short time intervals, planar motion may be parametrized by 3 DOFs: linear velocity (2 DOFs) and angular velocity (1 DOF). As the divergence and area metrics show in the Appendix A, planar motion is a well-posed warp. The resulting motion curves on the image plane do not lead to event collapse.

3.5.4. Similarity Transformation

The 1-DOF zoom in/out warp in Section 3.3 is a particular case of the 4-DOF warp in [20], which is an in-plane approximation to the motion induced by a freely moving camera. The same idea of combining translation, rotation, and scaling for CMax is expressed by the similarity transformation in [27]. Both 4-DOF warps enable event collapse because they allow for zoom-out motion curves. Formulas justifying it are given in the Appendix A.

3.6. Augmented Objective Function

We propose to augment previous objective functions (e.g., (5)) with penalties obtained from the metrics developed above for event collapse:

$$\theta^* = \underset{\theta}{\arg\min}\, J(\theta) = \underset{\theta}{\arg\min}(-G(\theta) + \lambda R(\theta)). \qquad (20)$$

We may interpret $G(\theta)$ (e.g., contrast or focus score [13]) as the data fidelity term and $R(\theta)$ as the regularizer, or, in Bayesian terms, the likelihood and the prior, respectively.

4. Experiments

We evaluate our method on publicly available datasets, whose details are described in Section 4.1. First, Section 4.2 shows that the proposed regularizers mitigate the overfitting issue on warps that enable collapse. For this purpose we use driving datasets (MVSEC [34], DSEC [39]). Next, Section 4.3 shows that the regularizers do not harm well-posed warps. To this end, we use the ECD dataset [40]. Finally, Section 4.4 conducts a sensitivity analysis of the regularizers.

4.1. Evaluation Datasets and Metrics

4.1.1. Datasets

The *MVSEC* dataset [34] is a widely used dataset for various vision tasks, such as optical flow estimation [16,18,19,41,42]. Its sequences are recorded on a drone (indoors) or on a car (outdoors), and comprise events, grayscale frames and IMU data from an mDAVIS346 [43] (346×260 pixels), as well as camera poses and LiDAR data. Ground truth optical flow is computed as the motion field [44], given the camera velocity and the depth of the scene (from the LiDAR). We select several excerpts from the *outdoor_day1* sequence with a forward motion. This motion is reasonably well approximated by collapse-enabled warps such as (6). In total, we evaluate 3.2 million events spanning 10 s.

The *DSEC* dataset [39] is a more recent driving dataset with a higher resolution event camera (Prophesee Gen3, 640×480 pixels). Ground truth optical flow is also computed as the motion field using the scene depth from a LiDAR [41]. We evaluate on the *zurich_city_11* sequence, using in total 380 million events spanning 40 s.

The *ECD* dataset [40] is the de facto standard to assess event camera ego-motion [5,8,28,45–48]. Each sequence provides events, frames, a calibration file, and IMU data (at 1kHz) from a DAVIS240C camera [49] (240×180 pixels), as well as ground-truth camera poses from a motion-capture system (at 200Hz). For rotational motion estimation (3DOF), we use the natural-looking *boxes_rotation* and *dynamic_rotation* sequences. We evaluate 43 million events (10 s) of the box sequence, and 15 million events (11 s) of the dynamic sequence.

The driving datasets (MVSEC, DSEC) and the selected sequences in the ECD dataset have different type of motions: forward (which enables event collapse) vs. rotational (which does not suffer from event collapse). Each sequence serves a different test purpose, as discussed in the next sections.

4.1.2. Metrics

The metrics used to assess optical flow accuracy (MVSEC and DSEC datasets) are the average endpoint error (AEE) and the percentage of pixels with AEE greater than N pixels (denoted by "NPE", for $N = \{3, 10, 20\}$). Both are measured over pixels with valid ground-truth values. We also use the FWL metric [50] to assess event alignment by means of the IWE sharpness (the FWL is the IWE variance relative to that of the identity warp).

Following previous works [13,27,28], rotational motion accuracy is assessed as the RMS error of angular velocity estimation. Angular velocity ω is assumed to be constant over a window of events, estimated and compared with the ground truth at the midpoint of the window. Additionally, we use the FWL metric to gauge event alignment [50].

The event time windows are as follows: the events in the time spanned by $dt = 4$ frames in MVSEC (standard in [16,18,41]), 500k events for DSEC, and 30k events for ECD [28]. The regularizer weights for divergence (λ_{div}) and deformation (λ_{def}) are as follows: $\lambda_{\text{div}} = 2$ and $\lambda_{\text{def}} = 5$ for MVSEC, $\lambda_{\text{div}} = 50$ and $\lambda_{\text{def}} = 100$ for DSEC, and $\lambda_{\text{div}} = 5$ and $\lambda_{\text{def}} = 10$ for ECD experiments.

4.2. Effect of the Regularizers on Collapse-Enabled Warps

Tables 1 and 2 report the results on the MVSEC and DSEC benchmarks, respectively, by using two different loss functions G: the IWE variance (4) and the squared magnitude of the IWE gradient, abbreviated "Gradient Magnitude" [13]. For MVSEC, we report the accuracy within the time interval of $dt = 4$ grayscale frame (at $\approx$45 Hz). The optimization algorithm is the Tree-Structured Parzen Estimator (TPE) sampler [51] for both experiments, with a number of sampling points equal to 300 (1 DOF) and 600 (4 DOF). The tables quantitatively capture the collapse phenomenon suffered by the original CMax framework [12] and the whitening technique [27]. Their high FWL values indicate that contrast is maximized; however, the AEE and NPE values are exceedingly high (e.g., >80 pixels, 20PE > 80%), indicating that the estimated flow is unrealistic.

Table 1. Results of MVSEC dataset [44].

		Variance					Gradient Magnitude				
		AEE ↓	3PE ↓	10PE ↓	20PE ↓	FWL ↑	AEE ↓	3PE ↓	10PE ↓	20PE ↓	FWL ↑
	Ground truth flow	–	–	–	–	1.05	–	–	–	–	1.05
	Identity warp	4.85	60.59	10.38	0.31	1.00	4.85	60.59	10.38	0.31	1.00
1 DOF	No regularizer	89.34	97.30	95.42	92.39	1.90	85.77	93.96	86.24	83.45	1.87
	Whitening [27]	89.58	97.18	96.77	93.76	1.90	81.10	90.86	89.04	86.20	1.85
	Divergence (Ours)	4.00	46.02	2.77	0.05	1.12	2.87	32.68	2.52	0.03	1.17
	Deformation (Ours)	4.47	52.60	5.16	0.13	1.08	3.97	48.79	3.21	0.07	1.09
	Div. + Def. (Ours)	3.30	33.09	2.61	0.48	1.20	2.85	32.34	2.44	0.03	1.17
4 DOF [20]	No regularizer	90.22	90.22	96.94	93.86	2.05	91.26	99.49	95.06	91.46	2.01
	Whitening [27]	90.82	99.11	98.04	95.04	2.04	88.38	98.87	92.41	88.66	2.00
	Divergence (Ours)	7.25	81.75	18.53	0.69	1.09	5.37	66.18	10.81	0.28	1.14
	Deformation (Ours)	8.13	87.46	18.53	1.09	1.03	5.25	64.79	13.18	0.37	1.15
	Div. + Def. (Ours)	5.14	65.61	10.75	0.38	1.16	5.41	66.01	13.19	0.54	1.14

Table 2. Results of DSEC dataset [39].

		Variance					Gradient Magnitude				
		AEE ↓	3PE ↓	10PE ↓	20PE ↓	FWL ↑	AEE ↓	3PE ↓	10PE ↓	20PE ↓	FWL ↑
	Ground truth flow	–	–	–	–	1.09	–	–	–	–	1.09
	Identity warp	5.84	60.45	16.65	3.40	1.00	5.84	60.45	16.65	3.40	1.00
1 DOF	No regularizer	156.13	99.88	99.33	98.18	2.58	156.08	99.93	99.40	98.11	2.58
	Whitening [27]	156.18	99.95	99.51	98.26	2.58	156.82	99.88	99.38	98.33	2.58
	Divergence (Ours)	12.49	69.86	20.78	6.66	1.43	5.47	63.48	14.66	1.35	1.34
	Deformation (Ours)	9.01	68.96	18.86	4.77	1.40	5.79	64.02	16.11	2.75	1.36
	Div. + Def. (Ours)	6.06	68.48	17.08	2.27	1.36	5.53	64.09	15.06	1.37	1.35
4 DOF [20]	No regularizer	157.54	99.97	99.64	98.67	2.64	157.34	99.94	99.53	98.44	2.62
	Whitening [27]	157.73	99.97	99.66	98.71	2.60	156.12	99.91	99.26	97.93	2.61
	Divergence (Ours)	14.35	90.84	41.62	10.82	1.35	10.43	91.38	41.63	9.43	1.21
	Deformation (Ours)	15.12	94.96	62.59	22.62	1.25	10.01	90.15	39.45	8.67	1.25
	Div. + Def. (Ours)	10.06	90.65	40.61	8.58	1.26	10.39	91.02	41.81	9.40	1.23

By contrast, our regularizers (Divergence and Deformation rows) work well to mitigate the collapse, as observed in smaller AEE and NPE values. Compared with the values of no regularizer or whitening [27], our regularizers achieve more than 90% improvement for AEE on average. The AEE values are high for optical flow standards (4–8 pix in MVSEC vs. 0.5–1 pixel [16], or 10–20 pix in DSEC vs. 2–5 pix [41]); however, this is due to the fact that the warps used have very few DOFs ($\leq$4) compared to the considerably higher DOFs ($2N_p$) of optical flow estimation algorithms. The same reason explains the high 3PE values (standard in [52]): using an end-point error threshold of 3 pix to consider that the flow is correctly estimated does not convey the intended goal of inlier/outlier classification for the low-DOF warps used. This is the reason why Tables 1 and 2 also report 10PE, 20PE metrics, and the values for the identity warp (zero flow). As expected, for the range of AEE values in the tables, the 10PE and 20PE figures demonstrate the large difference between methods suffering from collapse (20PE > 80%) and those that do not (20PE < 1.1% for MVSEC and <22.6% for DSEC).

The FWL values of our regularizers are moderately high ($\geq$1), indicating that event alignment is better than that of the identity warp. However, because the FWL depends on the number of events [50], it is not easy to establish a global threshold to classify each method as suffering from collapse or not. The AEE, 10PE, and 20PE are better for such a classification.

Tables 1 and 2 also include the results of the use of both regularizers simultaneously ("Div. + Def."). The results improve across all sequences if the data fidelity term is given by the variance loss, whereas they remain approximately the same for the gradient magnitude loss. Regardless of the choice of the proposed regularizer, the results in these tables clearly show the effectiveness of our proposal, i.e., the large improvements compared with prior works (rows "No regularizer" and [27]).

The collapse results are more visible in Figure 6, where we used the variance loss. Without a regularizer, the events collapse in the MVSEC and DSEC sequences. Our regularizers successfully mitigate overfitting, having a remarkable impact on the estimated motion.

4.3. Effect of the Regularizers on Well-Posed Warps

Table 3 shows the results on the ECD dataset for a well-posed warp (3-DOF rotational motion, in the benchmark). We use the variance loss and the Adam optimizer [53] with 100 iterations. All values in the table (RMS error and FWL, with and without regularization, are very similar, indicating that: (i) our regularizers do not affect the motion estimation algorithm, and (ii) results without regularization are good due to the well-posed warp. This is qualitatively shown in the bottom part of Figure 6. The fluctuations of the divergence and deformation values away from those of the identity warp (0 and 1, respectively) are at least one order of magnitude smaller than the collapse-enabled warps (e.g., 0.2 vs. 2).

Table 3. Results on ECD dataset [40].

	boxes_rot		dynamic_rot	
	RMS $\downarrow$	**FWL** $\uparrow$	**RMS** $\downarrow$	**FWL** $\uparrow$
Ground truth pose	–	1.559	–	1.414
No regularizer	8.858	1.562	4.823	1.420
Divergence (Ours)	9.237	1.558	4.826	1.420
Deformation (Ours)	8.664	1.561	4.822	1.420

4.4. Sensitivity Analysis

The landscapes of loss functions as well as sensitivity analysis of λ are shown in Figure 7, for the MVSEC experiments. Without regularizer ($\lambda = 0$), all objective functions tested (variance, gradient magnitude, and average timestamp [16]) suffer from event collapse, which is the undesired global minimum of (20). Reaching the desired local optimum depends on the optimizing algorithm and its initialization (e.g., starting gradient descent close enough to the local optimum). Our regularizers (divergence and deformation) change the landscape: the previously undesired global minimum becomes local, and the desired minimum becomes the new global one as λ increases.

Specifically, the larger the weight λ, the smaller the effect of the undesired minimum (at $h_z = 1$). However, this is true only within some reasonable range: a too large λ discards the data-fidelity part G in (20), which is unwanted because it would remove the desired local optimum (near $h_z \approx 0$). Minimizing (20) with only the regularizer is not sensible.

Observe that for completeness, we include the average timestamp loss in the last column. However, this loss also suffers from an undesired optimum in the expansion region ($h_z \approx -1$). Our regularizers could be modified to also remove this undesired optimum, but investigating this particular loss, which was proposed as an alternative to the original contrast loss, is outside the scope of this work.

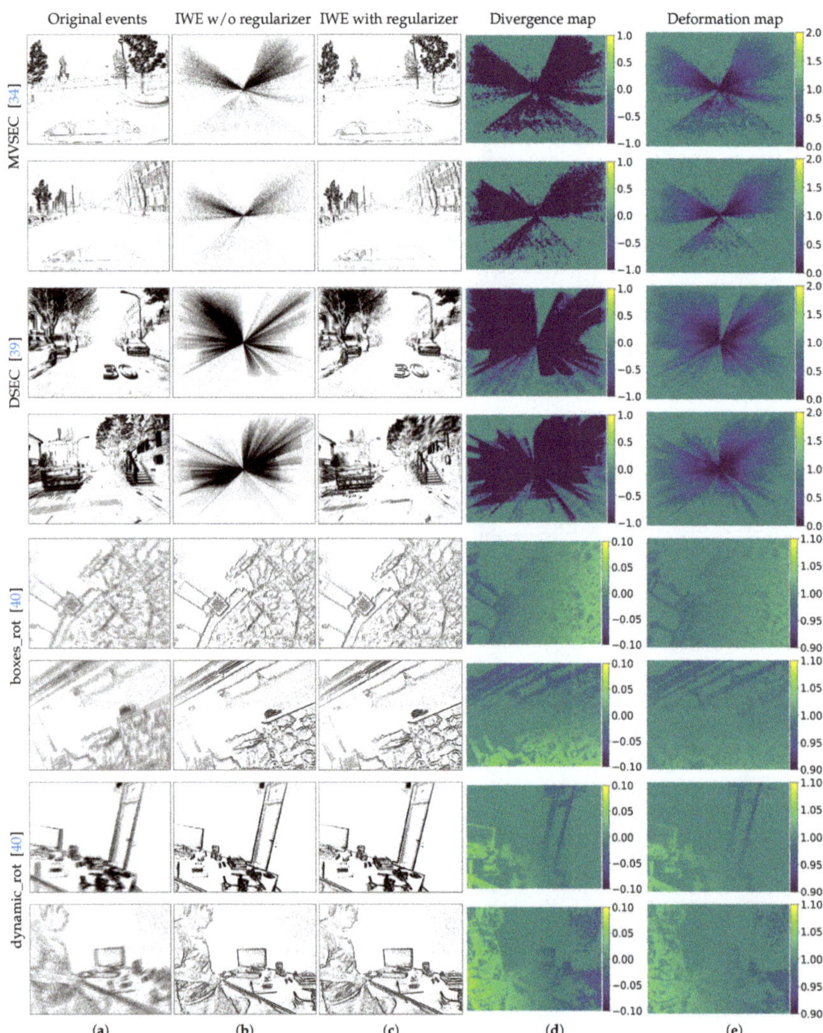

Figure 6. *Proposed regularizers and collapse analysis.* The scene motion is approximated by 1-DOF warp (zoom in/out) for MVSEC [34] and DSEC [39] sequences, and 3-DOF warp (rotation) for boxes and dynamic ECD sequences [40]. (**a**) Original events. (**b**) Best warp without regularization. Event collapse happens for 1-DOF warp. (**c**) Best warp with regularization. (**d**) Divergence map ((10) is zero-based). (**e**) Deformation map ((15), centered at 1). Our regularizers successfully penalize event collapse and do not damage non-collapsing scenarios.

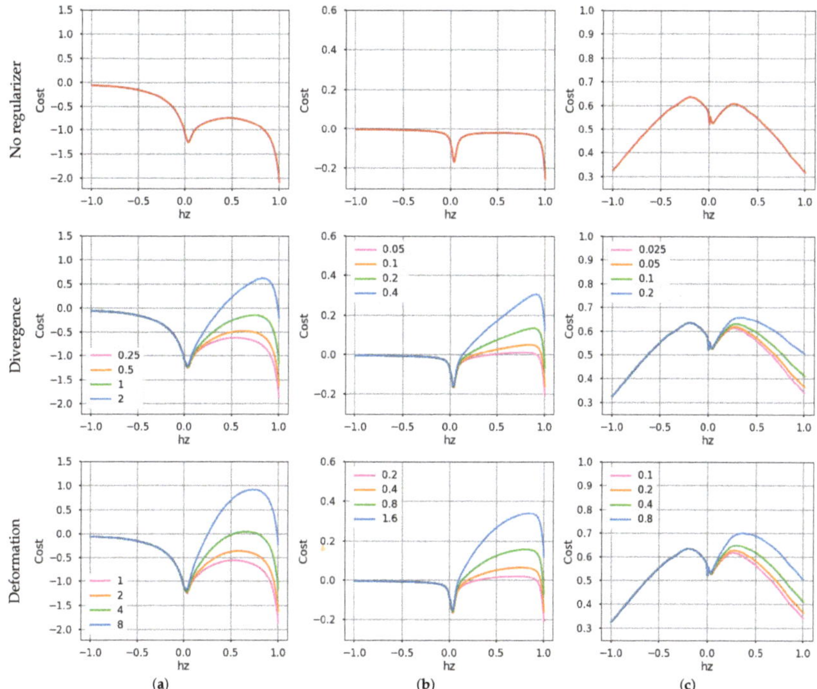

Figure 7. *Cost function landscapes over the warp parameter h_z for*: (**a**) Image variance [12], (**b**) gradient magnitude [13], and (**c**) mean square of average timestamp [16]. Data from MVSEC [34] with dominant forward motion. The legend weights denote λ in (20).

4.5. Computational Complexity

Computing the regularizer(s) requires more computation than the non-regularized objective. However, complexity is linear with the number of events and the number of pixels, which is an advantage, and the warped events are reutilized to compute the DIWE or IWA. Hence, the runtime is less than doubled (warping is the dominant runtime term [13] and is computed only once). The computational complexity of our regularized CMax framework is $O(N_e + N_p)$, the same as that of the non-regularized one.

4.6. Application to Motion Segmentation

Although most of the results on standard datasets comprise stationary scenes, we have also provided results on a dynamic scene (from dataset [40]). Because the time spanned by each set of events processed is small, the scene motion is also small (even for complicated objects like the person in the bottom row of Figure 6), hence often a single warp fits the scene reasonably well. In some scenarios, a single warp may not be enough to fit the event data because there are distinctive motions in the scene of equal importance. Our proposed regularizers can be extended to such more complex scene motions. To this end, we demonstrate it with an example in Figure 8.

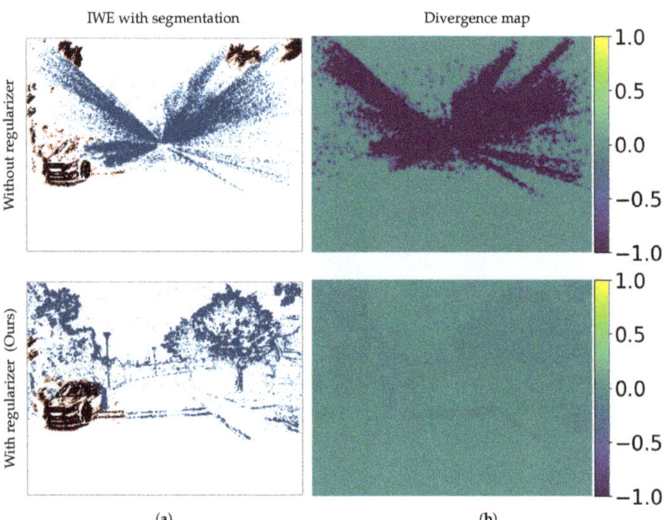

Figure 8. *Application to Motion Segmentation.* (**a**) Output IWE, whose colors (red and blue) represent different clusters of events (segmented according to motion). (**b**) Divergence map. The range of divergence values is larger in the presence of event collapse than in its absence. Our regularizer (divergence in this example) mitigates the event collapse for this complex motion, even with an independently moving object (IMO) in the scene.

Specifically, we use the MVSEC dataset, in a clip where the scene consists of two motions: the ego-motion (forward motion of the recording vehicle) and the motion of a car driving in the opposite direction in a nearby lane (an independently moving object—IMO). We model the scene by using the combination of two warps. Intuitively, the 1-DOF warp (6) describes the ego-motion, while the feature flow (2 DOF) describes the IMO. Then, we apply the contrast maximization approach (augmented with our regularizing terms) and the expectation-maximization scheme in [21] to segment the scene, to determine which events belong to each motion. The results in Figure 8 clearly show the effectiveness of our regularizer, even for such a commonplace and complex scene. Without regularizers, (i) event collapse appears in the ego-motion cluster of events and (ii) a considerable portion of the events that correspond to ego-motion are assigned to the second cluster (2-DOF warp), thus causing a segmentation failure. Our regularization approach mitigates event collapse (bottom row of Figure 8) and provides the correct segmentation: the 1-DOF warp fits the ego-motion and the feature flow (2-DOF warp) fits the IMO.

5. Conclusions

We have analyzed the event collapse phenomenon of the CMax framework and proposed collapse metrics using first principles of space-time deformation, inspired by differential geometry and physics. Our experimental results on publicly available datasets demonstrate that the proposed divergence and area-based metrics mitigate the phenomenon for collapse-enabled warps and do not harm well-posed warps. To the best of our knowledge, our regularizers are the only effective solution compared to the unregularized CMax framework and whitening. Our regularizers achieve, on average, more than 90% improvement on optical flow endpoint error calculation (AEE) on collapse-enabled warps.

This is the first work that focuses on the paramount phenomenon of event collapse. No prior work has analyzed this phenomenon in such detail or proposed new regularizers without additional data or reparameterizing the search space [14,16,27]. As we analyzed various warps from 1 DOF to 4 DOFs, we hope that the ideas presented here inspire further research to tackle more complex warp models. Our work shows how the divergence and area-based deformation can be computed for warps given by analytical formulas. For more complex warps, like those used in dense optical flow estimation [16,18], the divergence or area-based deformation could be approximated by using finite difference formulas.

Author Contributions: Conceptualization, S.S. and G.G.; methodology, G.G.; software, S.S.; validation, S.S.; formal analysis, S.S. and G.G.; investigation, S.S. and G.G.; resources, Y.A.; data curation, S.S.; writing—original draft preparation, S.S. and G.G.; writing—review and editing, S.S., Y.A. and G.G.; visualization, S.S. and G.G.; supervision, Y.A. and G.G.; project administration, S.S.; funding acquisition, S.S., Y.A. and G.G. All authors have read and agreed to the published version of the manuscript.

Funding: This research was funded by the German Academic Exchange Service (DAAD), Research Grant-Bi-nationally Supervised Doctoral Degrees/Cotutelle, 2021/22 (57552338). Ref. no.: 91803781.

Institutional Review Board Statement: Not applicable.

Informed Consent Statement: Not applicable.

Data Availability Statement: Data available in publicly accessible repositories. The data presented in this study are openly available in reference number [34,39,40].

Conflicts of Interest: The authors declare no conflict of interest.

Appendix A. Warp Models, Jacobians and Flow Divergence

Appendix A.1. Planar Motion — Euclidean Transformation on the Image Plane, SE(2)

If the point trajectories of an isometry are $\mathbf{x}(t)$, the warp is given by [27]

$$\begin{pmatrix} \mathbf{x}'_k \\ 1 \end{pmatrix} \sim \begin{pmatrix} \mathrm{R}(t_k\omega_Z) & t_k\mathbf{v} \\ \mathbf{0}^\top & 1 \end{pmatrix}^{-1} \begin{pmatrix} \mathbf{x}_k \\ 1 \end{pmatrix}, \tag{A1}$$

where $\mathbf{v}, \omega_Z$ comprise the 3 DOFs of a translation and an in-plane rotation. The in-plane rotation is

$$\mathrm{R}(\phi) = \begin{pmatrix} \cos\phi & -\sin\phi \\ \sin\phi & \cos\phi \end{pmatrix}. \tag{A2}$$

Since

$$\begin{pmatrix} \mathrm{A} & \mathbf{b} \\ \mathbf{0}^\top & 1 \end{pmatrix}^{-1} = \begin{pmatrix} \mathrm{A}^{-1} & -\mathrm{A}^{-1}\mathbf{b} \\ \mathbf{0}^\top & 1 \end{pmatrix} \tag{A3}$$

and $\mathrm{R}^{-1}(\phi) = \mathrm{R}(-\phi)$, we have

$$\begin{pmatrix} \mathbf{x}'_k \\ 1 \end{pmatrix} \sim \begin{pmatrix} \mathrm{R}(-t_k\omega_Z) & -\mathrm{R}(-t_k\omega_Z)(t_k\mathbf{v}) \\ \mathbf{0}^\top & 1 \end{pmatrix} \begin{pmatrix} \mathbf{x}_k \\ 1 \end{pmatrix}. \tag{A4}$$

Hence, in Euclidean coordinates the warp is

$$\mathbf{x}'_k = \mathrm{R}(-t_k\omega_Z)(\mathbf{x}_k - t_k\mathbf{v}). \tag{A5}$$

The Jacobian and its determinant are:

$$J_k = \frac{\partial \mathbf{x}'_k}{\partial \mathbf{x}_k} = \mathrm{R}(-t_k\omega_Z), \tag{A6}$$

$$\det(J_k) = 1. \tag{A7}$$

The flow corresponding to (A5) is:

$$\mathbf{f} = \frac{\partial \mathbf{x}'}{\partial t} = \mathrm{R}^\top\left(\frac{\pi}{2} + t\omega_Z\right)(\mathbf{x} - t\mathbf{v})\omega_Z - \mathrm{R}(-t\omega_Z)\mathbf{v}, \tag{A8}$$

whose divergence is

$$\nabla \cdot \mathbf{f} = -2\omega_Z \sin(t\omega_Z). \tag{A9}$$

Hence, for small angles $|t\omega_Z| \ll 1$, the divergence of the flow vanishes.

In short, this warp has the same determinant and approximate zero divergence as the 2-DOF feature flow warp (Section 3.5.1), which is well-behaved. Note, however, that the trajectories are not straight in space-time.

Appendix A.2. 3-DOF Camera Rotation, $SO(3)$

Using calibrated and homogeneous coordinates, the warp is given by [5,12]

$$\mathbf{x}_k^{h\prime} \sim R(t_k\boldsymbol{\omega})\, \mathbf{x}_k^h, \tag{A10}$$

where $\boldsymbol{\theta} = \boldsymbol{\omega} = (\omega_1, \omega_2, \omega_3)^\top$ is the angular velocity, and R (3×3 rotation matrix in space) is parametrized using exponential coordinates (Rodrigues rotation formula [35,36]).

By the chain rule, the Jacobian is:

$$J_k = \frac{\partial \mathbf{x}_k'}{\partial \mathbf{x}_k} = \frac{\partial \mathbf{x}_k'}{\partial \mathbf{x}_k^{h\prime}} \frac{\partial \mathbf{x}_k^{h\prime}}{\partial \mathbf{x}_k^h} \frac{\partial \mathbf{x}_k^h}{\partial \mathbf{x}_k} = \frac{1}{(\mathbf{x}_k^{h\prime})_3} \begin{pmatrix} 1 & 0 & -x_k' \\ 0 & 1 & -y_k' \end{pmatrix} R(t_k\boldsymbol{\omega}) \begin{pmatrix} 1 & 0 \\ 0 & 1 \\ 0 & 0 \end{pmatrix}. \tag{A11}$$

Letting $\mathbf{r}_{3,k}^\top$ be the third row of $R(t_k\boldsymbol{\omega})$, and using (32)–(34) in [38], gives

$$\det(J_k) = (\mathbf{r}_{3,k}^\top \mathbf{x}_k^h)^{-3}. \tag{A12}$$

Connection between Divergence and Deformation Maps

If the rotation angle $t_k\|\boldsymbol{\omega}\|$ is small, using the first two terms of the exponential map we approximate $R(t_k\boldsymbol{\omega}) \approx \mathrm{Id} + (t_k\boldsymbol{\omega})^\wedge$, where the hat operator $^\wedge$ in $SO(3)$ represents the cross product matrix [54]. Then, $\mathbf{r}_{3,k}^\top \mathbf{x}_k^h \approx (-t_k\omega_2, t_k\omega_1, 1)^\top (x_k, y_k, 1) = 1 + (y_k\omega_1 - x_k\omega_2)t_k$. Substituting this expression into (A12) and using the first two terms in Taylor's expansion around $z = 0$ of $(1+z)^{-3} \approx 1 - 3z + 6z^2$ (convergent for $|z| < 1$) gives $\det(J_k) \approx 1 + 3(x_k\omega_2 - y_k\omega_1)t_k$. Notably, the divergence (18) and the approximate amplification factor depend linearly on $3(x_k\omega_2 - y_k\omega_1)$. This resemblance is seen in the divergence and deformation maps of the bottom rows in Figure 6 (ECD dataset).

Appendix A.3. 4-DOF In-Plane Camera Motion Approximation

The warp presented in [20],

$$\mathbf{x}_k' = \mathbf{x}_k - t_k \left(\mathbf{v} + (h_z + 1)R(\phi)\mathbf{x}_k - \mathbf{x}_k \right) \tag{A13}$$

has 4 DOFs: $\boldsymbol{\theta} = (\mathbf{v}, \phi, h_z)^\top$. The Jacobian and its determinant are:

$$J_k = \frac{\partial \mathbf{x}_k'}{\partial \mathbf{x}_k} = (1 + t_k)\mathrm{Id} - (h_z + 1)t_k R(\phi), \tag{A14}$$

$$\det(J_k) = (1 + t_k)^2 - 2(1 + t_k)t_k(h_z + 1)\cos\phi + t_k^2(h_z + 1)^2. \tag{A15}$$

The flow corresponding to (A13) is given by

$$\mathbf{f} = \frac{\partial \mathbf{x}'}{\partial t} = -\left(\mathbf{v} + (h_z + 1)R(\phi)\mathbf{x} - \mathbf{x} \right), \tag{A16}$$

whose divergence is:

$$\nabla \cdot \mathbf{f} = -(h_z + 1)\nabla \cdot \left(R(\phi)\mathbf{x} \right) + \nabla \cdot \mathbf{x} \tag{A17}$$

$$= 2 - 2(h_z + 1)\cos(\phi). \tag{A18}$$

As particular cases of this warp, one can identify:

- 1-DOF Zoom in/out ($\mathbf{v} = \mathbf{0}, \phi = 0$). $\mathbf{x}'_k = (1 - t_k h_z)\mathbf{x}_k$.
- 2-DOF translation ($\phi = 0, h_z = 0$). $\mathbf{x}'_k = \mathbf{x}_k - t_k \mathbf{v}$.
- 1-DOF "rotation" ($\mathbf{v} = \mathbf{0}, h_z = 0$). $\mathbf{x}'_k = \mathbf{x}_k - t_k \left(\mathrm{R}(\phi)\mathbf{x}_k - \mathbf{x}_k\right)$.
  Using a couple of approximations of the exponential map in $SO(2)$, we obtain

$$
\begin{aligned}
\mathbf{x}'_k &= \mathbf{x}_k - t_k \left(\mathrm{R}(\phi) - \mathrm{Id}\right)\mathbf{x}_k & & \text{(A19)} \\
&\approx \mathbf{x}_k - t_k \phi^{\wedge}\mathbf{x}_k & \text{if } \phi \text{ is small} & \text{(A20)} \\
&= \left(\mathrm{Id} + (-t_k\phi)^{\wedge}\right)\mathbf{x}_k & & \text{(A21)} \\
&\approx \mathrm{R}(-t_k\phi)\mathbf{x}_k & \text{if } t_k\phi \text{ is small.} & \text{(A22)}
\end{aligned}
$$

Hence, ϕ plays the role of a small angular velocity ω_Z around the camera's optical axis Z, i.e., in-plane rotation.

- 3-DOF planar motion ("isometry") ($h_z = 0$). Using the previous result, the warp splits into translational and rotational components:

$$
\mathbf{x}'_k = \mathbf{x}_k - t_k \left(\mathbf{v} + \mathrm{R}(\phi)\mathbf{x}_k - \mathbf{x}_k\right) \tag{A23}
$$

$$
\overset{\text{(A22)}}{\approx} -t_k\mathbf{v} + \mathrm{R}(-t_k\phi)\mathbf{x}_k. \tag{A24}
$$

Appendix A.4. 4-DOF Similarity Transformation on the Image Plane, Sim(2)

Another 4-DOF warp is proposed in [27]. Its DOFs are the linear, angular and scaling velocities on the image plane: $\boldsymbol{\theta} = (\mathbf{v}, \omega_Z, s)^{\top}$.

Letting $\beta_k = 1 + t_k s$, the warp is:

$$
\begin{pmatrix} \mathbf{x}'_k \\ 1 \end{pmatrix} \sim \begin{pmatrix} \beta_k \mathrm{R}(t_k\omega_Z) & t_k\mathbf{v} \\ \mathbf{0}^{\top} & 1 \end{pmatrix}^{-1} \begin{pmatrix} \mathbf{x}_k \\ 1 \end{pmatrix}. \tag{A25}
$$

Using (A3) gives

$$
\begin{pmatrix} \mathbf{x}'_k \\ 1 \end{pmatrix} \sim \begin{pmatrix} \beta_k^{-1}\mathrm{R}(-t_k\omega_Z) & -\beta_k^{-1}\mathrm{R}(-t_k\omega_Z)(t_k\mathbf{v}) \\ \mathbf{0}^{\top} & 1 \end{pmatrix} \begin{pmatrix} \mathbf{x}_k \\ 1 \end{pmatrix}. \tag{A26}
$$

Hence, in Euclidean coordinates the warp is

$$
\mathbf{x}'_k = \beta_k^{-1}\mathrm{R}(-t_k\omega_Z)(\mathbf{x}_k - t_k\mathbf{v}). \tag{A27}
$$

The Jacobian and its determinant are:

$$
\mathrm{J}_k = \frac{\partial \mathbf{x}'_k}{\partial \mathbf{x}_k} = \beta_k^{-1}\mathrm{R}(-t_k\omega_Z), \tag{A28}
$$

$$
\det(\mathrm{J}_k) = \beta_k^{-2} = \frac{1}{(1 + t_k s)^2}. \tag{A29}
$$

The following result will be useful to simplify equations. For a 2D rotation $\mathrm{R}(\phi(t))$, it holds that:

$$
\frac{\partial \mathrm{R}(\phi(t))}{\partial t} = -\mathrm{R}^{\top}\left(\frac{\pi}{2} - \phi\right)\frac{\partial \phi}{\partial t}. \tag{A30}
$$

To compute the flow of (A27), there are three time-dependent factors. Hence, applying the product rule we obtain three terms, and substituting (A30) (with $\phi = -t\omega_Z$) gives:

$$
\mathbf{f}_k = \left(\frac{\partial \beta_k^{-1}}{\partial t_k}\mathrm{R}(-t_k\omega_Z) + \beta_k^{-1}\omega_Z\mathrm{R}^{\top}\left(\frac{\pi}{2} + t_k\omega_Z\right)\right)(\mathbf{x}_k - t_k\mathbf{v}) - \beta_k^{-1}\mathrm{R}(-t_k\omega_Z)\mathbf{v}, \tag{A31}
$$

where, by the chain rule,

$$\frac{\partial \beta_k^{-1}}{\partial t_k} = -\beta_k^{-2}\frac{\partial \beta_k}{\partial t_k} = -\beta_k^{-2}s = -\frac{s}{(1+t_k s)^2}. \tag{A32}$$

Hence, the divergence of the flow is:

$$\nabla \cdot \mathbf{f}_k = \frac{\partial \beta_k^{-1}}{\partial t_k}\nabla \cdot \left(\mathtt{R}(-t_k\omega_Z)\mathbf{x}_k\right) + \beta_k^{-1}\omega_Z\nabla \cdot \left(\mathtt{R}^\top\left(\frac{\pi}{2}+t_k\omega_Z\right)\mathbf{x}_k\right) \tag{A33}$$

$$= \frac{\partial \beta_k^{-1}}{\partial t_k}2\cos(t_k\omega_Z) + \beta_k^{-1}\omega_Z 2\sin(-t_k\omega_Z) \tag{A34}$$

The formulas for $SE(2)$ are obtained from the above ones with $s = 0$ (i.e., $\beta_k = 1$).

References

1. Delbruck, T. Frame-free dynamic digital vision. In Proceedings of the International Symposium on Secure-Life Electronics, Tokyo, Japan, 6–7 March 2008; pp. 21–26. [CrossRef]
2. Suh, Y.; Choi, S.; Ito, M.; Kim, J.; Lee, Y.; Seo, J.; Jung, H.; Yeo, D.H.; Namgung, S.; Bong, J.; et al. A 1280x960 Dynamic Vision Sensor with a 4.95-μm Pixel Pitch and Motion Artifact Minimization. In Proceedings of the IEEE International Symposium on Circuits and Systems (ISCAS), Seville, Spain, 12–14 October 2020. [CrossRef]
3. Finateu, T.; Niwa, A.; Matolin, D.; Tsuchimoto, K.; Mascheroni, A.; Reynaud, E.; Mostafalu, P.; Brady, F.; Chotard, L.; LeGoff, F.; et al. A 1280x720 Back-Illuminated Stacked Temporal Contrast Event-Based Vision Sensor with 4.86 μm Pixels, 1.066GEPS Readout, Programmable Event-Rate Controller and Compressive Data-Formatting Pipeline. In Proceedings of the IEEE International Solid- State Circuits Conference-(ISSCC), San Francisco, CA, USA, 16–20 February 2020; pp. 112–114. [CrossRef]
4. Gallego, G.; Delbruck, T.; Orchard, G.; Bartolozzi, C.; Taba, B.; Censi, A.; Leutenegger, S.; Davison, A.; Conradt, J.; Daniilidis, K.; et al. Event-based Vision: A Survey. *IEEE Trans. Pattern Anal. Mach. Intell.* **2020**, *44*, 154–180. [CrossRef] [PubMed]
5. Gallego, G.; Scaramuzza, D. Accurate Angular Velocity Estimation with an Event Camera. *IEEE Robot. Autom. Lett.* **2017**, *2*, 632–639. [CrossRef]
6. Kim, H.; Kim, H.J. Real-Time Rotational Motion Estimation with Contrast Maximization Over Globally Aligned Events. *IEEE Robot. Autom. Lett.* **2021**, *6*, 6016–6023. [CrossRef]
7. Zhu, A.Z.; Atanasov, N.; Daniilidis, K. Event-Based Feature Tracking with Probabilistic Data Association. In Proceedings of the IEEE International Conference on Robotics and Automation (ICRA), Singapore, 29 May–3 June 2017; pp. 4465–4470. [CrossRef]
8. Zhu, A.Z.; Atanasov, N.; Daniilidis, K. Event-based Visual Inertial Odometry. In Proceedings of the IEEE Conference on Computer Vision and Pattern Recognition (CVPR), Honolulu, HI, USA, 21–26 July 2017; pp. 5816–5824. [CrossRef]
9. Seok, H.; Lim, J. Robust Feature Tracking in DVS Event Stream using Bezier Mapping. In Proceedings of the IEEE Winter Conference on Applications of Computer Vision (WACV), Snowmass, CO, USA, 1–5 March 2020; pp. 1647–1656. [CrossRef]
10. Stoffregen, T.; Kleeman, L. Event Cameras, Contrast Maximization and Reward Functions: An Analysis. In Proceedings of the IEEE/CVF Conference on Computer Vision and Pattern Recognition (CVPR), Long Beach, CA, USA, 15–20 June 2019; pp. 12292–12300. [CrossRef]
11. Dardelet, L.; Benosman, R.; Ieng, S.H. An Event-by-Event Feature Detection and Tracking Invariant to Motion Direction and Velocity. *TechRxiv* **2021**. [CrossRef]
12. Gallego, G.; Rebecq, H.; Scaramuzza, D. A Unifying Contrast Maximization Framework for Event Cameras, with Applications to Motion, Depth, and Optical Flow Estimation. In Proceedings of the IEEE/CVF Conference on Computer Vision and Pattern Recognition, Salt Lake City, UT, USA, 18–23 June 2018; pp. 3867–3876. [CrossRef]
13. Gallego, G.; Gehrig, M.; Scaramuzza, D. Focus Is All You Need: Loss Functions For Event-based Vision. In Proceedings of the 2019 IEEE/CVF Conference on Computer Vision and Pattern Recognition (CVPR), Long Beach, CA, USA, 15–20 June 2019; pp. 12272–12281. [CrossRef]
14. Peng, X.; Gao, L.; Wang, Y.; Kneip, L. Globally-Optimal Contrast Maximisation for Event Cameras. *IEEE Trans. Pattern Anal. Mach. Intell.* **2021**, *44*, 3479–3495. [CrossRef]
15. Rebecq, H.; Gallego, G.; Mueggler, E.; Scaramuzza, D. EMVS: Event-based Multi-View Stereo—3D Reconstruction with an Event Camera in Real-Time. *Int. J. Comput. Vis.* **2018**, *126*, 1394–1414. [CrossRef]
16. Zhu, A.Z.; Yuan, L.; Chaney, K.; Daniilidis, K. Unsupervised Event-based Learning of Optical Flow, Depth, and Egomotion. In Proceedings of the IEEE/CVF Conference on Computer Vision and Pattern Recognition (CVPR), Long Beach, CA, USA, 15–20 June 2019; pp. 989–997. [CrossRef]
17. Paredes-Valles, F.; Scheper, K.Y.W.; de Croon, G.C.H.E. Unsupervised Learning of a Hierarchical Spiking Neural Network for Optical Flow Estimation: From Events to Global Motion Perception. *IEEE Trans. Pattern Anal. Mach. Intell.* **2019**, *42*, 2051–2064. [CrossRef]

18. Hagenaars, J.J.; Paredes-Valles, F.; de Croon, G.C.H.E. Self-Supervised Learning of Event-Based Optical Flow with Spiking Neural Networks. In Proceedings of the Advances in Neural Information Processing Systems (NeurIPS), Virtual-only Conference, 7–10 December 2021; Volume 34, pp. 7167–7179.
19. Shiba, S.; Aoki, Y.; Gallego, G. Secrets of Event-based Optical Flow. In Proceedings of the European Conference on Computer Vision (ECCV), Tel-Aviv, Israel, 23–27 October 2022.
20. Mitrokhin, A.; Fermuller, C.; Parameshwara, C.; Aloimonos, Y. Event-based Moving Object Detection and Tracking. In Proceedings of the IEEE/RSJ International Conference on Intelligent Robots and Systems (IROS), Madrid, Spain, 1–5 October 2018; pp. 1–9. [CrossRef]
21. Stoffregen, T.; Gallego, G.; Drummond, T.; Kleeman, L.; Scaramuzza, D. Event-Based Motion Segmentation by Motion Compensation. In Proceedings of the IEEE/CVF International Conference on Computer Vision (ICCV), Seoul, Korea, 27 October–2 November 2019; pp. 7243–7252. [CrossRef]
22. Zhou, Y.; Gallego, G.; Lu, X.; Liu, S.; Shen, S. Event-based Motion Segmentation with Spatio-Temporal Graph Cuts. *IEEE Trans. Neural Netw. Learn. Syst.* **2021**, 1–13. [CrossRef]
23. Parameshwara, C.M.; Sanket, N.J.; Singh, C.D.; Fermüller, C.; Aloimonos, Y. 0-MMS: Zero-shot multi-motion segmentation with a monocular event camera. In Proceedings of the IEEE International Conference on Robotics and Automation (ICRA), Xi'an, China, 30 May–5 June 2021. [CrossRef]
24. Lu, X.; Zhou, Y.; Shen, S. Event-based Motion Segmentation by Cascaded Two-Level Multi-Model Fitting. In Proceedings of the IEEE/RSJ International Conference on Intelligent Robots and Systems (IROS), Prague, Czech Republic, 27 September–1 October 2021; pp. 4445–4452. [CrossRef]
25. Duan, P.; Wang, Z.; Shi, B.; Cossairt, O.; Huang, T.; Katsaggelos, A. Guided Event Filtering: Synergy between Intensity Images and Neuromorphic Events for High Performance Imaging. *IEEE Trans. Pattern Anal. Mach. Intell.* **2021**. [CrossRef]
26. Zhang, Z.; Yezzi, A.; Gallego, G. Image Reconstruction from Events. Why learn it? *arXiv* **2021**, arXiv:2112.06242.
27. Nunes, U.M.; Demiris, Y. Robust Event-based Vision Model Estimation by Dispersion Minimisation. *IEEE Trans. Pattern Anal. Mach. Intell.* **2021**. [CrossRef] [PubMed]
28. Gu, C.; Learned-Miller, E.; Sheldon, D.; Gallego, G.; Bideau, P. The Spatio-Temporal Poisson Point Process: A Simple Model for the Alignment of Event Camera Data. In Proceedings of the IEEE/CVF International Conference on Computer Vision (ICCV), Montreal, QC, Canada, 10–17 October 2021; pp. 13495–13504. [CrossRef]
29. Liu, D.; Parra, A.; Chin, T.J. Globally Optimal Contrast Maximisation for Event-Based Motion Estimation. In Proceedings of the IEEE/CVF Conference on Computer Vision and Pattern Recognition (CVPR), Seattle, WA, USA, 13–19 June 2020; pp. 6348–6357. [CrossRef]
30. Stoffregen, T.; Kleeman, L. Simultaneous Optical Flow and Segmentation (SOFAS) using Dynamic Vision Sensor. In Proceedings of the Australasian Conference on Robotics and Automation (ACRA), Sydney, Australia, 11–13 December 2017.
31. Ozawa, T.; Sekikawa, Y.; Saito, H. Accuracy and Speed Improvement of Event Camera Motion Estimation Using a Bird's-Eye View Transformation. *Sensors* **2022**, *22*, 773. [CrossRef]
32. Lichtsteiner, P.; Posch, C.; Delbruck, T. A 128 × 128 120 dB 15 μs latency asynchronous temporal contrast vision sensor. *IEEE J. Solid-State Circuits* **2008**, *43*, 566–576. [CrossRef]
33. Ng, M.; Er, Z.M.; Soh, G.S.; Foong, S. Aggregation Functions For Simultaneous Attitude And Image Estimation with Event Cameras At High Angular Rates. *IEEE Robot. Autom. Lett.* **2022**, *7*, 4384–4391. [CrossRef]
34. Zhu, A.Z.; Thakur, D.; Ozaslan, T.; Pfrommer, B.; Kumar, V.; Daniilidis, K. The Multivehicle Stereo Event Camera Dataset: An Event Camera Dataset for 3D Perception. *IEEE Robot. Autom. Lett.* **2018**, *3*, 2032–2039. [CrossRef]
35. Murray, R.M.; Li, Z.; Sastry, S. *A Mathematical Introduction to Robotic Manipulation*; CRC Press: Boca Raton, FL, USA, 1994.
36. Gallego, G.; Yezzi, A. A Compact Formula for the Derivative of a 3-D Rotation in Exponential Coordinates. *J. Math. Imaging Vis.* **2014**, *51*, 378–384. [CrossRef]
37. Corke, P. *Robotics, Vision and Control: Fundamental Algorithms in MATLAB*; Springer Tracts in Advanced Robotics; Springer: Berlin/Heidelberg, Germany, 2017. [CrossRef]
38. Gallego, G.; Yezzi, A.; Fedele, F.; Benetazzo, A. A Variational Stereo Method for the Three-Dimensional Reconstruction of Ocean Waves. *IEEE Trans. Geosci. Remote Sens.* **2011**, *49*, 4445–4457. [CrossRef]
39. Gehrig, M.; Aarents, W.; Gehrig, D.; Scaramuzza, D. DSEC: A Stereo Event Camera Dataset for Driving Scenarios. *IEEE Robot. Autom. Lett.* **2021**, *6*, 4947–4954. [CrossRef]
40. Mueggler, E.; Rebecq, H.; Gallego, G.; Delbruck, T.; Scaramuzza, D. The Event-Camera Dataset and Simulator: Event-based Data for Pose Estimation, Visual Odometry, and SLAM. *Int. J. Robot. Res.* **2017**, *36*, 142–149. [CrossRef]
41. Gehrig, M.; Millhäusler, M.; Gehrig, D.; Scaramuzza, D. E-RAFT: Dense Optical Flow from Event Cameras. In Proceedings of the International Conference on 3D Vision (3DV), London, UK, 1–3 December 2021. [CrossRef]
42. Nagata, J.; Sekikawa, Y.; Aoki, Y. Optical Flow Estimation by Matching Time Surface with Event-Based Cameras. *Sensors* **2021**, *21*, 1150. [CrossRef]
43. Taverni, G.; Moeys, D.P.; Li, C.; Cavaco, C.; Motsnyi, V.; Bello, D.S.S.; Delbruck, T. Front and Back Illuminated Dynamic and Active Pixel Vision Sensors Comparison. *IEEE Trans. Circuits Syst. II* **2018**, *65*, 677–681. [CrossRef]
44. Zhu, A.Z.; Yuan, L.; Chaney, K.; Daniilidis, K. EV-FlowNet: Self-Supervised Optical Flow Estimation for Event-based Cameras. In Proceedings of the Robotics: Science and Systems (RSS), Pittsburgh, PA, USA, 26–30 June 2018. [CrossRef]

45. Rosinol Vidal, A.; Rebecq, H.; Horstschaefer, T.; Scaramuzza, D. Ultimate SLAM? Combining Events, Images, and IMU for Robust Visual SLAM in HDR and High Speed Scenarios. *IEEE Robot. Autom. Lett.* **2018**, *3*, 994–1001. [CrossRef]
46. Rebecq, H.; Horstschäfer, T.; Gallego, G.; Scaramuzza, D. EVO: A Geometric Approach to Event-based 6-DOF Parallel Tracking and Mapping in Real-Time. *IEEE Robot. Autom. Lett.* **2017**, *2*, 593–600. [CrossRef]
47. Mueggler, E.; Gallego, G.; Rebecq, H.; Scaramuzza, D. Continuous-Time Visual-Inertial Odometry for Event Cameras. *IEEE Trans. Robot.* **2018**, *34*, 1425–1440. [CrossRef]
48. Zhou, Y.; Gallego, G.; Shen, S. Event-based Stereo Visual Odometry. *IEEE Trans. Robot.* **2021**, *37*, 1433–1450. [CrossRef]
49. Brandli, C.; Berner, R.; Yang, M.; Liu, S.C.; Delbruck, T. A 240×180 130dB $3\mu s$ Latency Global Shutter Spatiotemporal Vision Sensor. *IEEE J. Solid-State Circuits* **2014**, *49*, 2333–2341. [CrossRef]
50. Stoffregen, T.; Scheerlinck, C.; Scaramuzza, D.; Drummond, T.; Barnes, N.; Kleeman, L.; Mahony, R. Reducing the Sim-to-Real Gap for Event Cameras. In Proceedings of the European Conference on Computer Vision (ECCV), Glasgow, UK, 23–28 August 2020.
51. Bergstra, J.; Bardenet, R.; Bengio, Y.; Kégl, B. Algorithms for Hyper-Parameter Optimization. In Proceedings of the Advances in Neural Information Processing Systems (NeurIPS), Granada, Spain, 12–15 December 2011; Volume 24.
52. Geiger, A.; Lenz, P.; Stiller, C.; Urtasun, R. Vision meets robotics: The KITTI dataset. *Int. J. Robot. Res.* **2013**, *32*, 1231–1237. [CrossRef]
53. Kingma, D.P.; Ba, J.L. Adam: A Method for Stochastic Optimization. In Proceedings of the International Conference on Learning Representations (ICLR), San Diego, CA, USA, 7–9 May 2015.
54. Barfoot, T.D. *State Estimation for Robotics—A Matrix Lie Group Approach*; Cambridge University Press: Cambridge, UK, 2015.

Article

Energy Efficient SWIPT Based Mobile Edge Computing Framework for WSN-Assisted IoT

Fangni Chen [1,2]**, Anding Wang** [3]**, Yu Zhang** [1]**, Zhengwei Ni** [3] **and Jingyu Hua** [3,]*****

[1] College of Information Engineering, Zhejiang University of Technology, Hangzhou 310012, China; chenfangni@zust.edu.cn (F.C.); yzhang@zjut.edu.cn (Y.Z.)

[2] School of Information and Electronic Engineering, Zhejiang University of Science and Technology, Hangzhou 310012, China

[3] School of Information and Electronic Engineering, Zhejiang Gongshang University, Hangzhou 310018, China; anding_704@zjgsu.edu.cn (A.W.); zhengwei.ni@zjgsu.edu.cn (Z.N.)

* Correspondence: eehjy@zjut.edu.cn

Abstract: With the increasing deployment of IoT devices and applications, a large number of devices that can sense and monitor the environment in IoT network are needed. This trend also brings great challenges, such as data explosion and energy insufficiency. This paper proposes a system that integrates mobile edge computing (MEC) technology and simultaneous wireless information and power transfer (SWIPT) technology to improve the service supply capability of WSN-assisted IoT applications. A novel optimization problem is formulated to minimize the total system energy consumption under the constraints of data transmission rate and transmitting power requirements by jointly considering power allocation, CPU frequency, offloading weight factor and energy harvest weight factor. Since the problem is non-convex, we propose a novel alternate group iteration optimization (AGIO) algorithm, which decomposes the original problem into three subproblems, and alternately optimizes each subproblem using the group interior point iterative algorithm. Numerical simulations validate that the energy consumption of our proposed design is much lower than the two benchmark algorithms. The relationship between system variables and energy consumption of the system is also discussed.

Keywords: mobile edge computaing; simultaneous wireless information and power transfer; energy minimization; 5G; wireless sensing network; IoT

Citation: Chen, F.; Wang, A.; Zhang, Y.; Ni, Z.; Hua, J. Energy Efficient SWIPT Based Mobile Edge Computing Framework for WSN-Assisted IoT. *Sensors* **2021**, *21*, 4798. https://doi.org/10.3390/s21144798

Academic Editor: Jing Tian

Received: 26 June 2021
Accepted: 12 July 2021
Published: 14 July 2021

Publisher's Note: MDPI stays neutral with regard to jurisdictional claims in published maps and institutional affiliations.

1. Introduction

1.1. Backgroud

The 5G enabled Internet of Things (5G-IoT) [1–4] connects the real world with the internet world and human civilization is currently transforming from informatization to intelligence. The four abilities of 5G communication system, namely massive capacity, ultra-low latency, high reliability and extensive connection, are the key driving forces for the development of IoT [5,6]. The integration of 5G and WSN-assisted IoT not only strengthens the connection between the real world and the internet world, but also widens the scope of IoT services such that IoT can not only serve the smart city [7] but also penetrate into agriculture [8], medical care [9], transportation [10], industry [11] and other fields [12]. However, WSN-assisted IoT is facing grand challenges and the huge amount data traffic brought by great number of IoT devices and sensors can impose an enormous burden on the network, resulting in higher service delays and reduced quality of service (QoS) [13].

Although the current terminal devices are equipped with high-performance hardware, it is still difficult to meet the needs of computing intensive tasks, especially in the case of ensuring low power consumption and low latency. Mobile Edge Computing (MEC) technology is considered as a crucial solution for 5G-IoT [14,15]. With the help of MEC, terminal devices can upload part of or all of the computing tasks to the edge computing platform for computing so as to reduce their own computing pressure and energy consumption, improve the computing efficiency and performance and bring better QoS [16–18]. The authors of [19] studied how the MEC enabled industrial verticals in 5G. Yang et al. [20] analyzed the main features of MEC in the context of 5G and IoT and presented several fundamental key technologies that enable MEC to be applied in 5G and IoT.

Despite the fact that MEC brings great benefits for IoT, it still faces an important problem: how can we effectively and conveniently extend the lifetime of IoT devices in the network? Simultaneous wireless information and power transfer (SWIPT) is considered as the key technology to solve this problem. The principle of SWIPT system is that RF signals can carry energy and information at the same time. It was first proposed by the author of [21] in 2008. Rui Zhang et.al. [22] proposed two practical SWIPT receivers, which are the time switching (TS) receiver and power splitting (PS) receiver. The TS receiver divides the time slot into two parts, the RF signal received in the first part of time is used for information demodulation and then the signal collected in the remaining time is used for energy harvesting. The PS receiver divides the received RF signal into two parts and then transmits them to the energy collector and information demodulator, respectively, so that the information demodulation and energy harvesting can be realized simultaneously. Since then, a lot of researches focus on the design and performance of the SWIPT systems [23,24]. The author of [25] proposes a more complex dynamic energy splitting receiver based on the above two receivers framework. The author of [26] studied the trade off between the harvested energy and forward data in SWIPT sensor networks. Tang et al. [27] proposed a TS receiver design to maximize the energy efficiency in MIMO channels for IoT.

1.2. Related Works

As both SWIPT and MEC technologies benefit the IoT system, MEC deployed WSN IoT network design with SWIPT has attracted increasing attention. In such new framework, the communication and computation resource allocation as well as wireless energy harvesting scheme are crucial for maximizing the system performance. The authors of [8] studied an energy efficiency optimization scheme for OFDM transmission WSN in smart agriculture. By jointly optimizing the power allocation and the pairing of subcarriers, the optimization scheme can help to solve the problem of energy deficiency. An achievable rate maximization problem was discussed in [28] for multiuser satellite IoT system with SWIPT and MEC to overcome the limitation in battery capacity and computing capability of IoT terminals. In [29], a UAV-enabled wireless powered MEC system was investigated, where the offloading modes were optimized to reach the maximum computation rate under the power constraint and the UAV speed constraint. The authors of [30] extended multi-access edge computing to support the long range (LoRa) system for IoT applications. The novel framework allowed dynamic IoT deployment at the edge and life cycle management.

With the development of artificial intelligence technology, reinforcement learning (RL) methods are used to solve various communication problems in 5G and IoT systems. Aiming at minimizing the difference between the distributed and demanded throughput for each user, ref. [31] presented a novel deep reinforcement learning (DRL) scheme, which satisfied the user requirements by power regulation. In [32], a RL based offloading scheme was studied to select the edge device and the offloading rate for IoT devices. The distinguished merit of this scheme is that the offloading policy can be optimized without knowledge required in traditional schemes. A hybrid-decision-based DRL approach is proposed in [33] to provide coordinated decisions of dynamic offloading scheme for multi-device multi-server MEC-IoT systems with energy harvesting devices.

1.3. Contributions

In this paper, we investigate energy consumption minimization for SWIPT based mobile edge computing in WSN assisted IoT System by using the optimization process. It is a continuation of our previous work [34] that focused on the cellular system. However, this study focus on the WSN assisted IoT network. Different from existing works, an in-depth research is carried to analyze the effects of computation task size, mobile node (including wireless sensor node) number, antenna number and energy harvest weight factor on the system energy consumption. The novelties of this work are summarized as follows.

(1) We design a novel WSN assisted IoT System, which integrates a MEC-deployed and FD-deployed anchor node (AN) and multiple SWIPT-equipped mobile nodes (MNs). The research problem of modeling is completely different from our previous work since the transmission conditions and requirements between cellular communication system and wireless sensor network are different. We aim to achieve the minimum energy consumption by optimizing CPU frequency, power allocation, offloading weight factor and SWIPT weight factor. Moreover, given the analysis of uplink, we provide the closed form expression of downlink rate and obtain the downlink transmission delay. Moreover, we optimize the SWIPT weight factor, which will affect the energy consumption of downlink harvesting and then affect the total system energy consumption. A more reasonable expression of harvesting energy is provided, which is based on the SWIPT weight factor and downlink time delay. In other words, the uplink and downlink parameters are jointly optimized.

(2) We formulate a more practical WSN energy minimization problem by jointly optimizing the key decision variables in the system. Since the multiple variables to be optimized are coupled and the original problem is non-convex, the optimization is quite challenging. An efficient algorithm called alternate group iteration optimization (AGIO) is proposed. We decompose the decision variables into three groups and divide the original problem into three subproblems. Then, we alternately optimize each subproblem using the interior point iteration method until the convergence.

The rest of the paper is organized as follows. Section 2 describes the system model and analyzes the transmission process. Section 3 formulates the energy minimization problem. Sections 4 and 5 present the algorithms to solve the problem and provides simulation results. Section 6 concludes the paper.

2. System Model

Let us consider a SWIPT-MEC enabled WSN assisted IoT system as demonstrated in Figure 1. There are N mobile nodes (MNs) including wireless sensor nodes denoted as $\{D_1, D_2, \ldots, D_N\}$ that are overwhelmed with computation tasks and one M-antenna Full Duplex enabled anchor node (AN). Each MN deploys single antenna and a power splitting (PS) SWIPT equipment to harvest energy. The PS receiver is capable of switching between energy harvesting (EH) state and information decoding (ID) state. The anchor node is equipped with a MEC server that can help MNs with the enormous amount of computation tasks.

Assume that each MN can divide its computation task into two parts and one is for local computing and the other is for offloading to the MEC enabled AN. The total computation task size of D_i is represented as L_i bits and the offloading computation task size is L_i^u bits, which satisfies $L_i^u = \alpha_i L_i$, where $0 \leq \alpha_i \leq 1$ is a offloading weight factor. Since MN can decide how much computation task will be offloaded to AN, α_i is a variable to be optimized to achieve better performance.

The operation processes of the system can be illustrated in the following steps.

(1) During the uplink process, MN D_i ($i \in \{1, \ldots, N\}$) offloads the computation task L_i^u to the MEC server at AN.

(2) When the MEC server receives the offloading task, it immediately implements the computation task. Due to the strong computation ability, MEC server can finish the offloading computation task in a short time, which can be ignored compared to the other operation times.

(3) Since AN deploys FD technology when MN D_i is offloading, AN can simultaneously download the computation result L_j^d from the MEC server to MN D_j ($j \in \{1, \ldots, N\}$, $j \neq i$), which shares the same frequency as the uplink MN D_i. The computation result satisfies $L_j^d = \beta_j L_i^u$, where $0 \leq \beta_j \leq 1$ is a weight factor.

(4) After D_j receives the computation result, the PS receiver will perform energy harvesting and information decoding according to the received RF signal.

(5) Mobile node D_i will perform local computing on the remaining computaion task when it finishes the offloading.

The above processes can be divided into three phases: the offloading phase, downloading phase and local computing phase. They are described in detail in the following analysis.

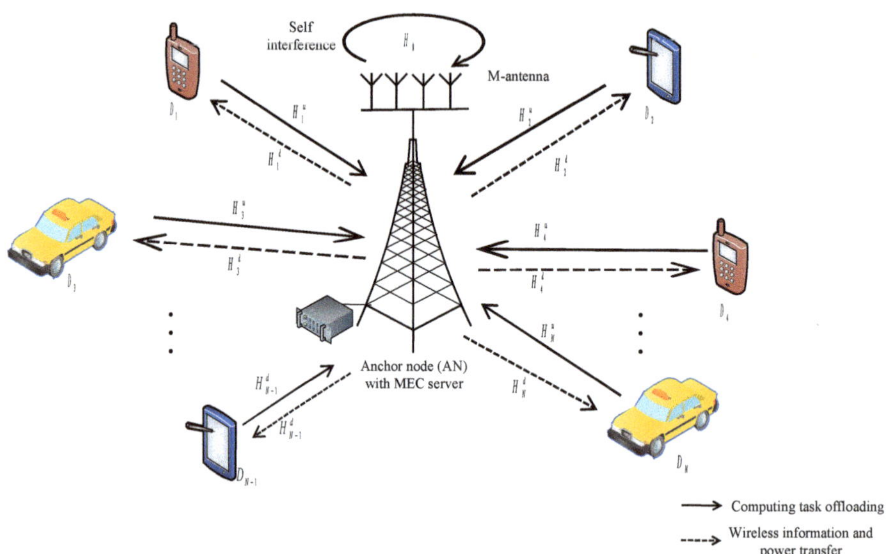

Figure 1. SWIPT-MEC enabled WSN assisted IoT System Model.

2.1. Offloading Phase

In the offloading phase, MNs $\{D_1, D_2, \ldots, D_N\}$ that are overwhelmed by the computation tasks will offload part of the tasks to AN. Without loss of generality, AN receives the offloading computation task L_i^u from MN D_i during the time interval t_i^u. Meanwhile, AN simultaneously downloads computation result L_j^d to D_j, where $i \in \{1, \ldots, N\}$, $j \in \{1, \ldots, N\}, i \neq j$. Thus, the received signal at AN is the following:

$$\mathbf{y}_i^u = \sqrt{p_i^u}\mathbf{H}_i^u s_i^u + \sqrt{\eta_j}\mathbf{H}_0(\sqrt{p_j^d}s_j^d) + \mathbf{n_{AN}}, \tag{1}$$

where p_i^u and s_i^u are the transmitted power and transmitted signal of D_i, p_j^d and s_j^d are the transmitted power and the computation result signal transmitted from AN to MN D_j. The two transmitted signals are assumed with normalized power, i.e., $|s_i^u|^2 = 1$ and $|s_j^d|^2 = 1$. $\mathbf{H}_i^u \in \mathbb{C}^{M \times 1}$ is the uplink channel from D_i to AN and $\mathbf{H}_0 \in \mathbb{C}^{M \times M}$ is the self-interference channel induced by FD transmission. η_j is the residual self-interference (RSI) coefficient. The received noise is $\mathbf{n_{AN}} \sim CN(0, \sigma_{AN}^2 \mathbf{I}_M)$.

According to the above expression, only the first part on the right side of Equation (1) contains the target offloading task s_i^u, the seond part is the RSI due to FD transmission of AN and the third part is the additive white Gaussian noise (AWGN) at AN. Thus, the signal to interference plus noise ratio (SINR) can be represented as the following:

$$\gamma_i^u = \frac{p_i^u tr\{\mathbf{H}_i^u(\mathbf{H}_i^u)^H\}}{\eta p_j^d tr\{\mathbf{H}_0\mathbf{H}_0^H\} + \sigma_{AN}^2}, \tag{2}$$

where $tr\{.\}$ represents the matrix trace. Then, we can obtain the transmission rate as the following:

$$R_i^u = Blog_2(1 + \gamma_i^u),\tag{3}$$

where B denotes the bandwidth allocated. Let R_{min}^u represent the minimum uplink transmission rate requirement. Therefore, we can obtain the first constraint condition in this system model shown as the following:

$$R_i^u \geq R_{min}^u.\tag{4}$$

Meanwhile, we can calculate the transmission time for offloading as follows.

$$t_i^u = \frac{L_i^u}{R_i^u}.\tag{5}$$

Thus, the energy consumed by D_i at the offloading phase can be expressed as the following:

$$E_i^{off} = p_i^u \frac{L_i^u}{R_i^u},\tag{6}$$

and the resulting transmission energy consumed by all MNs is provided by the following.

$$E^{off} = \sum_{i=1}^{N} E_i^{off}.\tag{7}$$

2.2. Downloading Phase

Since AN is equipped with FD technology, it can download the computation result to MN D_j and receive offloading task from other MN D_i simultaneously. Thus, the received signal at D_j is described as the following:

$$y_j^d = \sqrt{p_j^d}\mathbf{H}_j^d s_j^d + n_j^d,\tag{8}$$

where p_j^d is the transmitted power AN uses for downloading the computation result to D_j. n_j^d is the AWGN with power σ_j^2.

Here, we suppose that the co-channel interference can be canceled perfectly in the receiver for the sake of simplicity.

Then, the signal to interference plus noise ratio (SINR) and the transmission rate at MN D_j are the following.

$$\gamma_j^d = \frac{p_j^d tr\{\mathbf{H}_j^d (\mathbf{H}_j^d)^H\}}{\sigma_j^2},\tag{9}$$

$$R_j^d = Blog_2(1 + \gamma_j^d).\tag{10}$$

Let R_{min}^d represent the minimum downlink transmission rate requirement and we can obtain another constraint condition described as follows.

$$R_j^d \geq R_{min}^d.\tag{11}$$

Meanwhile, we can calculate the latency of the downlink transmission as follows.

$$t_j^d = \frac{L_j^d}{R_j^d}.\tag{12}$$

The PS receiver at D_j then divides the received RF signal into two parts; the θ_j ($0 \leq \theta \leq 1$) part is used for energy harvesting, while the rest $(1 - \theta_j)$ part is used for information decoding. We can obtain the harvest energy of D_j as follows.

$$E_j^{hav} = \theta_j(p_j^d tr\{\mathbf{H}_j^d(\mathbf{H}_j^d)^H\} + \sigma_j^2)t_j^d. \tag{13}$$

Meanwhile, the energy consumption of AN can be calculated as the following.

$$E^{AN} = \sum_{j=1}^{N} p_j^d t_j^d. \tag{14}$$

2.3. Local Computation Phase

After offloading, MN operates on the remaining computation task. Let f_i^n denote the CPU frequency needed for the n-th CPU cycle of D_i. The following constraint condition should be met:

$$0 \leq f_i^n \leq f_i^{max}, \forall i \tag{15}$$

where f_i^{max} is the maximum CPU frequency of D_i. Then, the time for local computation of D_i is the following:

$$t_i^{local} = \sum_{n=1}^{C(L_i-L_i^u)} \frac{1}{f_i^n}, \tag{16}$$

where C is the CPU cycles required for computing 1-bit of data. The energy consumption of local computation is given by the following:

$$E_i^{loc} = \sum_{n=1}^{C(L_i-L_i^u)} \kappa(f_i^n)^2, \tag{17}$$

where κ is the effective capacitance coefficient based on the chip architecture [35]. Thus, we can obtain the total local energy consumption as follows.

$$E^{loc} = \sum_{i=1}^{N} E_i^{loc}. \tag{18}$$

3. Problem Formulation

After we analyze the transmission process of the system, we can formulate an problem which can optimize the system performance. This paper aims at minimizing the total energy consumption of the system, while simultaneously ensuring the transmission requirements. The total energy consumption of the system contains AN energy consumption E^{AN}, the offloading energy consumption E^{off} and the energy consumption of local computation E^{loc}. In addition, we need to remove the harvest energy E^{hav} MNs can obtain. Based on Equations (7), (13), (14) and (18), we can write the total energy consumption of the system as the following.

$$\begin{aligned}
E^{total} &= E^{AN} + E^{off} + E^{loc} - E^{uh} \\
&= \sum_{j=1}^{N} p_j^d \frac{L_j^d}{R_j^d} + \sum_{i=1}^{N} p_i^u \frac{L_i^u}{R_i^u} + \sum_{i=1}^{N} \sum_{n=1}^{C(L_i-L_i^u)} \kappa(f_i^n)^2 \\
&\quad - \sum_{j=1}^{N} \theta_j(p_j^d tr\{\mathbf{H}_j^d(\mathbf{H}_j^d)^H\} + \sigma_j^2)\frac{L_j^d}{R_j^d}
\end{aligned} \tag{19}$$

Finally, the problem can be described as follows:

$$(P1) \quad \min_{\mathbf{f}, \mathbf{p}^u, \mathbf{p}^d, \boldsymbol{\alpha}, \boldsymbol{\theta}} \ E^{total}$$

$$s.t. \begin{cases} C1: 0 \leq \alpha, \theta \leq 1 \\ C2: R_i^u \geq R_{min}^u, \forall i \\ C3: R_j^d \geq R_{min}^d, \forall j \\ C4: p_i^u \leq P_{max}^u, \forall i \\ C5: p_j^d \leq P_{max}^d, \forall j \\ C6: 0 \leq t_i^{local} \leq t_i^u, \forall i \\ C7: 0 \leq f_i^n \leq f_i^{max}, \forall i \end{cases} \quad (20)$$

where $\mathbf{f} = [f_1, f_2, \ldots, f_N]$, $\mathbf{p}^u = [p_1^u, p_2^u, \ldots, p_N^u]$, $\mathbf{p}^d = [p_1^d, p_2^d, \ldots, p_N^d]$, $\boldsymbol{\alpha} = [\alpha_1, \alpha_2, \ldots, \alpha_N]$ and $\boldsymbol{\theta} = [\theta_1, \theta_2, \ldots, \theta_N]$.

In problem $P1$, $C1$ provides the constraints on weight factors. $C2$ and $C3$ imply that the offloading rate and downloading rate should not be less than the QoS requirements R_{min}^u and R_{min}^d, respectively. $C4$ and $C5$ indicate the uplink and downlink transmission power limits P_{max}^u and P_{max}^d. $C6$ denotes that the time of the local computation should be no more than the time of the offloading phase, otherwise it is better to offload all the tasks. $C7$ indicates the CPU frequency constraint according to Equation (15).

4. The Proposed Algorithm

In this section, we will solve the formulated problem in steps. The problem P1 is full of challenges since both the objective function and the constraints are non-convex. Although the variables that need to be optimized are all coupled in P1, we find that the CPU frequency $\mathbf{f}$ is the least relevant variable compared to other variables. We divide P1 into the following three subproblems. (i) Local computation optimization is as follows: In this subproblem, we obtain the optimal CPU frequency using the scheme in [36]. (ii) Power optimization is as follows: When $\mathbf{f}$ and the weight factors $\boldsymbol{\alpha}$, $\boldsymbol{\theta}$ are fixed, we can use the interior point algorithm to solve the problem. (iii) Weight factor optimization is as follows: After frequency and power optimization are completed, interior point algorithm can be used again to obtain the optimal weight factors. The three subproblems should be optimized alternately by the iteration method.

4.1. Local Computation Optimization

Inspired by [36], the optimal CPU frequency should satisfy the following.

$$f_i^1 = f_i^2 = \cdots = f_i^{C(L_i - L_i^u)} = \overline{f_i}. \quad (21)$$

The above equation reveals the CPU frequency should maintain the same in each cycle as $\overline{f_i}$. Suppose the other four variables have been optimized, the initial problem (P1) can be reformulated as follows:

$$(P2) \quad \min_{\overline{\mathbf{f}}} E + \sum_{i=1}^{N} C(L_i - L_i^u)\kappa(\overline{f_i})^2$$

$$s.t. \begin{cases} C6: 0 \leq \dfrac{C(L_i - L_i^u)}{\overline{f_i}} \leq t_i^u, \forall i \\ C7: 0 \leq \overline{f_i} \leq f_i^{max}, \forall i \end{cases} \quad (22)$$

where $E = E^{AN} + E^{off} - E^{hav}$, $\overline{\mathbf{f}} = [\overline{f_1}, \overline{f_2}, \ldots, \overline{f_N}]$. According to the optimization objective function, the energy consumption increases monotonically with $\overline{f_i}$. In other words, $\overline{f_i}$ has to be the smallest value to achieve the minimum energy consumption. Thus, from the constraint C6, we can obtain the following.

$$f_i^{opt} = \frac{C(L_i - L_i^u)}{t_i^u}. \quad (23)$$

Then, by replacing f_i^n with f_i^{opt} in Equation (16), we can obtain the local energy consumption of user D_i^u as follows.

$$E_i^{loc} = \frac{\kappa C^3 (L_i - L_i^u)^3}{(t_i^u)^2}. \tag{24}$$

According to Equation (16), we find that the effect of CPU frequency on energy consumption can be transformed to the effect of offloading ratio and transmitting power. Therefore, we only need to focus on the optimization of these two parameters in the following steps.

4.2. Power Optimization

After the CPU frequency has been optimized and with the assumption that the two weight factors have been optimized, problem P1 now can be rewritten as the following.

$$(P3) \quad \min_{\mathbf{p}^u, \mathbf{p}^d} \sum_{j=1}^N p_j^d \frac{L_j^d}{R_j^d} + \sum_{i=1}^N p_i^u \frac{L_i^u}{R_i^u} + \sum_{i=1}^N \frac{\kappa C^3 (L_i - L_i^u)^3}{(t_i^u)^2}$$
$$- \sum_{j=1}^N \theta_j (p_j^d tr\{\mathbf{H}_j^d (\mathbf{H}_j^d)^H\} + \sigma_j^2) \frac{L_j^d}{R_j^d} \tag{25}$$
$$s.t. \ \ C2, C3, C4, C5.$$

For problem P3, the second-order derivative of each variable, i.e., $p_1^u, p_2^u, \ldots, p_N^u$ and $p_1^d, p_2^d, \ldots, p_N^d$ of the objective function is zero. Moreover, all constraints are linear. Thus, the standard interior point algorithm can be applied to this convex problem and the optimal solution can be achieved.

4.3. Weight Factor Optimization

With the optimized variables $\mathbf{f}, \mathbf{p}^u, \mathbf{p}^d$, the problem can can be expressed as the following

$$(P4) \quad \min_{\alpha, \theta} \sum_{j=1}^N p_j^d \frac{\beta_j \alpha_i L_i}{R_j^d} + \sum_{i=1}^N p_i^u \frac{\alpha_i L_i}{R_i^u} + \sum_{i=1}^N \frac{\kappa C^3 (1 - \alpha_i)^3 (L_i)^3}{(t_i^u)^2}$$
$$- \sum_{j=1}^N \theta_j (P_j^d + \sigma_j^2) \frac{\beta_j \alpha_i L_i}{R_j^d} \tag{26}$$
$$s.t. \ \ C1 : 0 \leq \alpha, \theta \leq 1$$

where $P_j^d = p_j^d tr\{\mathbf{H}_j^d (\mathbf{H}_j^d)^H\}$. Similar to problem P3, problem P4 is a convex problem which can be solved by the interior point algorithm.

Based on the above discussion, the proposed AGIO algorithm can effectively solve the optimization problem P1, which is illustrated in Algorithm 1.

Algorithm 1 Alternate Group Iterative Optimization

Input: $N, M, B, \mathbf{H}^u, \mathbf{H}^d, \mathbf{H}_0, C, \kappa, \mathbf{L}, \sigma_j^2, \sigma_{AN}^2, \beta, \eta, P_{max}^u, P_{max}^d, R_{min}^u, R_{min}^d$

Output: optimal solutions $\mathbf{f}^*, \mathbf{p}^{(u)*}, \mathbf{p}^{(d)*}, \alpha^*, , \theta^*$

 1: Set iteration number $n = 1$.

 2: Set maximum iteration number I_{max}.

 3: Set the initial values: $\mathbf{p}^{u(0)}, \mathbf{p}^{d(0)}, \alpha^{(0)}, \theta^{(0)}$.

 4: **While** $n \leq I_{max}$

 5:     Solve problem P2 to obtain the optimal $\mathbf{f}$;

 6:     Solve problem P3 to obtain the optimal $(\mathbf{p}^{u(n)}, \mathbf{p}^{d(n)})$;

 7:     Solve problem P4 to obtain the optimal $(\alpha^{(n)}, \theta^{(n)})$

 8:     Set $n = n + 1$;

 9: **End While** .

---

5. Simulation and Analysis

5.1. Simulation Results

In this section, the research work above is simulated and the effects of different variables on the system performance are investigated. The simulation parameters are summarized as follows: the bandwidth is 5 MHz; the upper limits of transmitting power are $p^u_{1,max} = p^u_{2,max} = \cdots = p^u_{N,max} = 5$ W and $p^d_{1,max} = p^d_{2,max} = \cdots = p^d_{N,max} = 20$ W; the weight factors are $\beta_1 = \beta_2 = \cdots = \beta_N = 1$; and the noise power is $\sigma^2_{AN} = \sigma^2_j = -120$ dBm, $\forall j$. We also set the chip effective capacitance coefficient to $\kappa = 10^{-20}$ and the CPU cycles are $C = 10^3$ cycles/bit. We choose a random Rayleigh fading channel model for all the channel matrix in the simulation. Moreover, we apply two benchmark algorithms to compare with the proposed AGIO algorithm.

(1) The fixed-variable (FV) algorithm: The variables $\mathbf{p}^u, \mathbf{p}^d, \boldsymbol{\alpha}, \boldsymbol{\theta}$ are fixed at initial values.

(2) The full-offloading (FO) algorithm: MNs upload all computation tasks to the MEC server. Thus, the local computation task is zero, the CPU frequency of local computation is $f_1 = f_2 = \cdots = f_N = 0$ and the offloading weight factor is $\alpha_1 = \alpha_2 = \cdots = \alpha_N = 1$.

Figure 2 demonstrates that convergence performance of the proposed AGIO algorithm under different computation task size L (4 Mbits, 8 Mbits and 12 Mbits) with 6 MNs and the antenna number of AN is 6. As shown in the figure, the algorithm converges after five iterations under varying L. This proves the effectiveness of our algorithm.The convergence curves also indicate that the energy consumption increases with computation task size L.

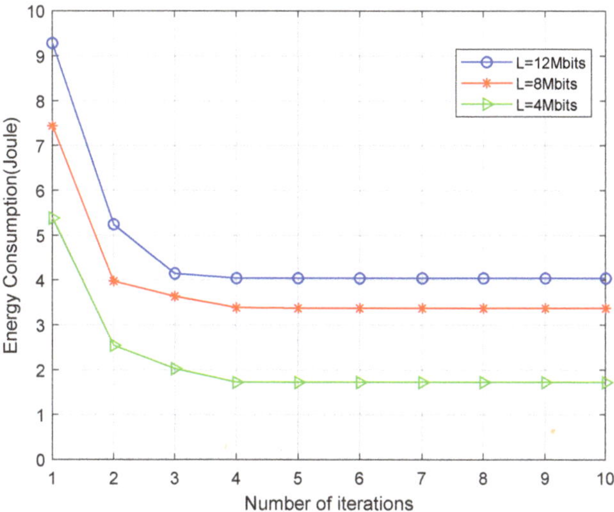

Figure 2. Convergence behavior of the proposed AGIO algorithm.

Figure 3 compares the system energy consumption of the three different algorithms under a different number of AN antennas M. The proposed AGIO has the minimum system energy consumption in all the scenarios, which shows the performance superiority of the AGIO algorithm. The number of antennas of the FV algorithm has the least impact on energy consumption. It increases gently with the number of AN antennas. On the other hand, for FO algorithm and the proposed AGIO algorithm, the energy consumption is greatly reduced with M. As a result, our algorithm indicates greater performance improvement in energy consumption when the number of the AN antennas increases. Thus, we can conclude that the proposed AGIO algorithm is suitable for multi-antenna AN.

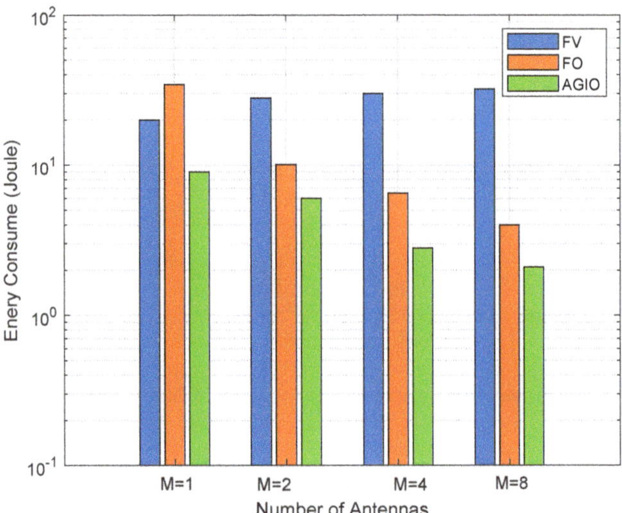

Figure 3. Comparison of energy consumption with a different number of antennas.

Figure 4 compares the three different algorithms under different number of MNs. The energy consumption increases with N which is attributed to the increased number in computation tasks when the number of users increases. Again, the proposed AGIO algorithm outperforms both FV and FO algorithms. The superiority increases with the number of MNs, which reveals the applicability of the proposed algorithm in multi-user senarios. In addition, it is clear that the FO scheme outperforms the FV scheme, which mainly benefits from the optimal power allocation process in FO scheme.

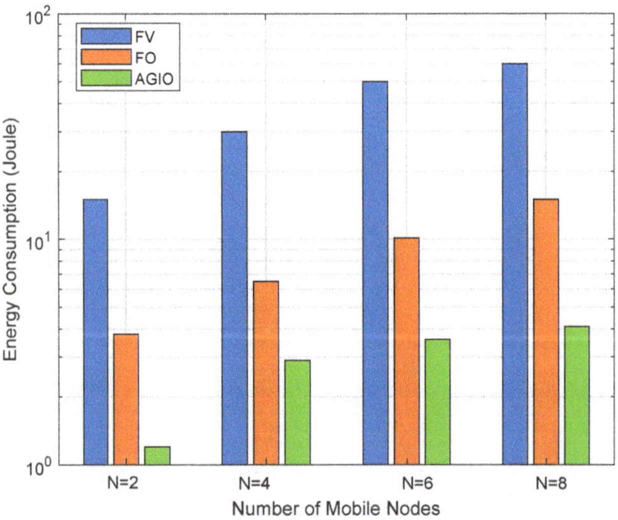

Figure 4. Comparison of energy consumption with a different number of mobile nodes.

Figure 5 shows the energy consumption of the three algorithms under different harvest weight factor θ with $N = 2$ and $N = 4$, respectively. For the three algorithms, the energy consumption decreases with the parameter θ. The FO and AGIO algorithms demonstrate increased significant decline than FV due to the reason that θ represents the capability of energy harvesting and the greater value of θ means more harvested energy and less total system energy consumption. Thus, we can conclude MN tends to offload a larger proportion of computation bits to AN in the view of energy efficiency.

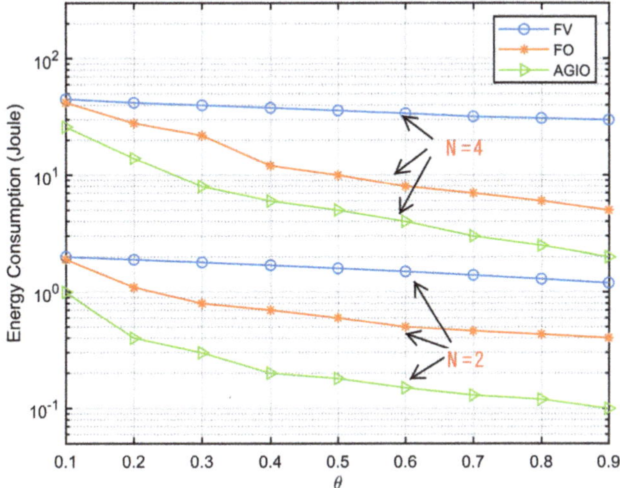

Figure 5. Comparison of energy consumption with different θ.

Figure 6 demonstrates the offloading delay of MNs in the uplink transmission. The FV scheme shows a slight advantage than the FO algorithm because we set the offloading weight factor $\alpha = 0.1$ for FV, which means smaller offloading task in FV than in FO. Compared with the two benchmark algorithms, AGIO algorithm displays better latency performance due to the excellent design of optimization steps. It also indicates the distinguished feature of our algorithm both in energy consumption performance and latency performance.

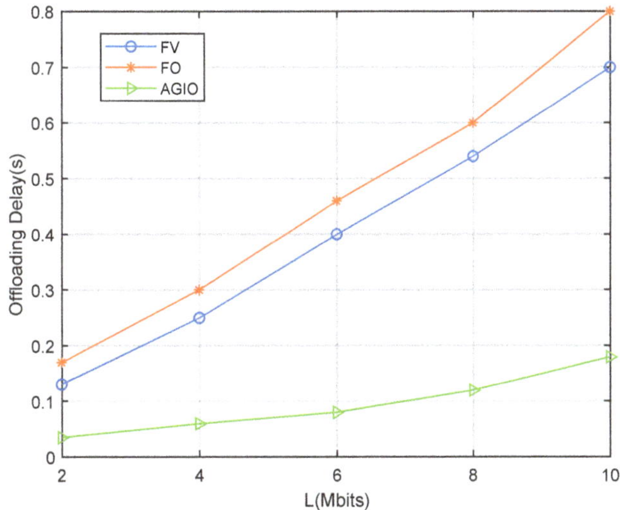

Figure 6. Comparison of offloading delay with different computation task size L.

5.2. Analysis and Discussion

Aiming at solving the data explosion and energy insufficiency challenges in WSN-assisted IoT system, we designed a novel framework that integrates MEC and SWIPT technologies into the IoT system. We formulate the energy consumption minimization problem and propose an AGIO algorithm to solve it. By jointly optimizing the CPU frequency, power allocation, offloading scheme and SWIPT scheme, we can achieve the minimum energy consumption.

Simulation results in Section 5.1 have verified our original intention. First, the proposed novel system shows great advantages in energy consumption and time delay, which are demonstrated in Figures 5 and 6. It can attribute the success to more reasonable expressions of several parameters, which affects the system performance. That is also achieved by the first contribution listed in Section 1.3. In addition, the convergence result displayed in Figure 2 confirms the effectiveness of the AGIO algorithm as we stated in the second contribution. Furthermore, the simulation result in Figure 3 shows that the energy consumption decreases with the number of AN antennas. It provides a clue that our system is suitable for multi-antenna system, which is more efficient in applications. The simulation results in Figure 4 shows the superiority of the proposed algorithm in a multi-user scenario, which is exactly the practical application of a WSN-assisted IoT system.

6. Conclusions

In this paper, we investigate the wireless information transmission and energy transfer of a novel SWIPT-MEC enabled WSN-assisted IoT System. We fomulate an optimization problem by jointly optimizing the CPU frequency, transmitted power, offloading weight factor and harvest weight factor to achieve the minimum system energy consumption. In order to render the problem solvable, we propose a novel alternate group iteration optimization (AGIO) algorithm, which decomposes the original problem into three subproblems and alternately optimizes each subproblem using the group interior point iterative optimization algorithm. Finally, numerical simulation of the proposed strategy is carried on to compare with the two other benchmark schemes. The results demonstrate that the proposed design presents the performance advantages both in energy consumption and latency.

Author Contributions: Conceptualization, F.C. and J.H.; methodology, F.C. and A.W.; software, F.C. and Y.Z.; validation, Y.Z. and Z.N.; writing—original draft preparation, F.C. and A.W.; writing—review and editing, all co-authors; supervision, J.H; funding acquisition, J.H. and F.C. All authors have read and agreed to the published version of the manuscript.

Funding: This work is funded by the National Natural Science Foundation of China (61601409), Zhejiang Natural Science Foundation (LQ21F010008), Zhejiang Province Science and technology Projects (2021C04004), the Scientific Research Project of the Department of Education of Zhejiang Province (Y202044549) and Zhejiang Postdoctoral Funding.

Data Availability Statement: The data used to support the findings of this study are available from the corresponding author upon request.

Conflicts of Interest: The authors declare that they have no conflict of interest.

References

1. Chettri, L.; Bera, R. A comprehensive survey on internet of things (IoT) toward 5G sireless systems. *IEEE Internet Things J.* **2020**, *7*, 16–32. [CrossRef]
2. Shafique, K.; Khawaja, B.A.; Sabir, F.; Qazi, S.; Mustaqim, M. Internet of things (IoT) for next-generation smart systems: A review of current challenges, future trends and prospects for emerging 5G-IoT scenarios. *IEEE Access* **2020**, *7*, 23022–23040. [CrossRef]
3. Wang, D.; Chen, D.; Song, B.; Guizani, N.; Yu, X.; Du, X. From IoT to 5G HoT: The next generation IoT-based intelligent algorithms and 5G technologies. *IEEE Commun. Mag.* **2018**, *56*, 114–120. [CrossRef]
4. Shahzadi, R.; Niaz, A.; Ali, M.; Naeem, M.; Rodrigues, J.J.; Qamar, F.; Anwar, S.M. Three tier fog networks: Enabling IoT/5G for latency sensitive applications. *China Commun.* **2019**, *16*, 1–11.
5. Wazid, M.; Das, A.K.; Shetty, S.; Gope, P.; Rodrigues, J.J. Security in 5G-Enabled internet of things communication: Issues, challenges, and future research roadmap. *IEEE Access* **2020**, *9*, 4466–4489. [CrossRef]

6. Ma, X.E.Z.; Yu, K. Energy-efficient computation offloading and resource allocation in SWIPT-based MEC Networks. *IEEE Access* **2020**. [CrossRef]

7. Dua, A.; Dutta, A.; Zaman, N.; Kumar, N. Blockchain-based E-waste Management in 5G Smart Communities. In Proceedings of the IEEE INFOCOM 2020—IEEE Conference on Computer Communications Workshops (INFOCOM WKSHPS), Toronto, ON, Canada, 6–9 July 2020; pp. 195–200.

8. Lu, W.; Xu, X.; Huang, G.; Li, B.; Wu, Y.; Zhao, N.; Yu, F.R. Energy efficiency optimization in SWIPT enabled WSNs for smart agriculture. *IEEE Trans. Ind. Inform.* **2020**, *17*, 4335–4344. [CrossRef]

9. Cui, J.; Zhang, Y.; Cao, M.; Wang, S.; Xu, Y. Thyroid tumor care risk based on medical IoT system. *Micropress. Microsyst.* **2021**, *82*, 103845. [CrossRef]

10. Zhang, Q.; Sun, H.; Wei, Z.; Feng, Z. Sensing and Communication Integrated System for Autonomous Driving Vehicles. In Proceedings of the IEEE INFOCOM 2020—IEEE Conference on Computer Communications Workshops (INFOCOM WKSHPS), Toronto, ON, Canada, 6–9 July 2020; pp. 1278–1279.

11. Lu, W.; Xu, X.; Ye, Q.; Li, B.; Peng, H.; Hu, S.; Gong, Y. Power optimisation in UAV-assisted wireless powered cooperative mobile edge computing systems. *IET Commun.* **2020**, *14*, 2516–2523. [CrossRef]

12. Jiang, D.; Wang, Z.; Lv, Z.; Li, W. Smart antenna-based multihop highly-energy-efficient DSA approach to drone-assisted backhaul networks for 5G. In Proceedings of the IEEE INFOCOM 2020—IEEE Conference on Computer Communications Workshops (INFOCOM WKSHPS), Toronto, ON, Canada, 6–9 July 2020; pp. 883–887.

13. Zhang, F.; Han, G.; Liu, L.; Martinez-Garcia, M.; Peng, Y. Joint Optimization of Cooperative Edge Caching and Radio Resource Allocation in 5G-Enabled Massive IoT Networks. *IEEE Internet Things J.* **2021**. [CrossRef]

14. Zhang, K.; Mao, Y.; Leng, S.; Zhao, Q.; Li, L.; Peng, X.; Zhang, Y. Energy-efficient offloading for mobile edge computing in 5G heterogeneous networks. *IEEE Access* **2016**, *4*, 5896–5907. [CrossRef]

15. Hewa, T.; Braeken, A.; Ylianttila, M.; Liyanage, M. Multi-Access Edge Computing and Blockchain-based Secure Telehealth System Connected with 5G and IoT. In Proceedings of the GLOBECOM 2020—2020 IEEE Global Communications Conference, Taipei, Taiwan, 7–11 December 2020.

16. Yang, X.; Fei, Z.; Zheng, J.; Zhang, N.; Anpalagan, A. Joint multi-user computation offloading and data caching for hybrid mobile cloud/Edge computing. *IEEE Trans. Veh. Technol.* **2019**, *68*, 11018–11030. [CrossRef]

17. Wu, Y.; Chen, J.; Qian, L.P.; Huang, J.; Shen, X.S. Energy-aware cooperative traffic offloading via device-to-device cooperations: An analytical approach. *IEEE Trans. Mob. Comput.* **2017**, *16*, 97–114. [CrossRef]

18. Liu, C.F.; Bennis, M.; Debbah, M.; Poor, H.V. Dynamic task offloading and resource allocation for ultra-reliable low-latency edge computing. *IEEE Trans. Commun.* **2019**, *67*, 4132–4150. [CrossRef]

19. Spinelli, F.; Mancuso, V. Toward Enabled Industrial Verticals in 5G: A Survey on MEC-Based Approaches to Provisioning and Flexibility. *IEEE Commun. Surv. Tutor.* **2021**, *23*, 596–630. [CrossRef]

20. Liu, Y.; Peng, M.; Shou, G.; Chen, Y.; Chen, S. Toward Edge Intelligence: Multiaccess Edge Computing for 5G and Internet of Things. *IEEE Internet Things J.* **2020**, *7*, 6722–6747. [CrossRef]

21. Varshney, L.R. Transporting information and energy simultaneously. In Proceedings of the 2008 IEEE International Symposium on Information Theory, Toronto, ON, Canada, 6–11 July 2008

22. Zhang, R.; Ho, C.K. MIMO Broadcasting for Simultaneous Wireless Information and Power Transfer. *IEEE Trans. Wirel. Commun.* **2013**, *12*, 1989–2001. [CrossRef]

23. Park, J.; Clerckx, B. Joint wireless information and energy transfer in a two-user mimo interference channel. *IEEE Trans. Wirel. Commun.* **2013**, *12*, 4210–4221. [CrossRef]

24. Nasir, A.A.; Zhou, X.; Durrani, S.; Kennedy, R.A. Relaying protocols for wireless energy harvesting and information processing. *IEEE Trans. Wirel. Commun.* **2013**, *12*, 3622–3636. [CrossRef]

25. Zhou, X.; Zhang, R.; Ho, C.K. Wireless information and power transfer: Architecture design and rate-energy tradeoff. *IEEE Trans. Commun.* **2013**, *61*, 4754–4767. [CrossRef]

26. Guo, S.; Shi, Y.; Yang, Y.; Xiao, B. Energy efficiency maximization in mobile wireless energy harvesting sensor networks. *IEEE Trans. Mob. Comput.* **2018**, *17*, 1524–1537. [CrossRef]

27. Tang, J.; So, D.; Zhao, N.; Shojaeifard, A.; Wang, K. Energy efficiency optimization with SWIPT in MIMO broadcast channels for internet of things. *IEEE Internet Things J.* **2018**, *5*, 2605–2619. [CrossRef]

28. Fu, J.; Hua, J.; Wen, J.; Zhou, K.; Li, J.; Sheng, B. Optimization of Achievable Rate in the Multiuser Satellite IoT System With SWIPT and MEC. *IEEE Trans. Ind. Inform.* **2021**, *17*, 2072–2080. [CrossRef]

29. Zhou, F.H.; Wu, Y.P.; Hu, R.Q.; Qian, Y. Computation rate maximization in UAV-enabled wireless-powered mobile-edge computing systems. *IEEE J. Sel. Areas Commun.* **2018**, *36*, 1927–1941. [CrossRef]

30. Ksentini, A.; Frangoudis, P.A. On Extending ETSI MEC to Support LoRa for Efficient IoT Application Deployment at the Edge. *IEEE Commun. Stand. Mag.* **2020**, *4*, 57–63. [CrossRef]

31. Giannopoulos, A.; Spantideas, S.; Tsinos, C.; Trakadas, P. Power Control in 5G Heterogeneous Cells Considering User Demands Using Deep Reinforcement Learning. *Int. Fed. Inf. Process.* **2021**, 95–105. [CrossRef]

32. Min, M.; Xiao, L.; Chen, Y.; Cheng, P.; Wu, D.; Zhuang, W. Learning-Based Computation Offloading for IoT Devices With Energy Harvesting. *IEEE Trans. Veh. Technol.* **2019**, *68*, 1930–1941. [CrossRef]

33. Zhang, J.; Du, J.; Shen, Y.; Wang, J. Dynamic Computation Offloading With Energy Harvesting Devices: A Hybrid-Decision-Based Deep Reinforcement Learning Approach. *IEEE Internet Things J.* **2020**, *7*, 9303–9317. [CrossRef]
34. Chen, F.; Fu, J.; Wang, Z.; Zhou, Y.; Qiu, W. Joint Communication and Computation Resource Optimization in FD-MEC Cellular Networks. *IEEE Access* **2019**, *7*, 168444–168454. [CrossRef]
35. Mao, Y.; Zhang, J.; Ben Letaief, K. Dynamic Computation Offloading for Mobile-Edge Computing With Energy Harvesting Devices. *IEEE J. Sel. Areas Commun.* **2016**, *34*, 3590–3605. [CrossRef]
36. Wang, F.; Xu, J.; Wang, X.; Cui, S. Joint offloading and computing optimization in wireless powered mobile-edge computing Systems. *IEEE Trans. Wirel. Commun.* **2018**, *17*, 1784–1797. [CrossRef]

Article

DFusion: Denoised TSDF Fusion of Multiple Depth Maps with Sensor Pose Noises

Zhaofeng Niu *, Yuichiro Fujimoto, Masayuki Kanbara, Taishi Sawabe and Hirokazu Kato

Nara Institute of Science and Technology (NAIST), Ikoma 630-0192, Nara, Japan; yfujimoto@is.naist.jp (Y.F.); kanbara@is.naist.jp (M.K.); t.sawabe@is.naist.jp (T.S.); kato@is.naist.jp (H.K.)
* Correspondence: niu.zhaofeng.mv7@is.naist.jp

Abstract: The truncated signed distance function (TSDF) fusion is one of the key operations in the 3D reconstruction process. However, existing TSDF fusion methods usually suffer from the inevitable sensor noises. In this paper, we propose a new TSDF fusion network, named DFusion, to minimize the influences from the two most common sensor noises, i.e., depth noises and pose noises. To the best of our knowledge, this is the first depth fusion for resolving both depth noises and pose noises. DFusion consists of a fusion module, which fuses depth maps together and generates a TSDF volume, as well as the following denoising module, which takes the TSDF volume as the input and removes both depth noises and pose noises. To utilize the 3D structural information of the TSDF volume, 3D convolutional layers are used in the encoder and decoder parts of the denoising module. In addition, a specially-designed loss function is adopted to improve the fusion performance in object and surface regions. The experiments are conducted on a synthetic dataset as well as a real-scene dataset. The results prove that our method outperforms existing methods.

Keywords: depth fusion; TSDF; sensor noises

Citation: Niu, Z.; Fujimoto, Y.; Kanbara, M.; Sawabe, T.; Kato, H. DFusion: Denoised TSDF Fusion of Multiple Depth Maps with Sensor Pose Noises. *Sensors* **2022**, *22*, 1631. https://doi.org/10.3390/s22041631

Academic Editor: Jing Tian

Received: 4 January 2022
Accepted: 2 February 2022
Published: 19 February 2022

Publisher's Note: MDPI stays neutral with regard to jurisdictional claims in published maps and institutional affiliations.

1. Introduction

Depth fusion is of great importance for many applications, such as augmented reality applications and autonomous driving. Many methods have been proposed in this area and truncated signed distance function (TSDF) [1] is one of the most famous. However, TSDF requires manual adjustment on its parameters, possibly leading to thick artifacts. To address this problem, some depth fusion methods have emerged with improved performance. Methods such as [2,3] use surfel-based or probabilistic approaches to generate 3D representations, which may be a voxel grid, a mesh or a point cloud. In addition, compared with these classical methods, convolutional neural network (CNN) based methods have shown advantages in the fusion performance. However, their results still suffer from noisy input, which results in missing surface details and incomplete geometry [4].

The data acquired by depth cameras inevitably contain a significant amount of noise. Although researchers have proposed many methods to remove the noise, most of the works only focus on removing the noise caused by depth maps but neglect the noise of camera poses (pose noises for simplicity). Figure 1 illustrates the two types of noises. Figure 1a shows the situation where there is no noise and a plane is in the sight of the camera. If there are depth noises, the noise may be outliers or missing data, as shown in Figure 1b, which leads to noisy TSDF volumes. As for the pose noise, Figure 1c provides an example when the camera has translation and rotation error compared with Figure 1a, which causes troubles when integrating the TSDF updates due to the inaccurate extrinsic data. Both types of noises may have adverse impacts on depth fusion results. However, there are only a few works that focus on removing noises for TSDF fusion, even given the fact that both types of noises are inevitable.

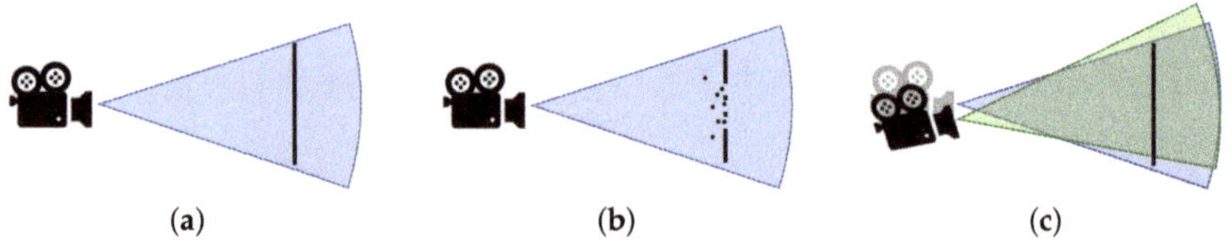

Figure 1. Illustration of the sensor noises. (**a**) Sensor without noises. (**b**) Depth noises. (**c**) Sensor pose noises.

RoutedFusion method [4], as an example, considers the depth noise and aims to obtain a robust TSDF volume against different levels of depth noise. It uses depth maps derived from synthetic datasets and puts random noises into the depth maps. However, in the fusion process, the camera pose they use is the ground-truth pose from the synthetic dataset, so that the results can only be robust against depth noise, but not against pose noise. In this paper, we propose a method named DFusion that considers not only depth noises but also pose noises, as shown in Figure 2. To the best of our knowledge, this is one of the earliest research that tries to avoid the performance drop caused by pose noises.

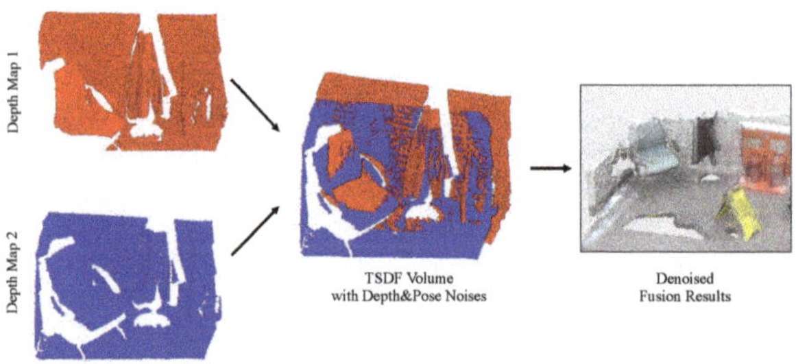

Figure 2. DFusion can minimize the influence of both types of noises.

Generally, depth fusion is conducted with 2D convolutional models. However, when considering the pose noise, it is better to remove the noise with the 3D representation because it is challenging to recognize and remove the surface shifts in the 2D space. Therefore, we firstly adopt a Fusion Module, as the first part of DFusion, with the same setting as the fusion network in the RoutedFusion method, to fuse the depth maps with camera poses into a TSDF volume. After gaining the integrated TSDF volume, we design a Denoising Module, an UNet-like neural network, as the second part of DFusion to denoise the TSDF volume. Since the input of the Denoising Module is a 3D volume, 3D convolutional layers are utilized to obtain the 3D features. Skip connections are used to avoid the vanishing gradient problem, which is prone to occur due to the small value of TSDF volume.

For training the networks, we utilize a synthetic dataset which can provide the ground-truth value of depth maps and camera poses. The model is trained in a supervised manner. In addition to the commonly-used fusion loss, several specially-designed loss functions are proposed, including a L_1 loss for all voxels in the whole scene and L_1 losses over the objects and surfaces for better fusion performance on these regions.

In sum, the contributions of this work are as follows:

- We propose a new fusion network named DFusion, which considers both depth noises and pose noises in the fusion process. DFusion can avoid the performance drops caused by both types of noises, and conduct accurate and robust depth fusion.
- We design new fusion loss functions that focus on all the voxels while emphasizing the object and surface regions, which can improve the overall performance.
- The experiments are conducted on a synthetic dataset as well as a real scene dataset, measuring the actual noise levels with the real-world setting and demonstrating the

denoising effects of the proposed method. The ablation study proves the effectiveness of the proposed loss function.

2. Related Works

2.1. Depth Fusion and Reconstruction

2.1.1. Classical Methods

TSDF fusion method [1] is one of the most important classical fusion methods that fuses depth maps with camera intrinsics and the corresponding viewpoints, i.e., camera poses, into a discretized signed distance function and weight function, thereby obtaining a volumetric representation. It has been adopted as the fundamental in the majority of depth map fusion based 3D reconstruction, including KinectFusion [5], BundleFusion [6], and voxel hashing [7,8]. However, the depth maps always involve noises but all these methods update a wider band to deal with the noise, as a result, there are noise artifacts, especially outlier blobs and thickening surfaces, on the results.

In contrast to the voxel-based method, there are some reconstruction approaches that update the results in different ways. For example, Zienkiewicz et al. [9] introduce a scalable method that fuses depth maps into a multi-resolution mesh instead of a voxel grid. Keller et al. [10] design a flat point-based representation method [2], which utilizes the input from the depth sensor directly without converting representations, thereby saving the memory and increasing the speed. In addition, the surfel-based approach that approximates the surface with local points is adopted for reconstruction [2,11]. The unstructured neighborhood relationship can be built by this approach, although it usually tends to miss connectivity information among surfels. MRSMap [12], as an example, integrates depth maps into a multi-resolution surfel map for objects and indoor scenes.

Some researchers also regard the depth map fusion process as a probabilistic density problem [3,12–14], considering various ray directions. Yong et al. [15] estimate the probability density function based on the original point cloud instead of the depth map and use a mathematical expectation method to decrease the complexity of computation. In [16], the marginal distribution of each voxel's occupancy and appearance is calculated by a Markov random field along with the camera rays. However, all these classical methods have limitations to balance reconstruction quality, scene assumptions, speed and spatial scale due to the large and complex computation but limited memory.

2.1.2. Learning-Based Methods

Along with the development of deep learning methods, there exist lots of proposals that fuse and improve the performance of classical 3D reconstruction [17]. For example, ScanComplete [18] method completes and refines the 3D scan with a CNN model, which can deal with the large-scale input and obtain the high-resolution output. RayNet [19], which combines a CNN model with Markov random fields method, considers both local information and global information of the multi-view images. It can cope with large surfaces and solve the occlusion problem. Based on Mask R-CNN method [20], Mesh R-CNN [21] detects objects in an image, then builds meshes with a mesh prediction model and refines the meshes with a mesh refinement model.

Specifically, in many learning-based approaches, TSDF fusion is still one of the important steps [22]. OctNetFusion [23] fuses the depth maps with TSDF fusion and subsequently utilizes a 3D CNN model to deal with the occluded regions and refines the surfaces. Leroy et al. [24] propose a deep learning-based method to achieve multi-view photoconsistency, which focuses on matching features among viewpoints for obtaining the depth information. Similarly, the depth maps are finally fused by TSDF fusion. RoutedFusion [4] also fuses the depth maps based on the standard TSDF fusion. Different from other methods, it reproduces TSDF fusion by a CNN model, which predicts the parameters of volume and weight, then the volumetric representation can be updated with new volume and weight sequentially.

Compared with the classical method, deep learning-based methods show advantages in handling thickening artifacts and increasing diversity and efficiency. In addition, existing methods pay little attention to the noise problem during the fusion process. Our method adopts a part of RoutedFusion models to fuse the 3D volume firstly, then combines a special-designed neural network to remove the noise, thereby improving the performance of the depth fusion.

2.2. Denoising/Noise Reduction

Most of the works consider the noise as the depth noise and try to remove the noise at the beginning of the fusion process. The authors in [3,25] adopt Gaussian noise to mimic the real depth noise derived from the depth sensors, then achieve the scene reconstruction. Cherabier et al. [26] also remove some regions of random shapes, such as circles and triangles, to simulate the missing data. In RoutedFusion [4], the authors add random noise to the depth maps and propose a routing network that can remove the random noise, then use a fusion network to fuse the denoised depth maps into a TSDF volume. The experiments prove that the routing network has a significant effect on improving accuracy.

Another way to cope with the noise is to refine the 3D representation directly. NPD [27] trains the network by utilizing a reference plane from the noiseless point cloud as well as the normal vector of each point while PointCleanNet [28] removes the outlier firstly then denoises the remaining points by estimating normal vectors. Han et al. [29] propose a local 3D network to refine the patch-level surface but it needs to obtain the global structure from the depth images firstly, which is inconvenient and time-consuming. Zollhöfer et al. [25] propose a method that utilizes the details, such as shading cues, of the color image to refine the fused TSDF volume since the color image typically has a higher resolution. A 3D-CFCN model [30], which is a cascaded fully convolutional network, combines the feature of low-resolution input TSDF volume and high-resolution input TSDF volume to remove the noise and refine the surface. However, all these methods only consider either the outliers of the 3D representation or the noises caused by depth maps. In our method, we design a denoising network with 3D convolutional layers, which remove the noise for the TSDF volume without any other additional information. In addition, we take the noise of both depth maps and camera poses into account; thus, the network is robust against not only depth noises but also pose noises.

3. Methodology

3.1. TSDF Fusion

Standard TSDF fusion, which is proposed by Curless and Levoy [1], integrates a depth map D_i with the camera pose and camera intrinsic into a signed distance function $V_i \in R^{X \times Y \times Z}$ and weight function $W_i \in R^{X \times Y \times Z}$. For location x, the integration process can be expressed as follows:

$$V_i(x) = \frac{W_{i-1}(x)V_{i-1}(x) + w_i(x)v_i(x)}{W_{i-1}(x) + w_i(x)} \tag{1}$$

$$W_i(x) = W_{i-1}(x) + w_i(x) \tag{2}$$

It is an incremental process, and V_0 and W_0 are initially set as zero volumes. In each time step i, the signed distance v_i and its weight w_i are estimated according to the depth map of the current ray, then are integrated into a cumulative signed distance function $V_i(x)$ and a cumulative weight $W_i(x)$.

However, in the traditional way, the parameters are tuned manually, so that it is a heavy task and difficult to exclude artifacts and maintain high performance. In RoutedFusion [4], the TSDF fusion process has been conducted in a convolutional network, named depth fusion network, which is trained to tune the parameters automatically. The input of the fusion network is depth maps, camera intrinsics and camera poses. The depth map is fused into the previous TSDF volume with the camera intrinsic and camera pose incrementally. The main purpose of RoutedFusion method is to deal with the noise of the TSDF volume caused by the noise on depth maps. To remove the depth noise, the authors firstly adopt the depth maps with random noises for training, then use a routing network to denoise the depth maps before fusing them with the fusion network.

In a real application, however, the pose noise is also inevitable. Therefore, in our method, the inputs include noised depth maps and noised camera poses.

3.2. Network Architecture

The proposed DFusion method mainly includes two parts: a Fusion Module for fusing depth maps and a Denoising Module for removing the depth noises and pose noises. These two modules are trained independently, with different loss functions.

Fusion Module. The Fusion Module follows the design of the fusion network proposed in the RoutedFusion method [4]. It fuses depth maps incrementally with a learned TSDF updating function, using the information of camera intrinsics and camera poses. Then the TSDF update will be integrated to form a TSDF volume for the whole scene. The process of the Fusion Module is illustrated in the upper part of Figure 3. Although RoutedFusion can remove the depth noise, its denoising process is implemented as a pre-processing network, i.e., the routing network as metioned in Section 3.1, rather than the Fusion Module which is used in our method. Also, different from the RoutedFusion method, we consider not only the depth noise but also the pose noise, the latter of which is much more obvious when fusion is finished than before/during fusion. Therefore, we add a post-processing module to deal with both of these two types of noises.

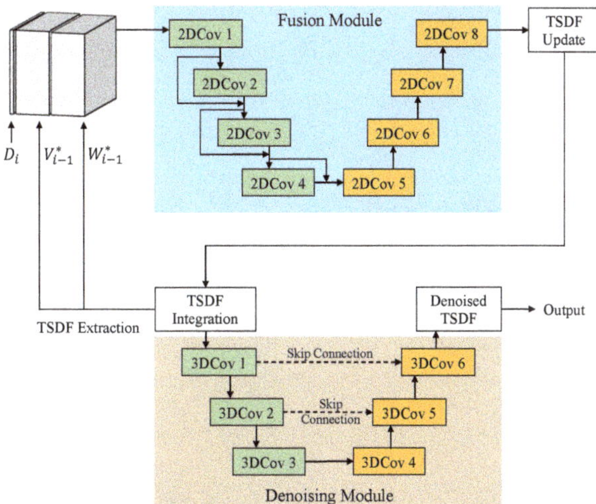

Figure 3. The DFusion model.

Denoising Module. After obtaining the TSDF volume, the Denoising Module is designed to remove the noise of the TSDF volume. The input of the Denoising Module, which is also the output of the Fusion Module, is a TSDF volume with depth noises and pose noises. Since it deals with a 3D volume, we adopt 3D convolutional layers instead of 2D convolutional layers, aiming to capture more 3D features to remove the noise (as using 3D convolutional layers is a natural choice for tasks such as 3D reconstruction [30] and recognizing 3D shifts are extremely difficult for 2D convolutions). As shown in Figure 3, the Denoising Module is implemented as an UNet-like network, which downsamples the features in the encoder part and upsamples them back to the original size in the decoder part. Skip connections are added among encoder layers and decoder layers.

In the training phase, to mimic the noises of real-world applications, we add random noises to the ground-truth depth maps and camera poses of the dataset. Therefore, the output of the Fusion Module, as well as the input of the Denoising Module, is noisy and needs to be fixed. For the depth noise, we add the noises B_d that follow a normal distribution to all pixels P in the depth maps (following the solutions in [4,23]). This process can be represented as

$$P' := P + B_d, \tag{3}$$

and

$$B_d \sim N[0, \sigma_d], \tag{4}$$

where σ_d is the pre-defined scale parameter. This parameter should be set to reflect the actual noise levels of the applications. We set $\sigma_d = 0.005$ following [4,23].

As for pose noises, we add the noise to translation matrix T and rotation matrix R, respectively. Firstly, given a random translation error B_t, a random rotation error B_r, two random unit vectors $n_t = (n_1, n_2, n_3)$ and $n_r = (n_4, n_5, n_6)$ (respectively, for translation and rotation errors), the noised translation matrix and rotation matrix are calculated as follows.

$$T' := T + n_t \cdot B_t$$
$$R' := R + \mathrm{Rodri}(n_r, B_r), \tag{5}$$

where $\mathrm{Rodri}(n_r, B_r)$ follows Rodrigues's rotation formula and it can be represented as:

$$\begin{pmatrix} n_4^2(1 - cosB_r) + cosB_r & n_4n_5(1 - cosB_r) - n_6sinB_r & n_4n_6(1 - cosB_r) + n_5sinB_r \\ n_4n_5(1 - cosB_r) + n_6sinB_r & n_5^2(1 - cosB_r) + cosB_r & n_5n_6(1 - cosB_r) - n_4sinB_r \\ n_4n_6(1 - cosB_r) - n_5sinB_r & n_5n_6(1 - cosB_r) + n_4sinB_r & n_6^2(1 - cosB_r) + cosB_r \end{pmatrix} \tag{6}$$

In addition, B_t and B_r also follow the normal distribution.

$$B_t \sim N[\mu_t, \sigma_t]$$
$$B_r \sim N[\mu_r, \sigma_r] \tag{7}$$

Since there is no existing method that adds artificial pose noises to improve the denoising performance, the value of μ and σ is decided based on a real scene dataset. More details are given in Section 4.2.

3.3. Loss Functions

Since there are two modules in the network, i.e., Fusion module and Denoising module, the total loss function involves two parts as follows.

Fusion Loss. The loss function of the Fusion Module is expressed as follows:

$$L_F = \sum_a \lambda_1^F L_1(V_{local,a}, V'_{local,a}) + \lambda_2^F L_C(V_{local,a}, V'_{local,a}), \tag{8}$$

where V_{local} and V'_{local} are two local volumes along ray a, respectively, from the the network output and from the ground-truth. L_1 is the L1 loss and can be represented as

$$L_1(V, V') = \frac{\sum_{v_m \in V, v'_m \in V'} |v_m - v'_m|}{|V|} \tag{9}$$

In addition, we use the cosine distance loss L_C (on the signs of the output volume and ground-truth volume) to ensure the fusion accuracy of the surface, following the setting in [4], which can be represented as

$$L_C(V, V') = 1 - cos(sign(V), sign(V')), \tag{10}$$

where $sign()$ is to get the signs of the inputs and $cos()$ is to get the cosine values of the angles between the input vectors.

In addition, λ_1^F and λ_2^F are the weigths for the loss terms and are emperically decided as 1 and 0.1 [4], respectively.

Denoising Loss. The Denoising Module is also trained in a supervised manner, considering the fusion accuracy on the whole scene, objects, and surface regions. The loss function is defined as follows:

$$L_D = \lambda_1^D L_{SPACE} + \lambda_2^D L_{OBJECT} + \lambda_3^D L_{SURFACE}, \tag{11}$$

where L_{SPACE}, L_{OBJECT}, and $L_{SURFACE}$ are, respectively, for the losses of the whole scene, objects, and the surface regions (as shown in Figure 4). λ_1^D, λ_2^D, and λ_3^D are the weights to adjust their relative importance.

L_{SPACE} is defined as

$$L_{SPACE} = L_1(V, V'), \tag{12}$$

where V is the predicted scene volume while V' is the ground-truth volume.

Let $V_{OBJECT} \subseteq V$, and for each $v_m \in V_{OBJECT}$, $v'_m \leq 0$, then

$$L_{OBJECT} = L_1(V_{OBJECT}, V'_{OBJECT}) \tag{13}$$

Similarly, let $V_{SURFACE} \subseteq V$, and for each v_m in $V_{SURFACE}$, $-S \leq v'_m \leq S$, where S is a threshold of the surface range (we set S to 0.02), then

$$L_{SURFACE} = L_1(V_{SURFACE}, V'_{SURFACE}) \tag{14}$$

We set the values of hyperparameter λ_1^D, λ_2^D, and λ_3^D to 0.5, 0.25, and 0.25, respectively. The effects of object loss and surface loss are explored in the ablation study.

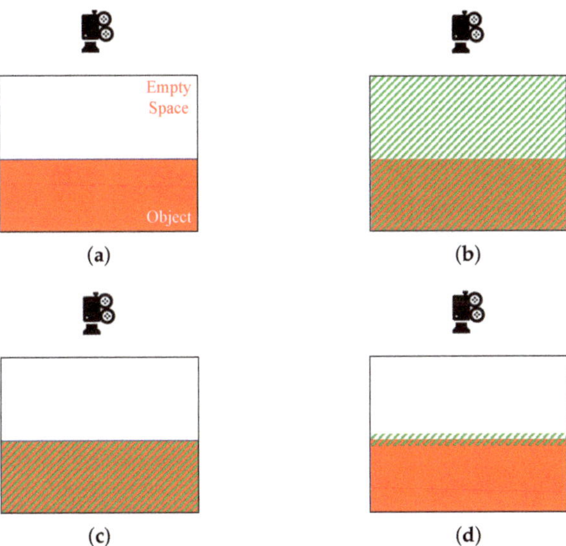

Figure 4. The focus regions of the loss functions (green masks for the focus regions). (**a**) The illustration of the example scene, where one object exists. (**b**) The scene loss. (**c**) The object loss. (**d**) The surface loss.

4. Experiments

In this section, we first explain the details of the experimental setup. Then we introduce the adopted datasets, with which both quantitative and qualitative results prove that our proposed method outperforms existing methods.

4.1. Experimental Setup

All the network models are implemented in PyTorch and trained with NVIDIA P100 GPU. The RMSprop optimization algorithm [31] is adopted with an initial learning rate of 10^{-4} and the momentum of 0.9, for both the fusion network and denoising network. The networks are trained sequentially, that is, the fusion network is pre-trained before the training of the denoising network. 10K frames sampled from ShapeNet dataset [32] are utilized for training the network.

4.2. Dataset and Noise Simulation

ShapeNet dataset [32] includes a large scale of synthetic 3D shapes, such as the plane, sofa and car. The ground-truth data, including depth maps, camera intrinsics and camera poses, can be obtained from the 3D shapes. Similar to RoutedFusion [4], we use the ShapeNet dataset to train the networks. To simulate the realistic noisy situation, not only depth maps but also camera poses are added random noises in the training process.

CoRBS dataset [33], a comprehensive RGB-D benchmark for SLAM, provides (i) real depth data and (ii) real color data, which are captured with a Kinect v2, (iii) a ground-truth trajectory of the camera that is obtained with an external motion capture system, and (iv) a ground-truth 3D model of the scene that is generated via an external 3D scanner. Totally, the dataset involves 20 image sequences of 4 different scenes.

Noise Simulation. As introduced in Section 3.2, we need the μ_t, σ_t, μ_r, and σ_r parameters to mimic the real sensor noises. Since the CoRBS dataset provides not only real-scene data but also the ground-truth data, we adopt it to obtain the realistic pose noise for simulation. In order to measure the pose noise, we follow the calculation process of the commonly-used relative pose error (RPE) [34]. RPE is defined as the drift of the trajectory over a fixed time interval Δ. For a sequence of n frames, firstly, the relative pose error at time step i is calculated as follows:

$$E_i = (I_i^{-1} I_{i+\Delta})^{-1}(J_i^{-1} J_{i+\Delta}), \tag{15}$$

where I is the ground-truth trajectory and J is the estimated trajectory. Then $m = n - \Delta$ individual relative pose error matrices can be obtained along the sequence. Generally, the RPE is considered as two components, i.e., RPE for translation matrix ($T = trans(E_i)$) and RPE for rotation matrix ($R = rot(E_i)$). We use the following formulas for obtaining the μ and σ parameters for the normal distribution.

$$\mu_t = \frac{1}{m} \sum_{i=1}^{m} \| trans(E_i) \| \tag{16}$$

$$\sigma_t = \sqrt{\frac{1}{m} \sum_{i=1}^{m} (\| trans(E_i) \| - \mu_t)^2} \tag{17}$$

$$\mu_r = \frac{1}{m} \sum_{i=1}^{m} \angle rot(E_i) \tag{18}$$

$$\sigma_r = \sqrt{\frac{1}{m} \sum_{i=1}^{m} (\angle rot(E_i) - \mu_r)^2}, \tag{19}$$

where $\angle rot(E_i) = arccos(\frac{Tr(R)-1}{2})$ and $Tr(R)$ represents the sum of the diagonal elements of the rotation matrix R.

For the translation error, μ_t is 0.006 and σ_t is 0.004, while for the rotation error, μ_r is 0.094 and σ_r is 0.068, which are used in the noise simulation for our experiments. These parameters are also preferable in the training of DFusion model for actual uses, while they can also be increased a bit (better keeping μ_t and σ_t no larger than 0.02, μ_r and σ_r no larger than 0.2, with which the DFusion model can give good fusion results) if strong sensor noises are expected.

4.3. Evaluation Results

The experiments are conducted on ShapeNet and CoRBS datasets. For ShapeNet dataset, which involves the synthetic data, we add only depth noises and both depth noises and pose noises, respectively. The results are shown in Tables 1 and 2. To compare with state-of-the-art methods, our method is evaluated with four metrics, i.e., the mean squared error (MSE), the mean absolute distance (MAD), intersection over union (IoU) and accuracy (ACC). MSE and MAD mainly focus on the distance between the estimated TSDF and the ground truth, while IoU and ACC quantify the occupancy of the estimation. According to the results, our method outperforms the state-of-the-art methods on all metrics for both scenarios. Especially when there exist both depth noises and pose noises, our method shows a significant advantage over other methods. When only depth noises exist, the RoutedFusion method and the proposed DFusion method have similar performance, while the latter shows a slight advantage due to the post-processing of the Denoising Module. Figures 5 and 6 illustrate the fusion results on the ShapeNet dataset with depth noises or pose noises, respectively, which is more intuitive to show the advantages of DFusion method. Consistent with the metric results, we can see that DFusion can give clean and precise fusion for all these objects. Due to the use of deep learning models, RoutedFusion and DFusion both have satisfactory outputs when depth noises are added, as shown in Figure 5. However, when pose noises exist (as shown in Figure 6), the fusion results of RoutedFusion deteriorate a lot, while our DFusion model can still have a precise output.

Table 1. Comparison results on ShapeNet (with only depth noise).

Methods	MSE	MAD	ACC	IoU
DeepSDF [35]	412.0	0.049	68.11	0.541
OccupacyNetworks [23]	47.5	0.016	86.38	0.509
TSDF Fusion [1]	10.9	0.008	88.07	0.659
RoutedFusion [4]	5.4	0.005	95.29	0.816
DFusion (Ours)	**3.5**	**0.003**	**96.12**	**0.847**

Table 2. Comparison results on ShapeNet (with depth noise and pose noise).

Methods	MSE	MAD	ACC	IoU
DeepSDF [35]	420.3	0.052	66.90	0.476
OccupacyNetworks [23]	108.6	0.037	77.34	0.453
TSDF Fusion [1]	43.4	0.020	80.45	0.582
RoutedFusion [4]	20.8	0.017	88.19	0.729
DFusion (Ours)	**6.1**	**0.006**	**95.08**	**0.801**

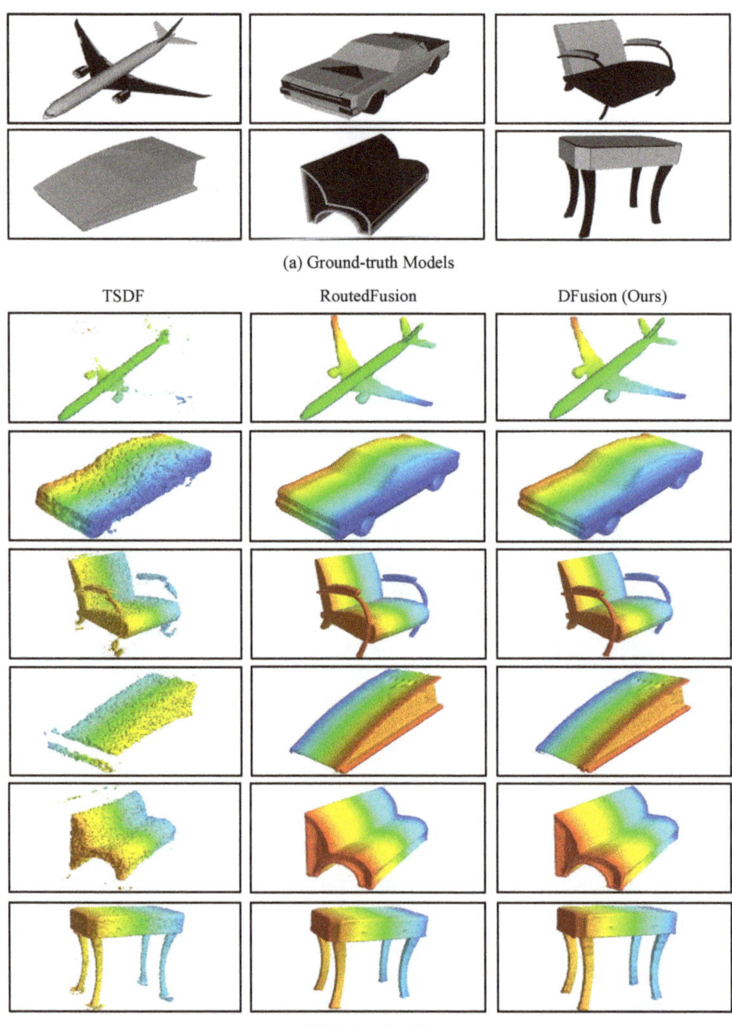

(a) Ground-truth Models

TSDF RoutedFusion DFusion (Ours)

(b) Fusion Results

Figure 5. Fusion results on the ShapeNet dataset with depth noise added.

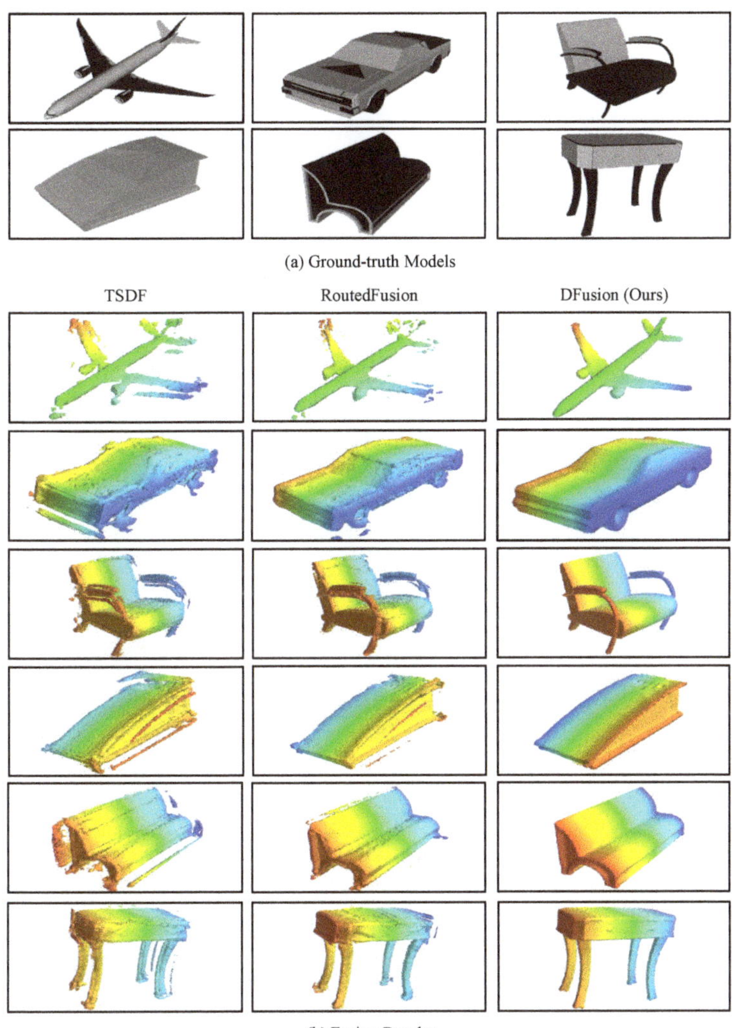

(a) Ground-truth Models

| TSDF | RoutedFusion | DFusion (Ours) |

(b) Fusion Results

Figure 6. Fusion results on the ShapeNet dataset with pose noise added.

For the CoRBS dataset, we choose four real scenes to perform the comparison with KinectFusion and RoutedFusion method. However, the pose information needs to be calculated before fusing the depth maps. KinectFusion method involves the process of calculating the pose information, which is the iterative closest point (ICP) algorithm [36]. Hence, to generate the TSDF volume, we use the ICP algorithm to obtain pose information for RoutedFusion and DFusion method, then compare the results on the MAD metric. The results are shown in Table 3. For all the scenes, our method achieves the best result. We also show some visualization results in Figure 7, which proves that our method can denoise the TSDF volume effectively and obtain more complete and smooth object models (note the cabinet edges, desk legs, and the human model arms).

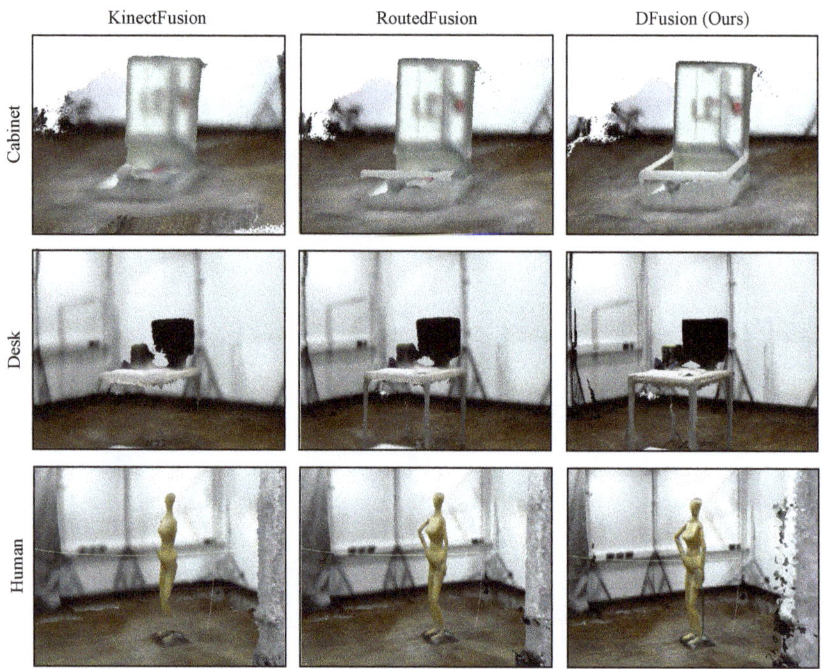

Figure 7. Fusion results on the CoRBS dataset. ICP algorithm [36] is used to obtain the sensor trajectory for RoutedFusion and DFusion.

Table 3. Quantitative results (MAD) on the CoRBS dataset.

Methods	Human	Desk	Cabinet	Car
KinectFusion [5]	0.015	0.005	0.009	0.009
ICP + RoutedFusion [4]	0.014	0.005	0.008	0.009
ICP + DFusion (Ours)	**0.012**	**0.004**	**0.006**	**0.007**

4.4. Ablation Study

To verify the effectiveness of the proposed loss function, we perform an ablation study, which compares the results with other three variants of the loss function, i.e., the loss function without object loss, the loss function without surface loss and the loss function without both object and surface loss. The original loss is our default setting which involves space loss, object loss and surface loss. For all variants, the experiment is conducted on the ShapeNet dataset with both depth noises and pose noises added. The results are shown in Table 4. It can be seen that the original setting can achieve the best performance for all metrics, which demonstrates the effectiveness of the proposed loss functions.

Table 4. Variants of the proposed method (with depth noise and pose noise).

Methods	MSE	MAD	ACC	IoU
Without object loss	8.3	0.007	92.11	0.744
Without surface loss	7.5	**0.006**	91.83	0.769
Without object&surface loss	16.3	0.015	90.87	0.740
Original	**6.1**	**0.006**	**95.08**	**0.801**

5. Conclusions

In this paper, we propose a new depth fusion network, considering not only depth noises but also pose noises of depth sensors, which is more realistic in 3D reconstruction. To improve the fusion quality, a new CNN model is proposed after fusing the depth maps. A synthetic dataset and a real-scene dataset are adopted to verify the effectiveness of our method. It has been proved that our method outperforms existing depth fusion methods for both quantitative results and qualitative results.

One limitation of our proposed method is that it can only be used after all depth sequences have been obtained. Therefore, it cannot be deployed in systems that require real-time fusion. A possible solution is to involve incomplete depth sequences in the training process, where we may need to redesign the noise generation and model optimization methods, which can be one of the future objectives. In addition, DFusion may have some performance issues if it is only trained on a small dataset, as the Denoising Module requires enough training samples. Therefore, more works are needed to lower its data requirements.

Author Contributions: Conceptualization, Z.N. and H.K.; methodology, Z.N.; software, Z.N.; validation, Y.F., M.K. and H.K.; formal analysis, T.S. and H.K.; investigation, Z.N.; resources, Z.N.; data curation, Z.N.; writing—original draft preparation, Z.N.; writing—review and editing, Y.F., M.K. and T.S.; visualization, Z.N.; supervision, H.K.; project administration, H.K. All authors have read and agreed to the published version of the manuscript.

Funding: This research received no external funding.

Institutional Review Board Statement: Not applicable.

Informed Consent Statement: Not applicable.

Data Availability Statement: The data is from the public datasets.

Conflicts of Interest: The authors declare no conflict of interest.

References

1. Curless, B.; Levoy, M. A volumetric method for building complex models from range images. In Proceedings of the 23rd annual conference on Computer Graphics and Interactive Techniques, New Orleans, LA, USA, 4–9 August 1996; pp. 303–312.
2. Lefloch, D.; Weyrich, T.; Kolb, A. Anisotropic point-based fusion. In Proceedings of the 2015 18th International Conference on Information Fusion (Fusion), Washington, DC, USA, 6–9 July 2015; pp. 2121–2128.
3. Dong, W.; Wang, Q.; Wang, X.; Zha, H. PSDF fusion: Probabilistic signed distance function for on-the-fly 3D data fusion and scene reconstruction. In Proceedings of the European Conference on Computer Vision (ECCV), Munich, Germany, 8–14 September 2018; pp. 701–717.
4. Weder, S.; Schonberger, J.; Pollefeys, M.; Oswald, M.R. RoutedFusion: Learning real-time depth map fusion. In Proceedings of the IEEE/CVF Conference on Computer Vision and Pattern Recognition, Seattle, WA, USA, 13–19 June 2020; pp. 4887–4897.
5. Newcombe, R.A.; Izadi, S.; Hilliges, O.; Molyneaux, D.; Kim, D.; Davison, A.J.; Kohi, P.; Shotton, J.; Hodges, S.; Fitzgibbon, A. KinectFusion: Real-time dense surface mapping and tracking. In Proceedings of the 2011 10th IEEE International Symposium on Mixed and Augmented Reality, Basel, Switzerland, 26–29 October 2011; pp. 127–136.
6. Dai, A.; Nießner, M.; Zollhöfer, M.; Izadi, S.; Theobalt, C. BundleFusion: Real-time globally consistent 3D reconstruction using on-the-fly surface reintegration. *ACM Trans. Graph. (ToG)* **2017**, *36*, 1. [CrossRef]
7. Nießner, M.; Zollhöfer, M.; Izadi, S.; Stamminger, M. Real-time 3D reconstruction at scale using voxel hashing. *ACM Trans. Graph. (ToG)* **2013**, *32*, 1–11. [CrossRef]
8. Marniok, N.; Goldluecke, B. Real-time variational range image fusion and visualization for large-scale scenes using GPU hash tables. In Proceedings of the 2018 IEEE Winter Conference on Applications of Computer Vision (WACV), Lake Tahoe, NV, USA, 12–15 March 2018; pp. 912–920.
9. Zienkiewicz, J.; Tsiotsios, A.; Davison, A.; Leutenegger, S. Monocular, real-time surface reconstruction using dynamic level of detail. In Proceedings of the 2016 Fourth International Conference on 3D Vision (3DV), Stanford, CA, USA, 25–28 October 2016; pp. 37–46.
10. Keller, M.; Lefloch, D.; Lambers, M.; Izadi, S.; Weyrich, T.; Kolb, A. Real-time 3D reconstruction in dynamic scenes using point-based fusion. In Proceedings of the 2013 International Conference on 3D Vision-3DV 2013, Seattle, WA, USA, 29 June–1 July 2013; pp. 1–8.
11. Schöps, T.; Sattler, T.; Pollefeys, M. SurfelMeshing: Online surfel-based mesh reconstruction. *IEEE Trans. Pattern Anal. Mach. Intell.* **2019**, *42*, 2494–2507. [CrossRef] [PubMed]

12. Stückler, J.; Behnke, S. Multi-resolution surfel maps for efficient dense 3D modeling and tracking. *J. Vis. Commun. Image Represent.* **2014**, *25*, 137–147. [CrossRef]
13. Woodford, O.J.; Vogiatzis, G. A generative model for online depth fusion. In *European Conference on Computer Vision*; Springer: Berlin/Heidelberg, Germany, 2012; pp. 144–157.
14. Ulusoy, A.O.; Black, M.J.; Geiger, A. Patches, planes and probabilities: A non-local prior for volumetric 3D reconstruction. In Proceedings of the IEEE Conference on Computer Vision and Pattern Recognition, Las Vegas, NV, USA, 27–30 June 2016; pp. 3280–3289.
15. Yong, D.; Mingtao, P.; Yunde, J. Probabilistic depth map fusion for real-time multi-view stereo. In Proceedings of the 21st International Conference on Pattern Recognition (ICPR2012), Tsukuba Science City, Japan, 11–15 November 2012; pp. 368–371.
16. Ulusoy, A.O.; Geiger, A.; Black, M.J. Towards probabilistic volumetric reconstruction using ray potentials. In Proceedings of the 2015 International Conference on 3D Vision (3DV), Lyon, France, 19–22 October 2015; pp. 10–18.
17. Dai, A.; Nießner, M. 3DMV: Joint 3D-multi-view prediction for 3D semantic scene segmentation. In Proceedings of the European Conference on Computer Vision (ECCV), Munich, Germany, 8–14 October 2018; pp. 452–468.
18. Dai, A.; Ritchie, D.; Bokeloh, M.; Reed, S.; Sturm, J.; Nießner, M. Scancomplete: Large-scale scene completion and semantic segmentation for 3d scans. In Proceedings of the IEEE Conference on Computer Vision and Pattern Recognition, Salt Lake City, UT, USA, 18–23 June 2018; pp. 4578–4587.
19. Paschalidou, D.; Ulusoy, O.; Schmitt, C.; Van Gool, L.; Geiger, A. RayNet: Learning volumetric 3D reconstruction with ray potentials. In Proceedings of the IEEE Conference on Computer Vision and Pattern Recognition, Salt Lake City, UT, USA, 18–23 June 2018; pp. 3897–3906.
20. He, K.; Gkioxari, G.; Dollár, P.; Girshick, R. Mask r-cnn. In Proceedings of the IEEE International Conference on Computer Vision, Venice, Italy, 22–29 October 2017; pp. 2961–2969.
21. Gkioxari, G.; Malik, J.; Johnson, J. Mesh R-CNN. In Proceedings of the IEEE/CVF International Conference on Computer Vision, Seoul, Korea, 27–28 October 2019; pp. 9785–9795.
22. Murez, Z.; van As, T.; Bartolozzi, J.; Sinha, A.; Badrinarayanan, V.; Rabinovich, A. Atlas: End-to-end 3D scene reconstruction from posed images. In Proceedings of the Computer Vision–ECCV 2020: 16th European Conference, Glasgow, UK, 23–28 August 2020; Part VII 16; Springer: Berlin/Heidelberg, Germany, 2020; pp. 414–431.
23. Riegler, G.; Ulusoy, A.O.; Bischof, H.; Geiger, A. OctnetFusion: Learning depth fusion from data. In Proceedings of the 2017 International Conference on 3D Vision (3DV), Qingdao, China, 10–12 October 2017; pp. 57–66.
24. Leroy, V.; Franco, J.S.; Boyer, E. Shape reconstruction using volume sweeping and learned photoconsistency. In Proceedings of the European Conference on Computer Vision (ECCV), Munich, Germany, 8–14 September 2018; pp. 781–796.
25. Zollhöfer, M.; Dai, A.; Innmann, M.; Wu, C.; Stamminger, M.; Theobalt, C.; Nießner, M. Shading-based refinement on volumetric signed distance functions. *ACM Trans. Graph. (TOG)* **2015**, *34*, 1–14. [CrossRef]
26. Cherabier, I.; Schonberger, J.L.; Oswald, M.R.; Pollefeys, M.; Geiger, A. Learning priors for semantic 3D reconstruction. In Proceedings of the European Conference on Computer Vision (ECCV), Munich, Germany, 8–14 September 2018; pp. 314–330.
27. Duan, C.; Chen, S.; Kovacevic, J. 3D point cloud denoising via deep neural network based local surface estimation. In Proceedings of the ICASSP 2019-2019 IEEE International Conference on Acoustics, Speech and Signal Processing (ICASSP), Brighton, UK, 12–17 March 2019; pp. 8553–8557.
28. Rakotosaona, M.J.; La Barbera, V.; Guerrero, P.; Mitra, N.J.; Ovsjanikov, M. PointCleanNet: Learning to denoise and remove outliers from dense point clouds. In *Computer Graphics Forum*; Wiley Online Library: Hoboken, NJ, USA, 2020; Volume 39, pp. 185–203.
29. Han, X.; Li, Z.; Huang, H.; Kalogerakis, E.; Yu, Y. High-resolution shape completion using deep neural networks for global structure and local geometry inference. In Proceedings of the IEEE International Conference on Computer Vision, Venice, Italy, 22–29 October 2017; pp. 85–93.
30. Cao, Y.P.; Liu, Z.N.; Kuang, Z.F.; Kobbelt, L.; Hu, S.M. Learning to reconstruct high-quality 3D shapes with cascaded fully convolutional networks. In Proceedings of the European Conference on Computer Vision (ECCV), Munich, Germany, 8–14 September 2018; pp. 616–633.
31. Graves, A. Generating sequences with recurrent neural networks. *arXiv* **2013**, arXiv:1308.0850.
32. Chang, A.X.; Funkhouser, T.; Guibas, L.; Hanrahan, P.; Huang, Q.; Li, Z.; Savarese, S.; Savva, M.; Song, S.; Su, H.; et al. ShapeNet: An information-rich 3D model repository. *arXiv* **2015**, arXiv:1512.03012.
33. Wasenmüller, O.; Meyer, M.; Stricker, D. CoRBS: Comprehensive RGB-D benchmark for SLAM using Kinect v2. In Proceedings of the 2016 IEEE Winter Conference on Applications of Computer Vision (WACV), Lake Placid, NY, USA, 7–10 March 2016; pp. 1–7.
34. Sturm, J.; Engelhard, N.; Endres, F.; Burgard, W.; Cremers, D. A benchmark for the evaluation of RGB-D SLAM systems. In Proceedings of the 2012 IEEE/RSJ International Conference on Intelligent Robots and Systems, Vilamoura, Portugal, 7–12 October 2012; pp. 573–580.
35. Park, J.J.; Florence, P.; Straub, J.; Newcombe, R.; Lovegrove, S. DeepSDF: Learning continuous signed distance functions for shape representation. In Proceedings of the IEEE/CVF Conference on Computer Vision and Pattern Recognition, Long Beach, CA, USA, 15–20 June 2019; pp. 165–174.
36. Besl, P.J.; McKay, N.D. Method for registration of 3-D shapes. In Proceedings of the Sensor Fusion IV: Control Paradigms and Data Structures, International Society for Optics and Photonics, Munich, Germany, 12–15 November 1992; Volume 1611; pp. 586–606.

Article

Content-Aware SLIC Super-Pixels for Semi-Dark Images (SLIC++)

Manzoor Ahmed Hashmani [1], Mehak Maqbool Memon [1,*], Kamran Raza [2], Syed Hasan Adil [2], Syed Sajjad Rizvi [3] and Muhammad Umair [1]

[1] High Performance Cloud Computing Center (HPC3), Department of Computer and Information Sciences, Universiti Teknologi PETRONAS, Seri Iskandar 32610, Malaysia; manzoor.hashmani@utp.edu.my (M.A.H.); muhammad_17008606@utp.edu.my (M.U.)
[2] Faculty of Engineering Science and Technology, Iqra University, Karachi 75600, Pakistan; kraza@iqra.edu.pk (K.R.); hasan.adil@iqra.edu.pk (S.H.A.)
[3] Department of Computer Science, Shaheed Zulfiqar Ali Bhutto Institute of Science and Technology, Karachi 75600, Pakistan; sshussainr@gmail.com
* Correspondence: mehak_19001057@utp.edu.my

Abstract: Super-pixels represent perceptually similar visual feature vectors of the image. Super-pixels are the meaningful group of pixels of the image, bunched together based on the color and proximity of singular pixel. Computation of super-pixels is highly affected in terms of accuracy if the image has high pixel intensities, i.e., a semi-dark image is observed. For computation of super-pixels, a widely used method is SLIC (Simple Linear Iterative Clustering), due to its simplistic approach. The SLIC is considerably faster than other state-of-the-art methods. However, it lacks in functionality to retain the content-aware information of the image due to constrained underlying clustering technique. Moreover, the efficiency of SLIC on semi-dark images is lower than bright images. We extend the functionality of SLIC to several computational distance measures to identify potential substitutes resulting in regular and accurate image segments. We propose a novel SLIC extension, namely, SLIC++ based on hybrid distance measure to retain content-aware information (lacking in SLIC). This makes SLIC++ more efficient than SLIC. The proposed SLIC++ does not only hold efficiency for normal images but also for semi-dark images. The hybrid content-aware distance measure effectively integrates the Euclidean super-pixel calculation features with Geodesic distance calculations to retain the angular movements of the components present in the visual image exclusively targeting semi-dark images. The proposed method is quantitively and qualitatively analyzed using the Berkeley dataset. We not only visually illustrate the benchmarking results, but also report on the associated accuracies against the ground-truth image segments in terms of boundary precision. SLIC++ attains high accuracy and creates content-aware super-pixels even if the images are semi-dark in nature. Our findings show that SLIC++ achieves precision of 39.7%, outperforming the precision of SLIC by a substantial margin of up to 8.1%.

Keywords: clustering; similarity measure; geodesic measure; Euclidean measure

Citation: Hashmani, M.A.; Memon, M.M.; Raza, K.; Adil, S.H.; Rizvi, S.S.; Umair, M. Content-Aware SLIC Super-Pixels for Semi-Dark Images (SLIC++). *Sensors* **2022**, *22*, 906. https://doi.org/10.3390/s22030906

Academic Editor: Jing Tian

Received: 27 September 2021
Accepted: 17 November 2021
Published: 25 January 2022

Publisher's Note: MDPI stays neutral with regard to jurisdictional claims in published maps and institutional affiliations.

1. Introduction

Image segmentation has potential to reduce the image complexities associated with processing of singular image primitives. Low-level segmentation of an image in non-overlapping set of regions called super-pixels helps in pre-processing and speeding up further high-level computational tasks related to visual images. The coherence feature of super-pixels allows faster architectural functionalities of many visual applications including object localization [1], tracking [2], posture estimation [3], recognition [4,5], semantic segmentation [6], instance segmentation [7], and segmentation of medical imagery [8,9]. These applications will be aided by super-pixels in terms of boosted performances, as the super-pixels put forward only the discriminating visual information [10].

Low-level segmentation tends to result in incorrect segmentations if the visual image has high pixel intensities; these high pixeled values are usually the biproduct of visual scenes captured in low lighting conditions, i.e., semi-dark images and dark images. The obtained incorrect super-pixels are attributed to the underlying approach used for the final segment creation. The existing super-pixel methods fail due to incorrect pixel manipulations for base operational functionality. The currently employed pixel manipulation relies on straight line differences for super-pixel creation. These straight-line difference manipulations fail to retain the content-aware information of the image. The results are further degraded if the image has low contrasted values, which result in no clear discrimination among the objects present. The existing super-pixel creation methods are divided into two categories based on the implemented workflow. The two categories are graph-based and gradient ascent based [10,11]. The former, focuses on minimization of cost function by grouping and treating each pixel of the image as a graph node. The later, iteratively processes each image pixel using clustering techniques until convergence [12]. One of the typical features observed for identification of super-pixel accuracy is regularity, i.e., to what extent the super-pixel is close to the actual object boundary of the image. All the graph-based methods for super-pixel calculations suffer from poorly adhered super-pixels which result in irregularity of segments presented by super-pixels [11]. Additionally, graph-based methods are constrained by the excessive computational complexity and other initialization parameters. Whereas, the gradient-ascent methods are simplistic in nature and are recommended in the literature due to resultant high performance and accuracy [13,14]. However, there are some issues associated with content irrelevant manipulation of singular pixels to form resultant super-pixels. Some of the key features of using super-pixel segmentations are:

- Super-pixels abstraction potentially decreases the overhead of processing each pixel at a time.
- Gray-level or color images can be processed by a single algorithm implementation.
- Integrated user control provided for performance tuning.

With these advantages of super-pixels, they are highly preferred. However, super-pixel abstraction methods backed by gradient-ascent workflows are also limited in their working functionality to retain the contextual information of the given image [15]. The contextual information retainment is required to achieve the richer details of the visual image. This loss of contextual information is caused by flawed pixel clusters created based on Euclidean distance measure [16]. As the Euclidean distance measures calculates straight line differences among pixels which ends up in irregular and lousy super-pixels. Moreover, further degradation can be expected to process the semi-dark images where high pixel intensities along with no clear boundaries are observed. In such scenarios, the propagation of inaccurate super-pixels will affect the overall functionality of automated solutions [11]. To overcome this problem of information loss for creation of compact and uniform super-pixels, we propose content-aware distance measures for image pixel cluster creation. The content-aware distance measure as the core foundational component of gradient-ascent methods for super-pixel creation will not only help in alleviation of information loss, but it will also help in preserving less observant/perceptually visible information of semi-dark images. The state-of-the-art methods for super-pixels creation have not been analyzed exclusively on the semi-dark images which further raises the concerns related to segmentation accuracy. In nutshell, the problems in existing segmentation algorithms are:

- Absence of classification of state-of-the-art methods based on low level manipulations.
- Inherited discontinuity in super-pixel segmentation due to inconsistent manipulations.
- Unknown pixel grouping criteria in terms of distance measure to retain fine grained details.
- Unknown effect of semi-dark images on final super-pixel segmentations.

To resolve these issues, the presented research exclusively presents multifaceted study offering following features:

1. Classification of Literary Studies w.r.t Singular Pixel Manipulation Strategies:

The categorization of the existing studies is based on entire image taken as one entity. The image entity can represent either graph or a feature space to be clustered, i.e., graph-based or gradient-ascent based methodology for pixel grouping. To the best of our knowledge, there has been no study that categorizes existing studies based on the manipulation strategy performed over each pixel. We present the detailed comparative analysis of existing research highlighting their core functionality as the basis for classification.

2.   Investigation of Appropriate Pixel Grouping Scheme:

The grouping scheme backing the image segmentation module being crucial component can highly affect the accuracy of the entire model. For this reason, to propose a novel extension as a generalized solution of all types of images including semi-dark images, we present a detailed qualitative investigation of up to seven distance measures for grouping pixels to create super-pixels. The investigation resulted in shortlisted pixel grouping measures to retain fine grained details of the visual image.

3.   Novel Hybrid Content-Aware Extension of SLIC—SLIC++:

SLIC, being the simplest and fastest solution for the pixel grouping, remains the inspiration, and we enhance the performance of SLIC by adding content-aware feature in its discourse. The proposed extension holds the fundamental functionality with improved features to preserve content-aware information. The enhancement results in better segmentation accuracy by extracting regular and continuous super-pixels for all scenarios including semi-dark scenarios.

4.   Comprehensive Perceptual Results focusing Semi-dark Images:

To assess the performance of the proposed extension SLIC++ for extraction of the richer information of the visual scene, we conduct experiments over semi-dark images. The experimental analysis is benchmarked against the standard super-pixel creation methods to verify that incorporating content-aware hybrid distance measure leads to improved performance. The perceptual results further conform better performance, the scalability, and generalizability of results produced by SLIC++.

Paper Organization

The remainder of paper is organized as follows: Section 2 presents the prior super-pixel creation research and its relevance to semi-dark technology. In this section we also present critical analysis of studies proposed over period of two decades and their possible applicability for semi-dark imagery. We also critically analyze two closely related studies and highlight the difference among them w.r.t SLIC++. Section 3 describes the extension hypothesis and the final detailed proposal for super-pixel segmentation of semi-dark images. Section 4 presents the detailed quantitative and qualitative analysis of SLIC extension against state-of-the-art algorithms validating the proposal. Section 5 discusses the applicability of the proposed algorithm in the domain of computer vision. Finally, Section 6 concludes the presented research and points out some future directions of research.

2. Literature Review

2.1. Limited Semi-Dark Image Centric Research Focusing Gradient-Ascent Methods

The gradient ascent methods are also called clustering-based methods. These methods take the input image and rasterize the image. Then, based on the local image cues such as color and spatial information the pixels are clustered iteratively. After each iteration, gradients are calculated to refine the new clusters from the previously created clusters [14]. The iterative process continues till the algorithm converges after the gradients stop changing, thus named gradient-ascent methods. A lot of research has been already done in the domain of gradient-ascent methods; the list of these methods is presented in Table 1.

Table 1. Critical analysis of gradient-ascent super-pixel creation algorithms.

Method	Complexity	Pixel Manipulation Strategy	Distance Measure	Dataset	Semi-Dark Image Mentions	Year	Ref.
Meanshift	$O(N^2)$	Mode seeking to locate local maxima	Euclidean	Not mentioned	✗	2002	[17]
Medoidshift	$O(N^2)$	Approximates local gradient using weighted estimates of medoids.	Euclidean	Not mentioned	✗	2007	[18]
Quickshift	$O(dN^2)$	Parzen's density estimation for pixel values	Non-Euclidean	Caltech-4	✗	2008	[19]
TurboPixel	$O(N)$	Geometric flows for limited pixels	Gradient calculation for boundary pixels only	Berkeley Dataset	✗	2009	[20]
Scene Shape Super-pixel (SSP)	$O\left(N^{\frac{3}{2}}logN\right)$	Shortest path manipulation with prior information of boundary.	Probabilistic modeling plus Euclidean space manipulation	Dynamic road scenes. No explicit mentions of semi-dark images but we suspect presence of semi	✔	2009	[21]
Compact Super-pixels (CS)	-	Approximation of distance between pixels and further optimization with graph cut methods	Euclidean	3D images	✗	2010	[22]
Compact Intensity Super-pixels	-	Same as CS, With added color constant information.	Euclidean	3D images	✗	2010	[22]
SLIC	$O(N)$	Gradient optimization after every iteration	Euclidean	Berkeley Dataset	✗	2012	[11]
SEEDS	-	Energy optimization for super-pixel is based on hill-climbing.	Euclidean	Berkeley Dataset	✗	2012	[23]
Structure Sensitive Super-pixels	$O(N)$	Super-pixel densities are checked, and energy minimization is conformed.	Geodesic	Berkeley Dataset	✗	2013	[16]
Depth-adaptive super-pixels	$O(N)$	Super-pixel density identification, followed by sampling and finally k-means to create final clusters	Euclidean	RGB-D dataset consisting of 11 images	✗	2013	[24]
Contour Relaxed Super-pixels	$O(N)$	Uses pre-segmentation technique to create homogeneity constraint	Not mentioned	Not mentioned	✗	2013	[25]
Saliency-based super-pixel	$O(N)$	Super-pixel creation followed by merging operator based on saliency.	Euclidean	Not mentioned	✗	2014	[26]

Table 1. *Cont.*

Method	Complexity	Pixel Manipulation Strategy	Distance Measure	Dataset	Semi-Dark Image Mentions	Year Ref.
Linear Spectral Clustering	$O(N)$	Two-fold pixel manipulation strategy of optimization based on graph and clustering based algorithms.	Euclidean	Berkeley Dataset	✗	2015 [27]
Manifold SLIC	$O(N)$	Same as SLIC but with mapping over manifold.	Euclidean	Berkeley Dataset (Random)	✗	2016 [15]
BASS (Boundary-Aware Super-pixel Segmentation)	$O(NlogN)$	Extension of SLIC initially creates boundary then uses SLIC with different distance measures, along with optimization of initialization parameters.	Euclidean + Geodesic	Fashionista, Berkeley Segmentation Dataset (BSD), HorseSeg, DogSeg, MSRA Salient Object Database, Complex Scene Saliency Dataset (CSSD) and Extended CSSD (ECSSD)	✔	2016 [28]
BSLIC	$O(6N)$	Extension of SLIC initializes seed within a hexagonal space rather than square	Euclidean	Berkeley Dataset	✗	2017 [29]
Intrinsic Manifold SLIC	$O(N)$	Extension of Manifold SLIC with geodesic distance measure	Geodesic	Berkeley Dataset (Random)	✗	2017 [30]
Similarity Ratio based Super-pixels	$O(N)$	Extension of SLIC. Proposes automatic scaling of coordinate axes.	Mahanlanobis	SAR Image dataset	✔	2017 [31]
Scalable SLIC	$O(N)$	Optimized initialization parameters such as 'n' number of super-pixels, focused research to parallelization of sequential implementation.	Euclidean	cyrosection Visible Human Male dataset	✗	2018 [32]
Content adaptive super-pixel segmentation	$O(N^2)$	Work on prior transformation of image with highlighted edges created by edge filters	Euclidean (with graph-based transformation)	Berkeley Dataset	✗	2019 [33]
BASS (Bayesian Adaptive Super-Pixel Segmentation)	$O(N)$	Uses probabilistic methods to intelligently initialize the super-pixel seeds.	Euclidean	Berkeley Dataset	✗	2019 [34]
Super-pixel segmentation with fully convolutional networks	$O\left(N^{No.\ of\ layers}\right)$	Attempts to use neural networks for automatic seed initialization over grid.	Euclidean	Berkeley Dataset, SceneFlow Dataset	✗	2020 [35]

Table 1. *Cont.*

Method	Complexity	Pixel Manipulation Strategy	Distance Measure	Dataset	Semi-Dark Image Mentions	Year Ref.
Texture-aware and structure preserving Super-pixels	$O(N)$	The seed initialization takes place in circular grid.	Three different distance measure (without explicit details)	Berkeley Dataset	✗	2021 [36]
Efficient Image-Warping Framework for Content-Adaptive Super-pixels Generation	$O(N)$	Warping transform is used along with SLIC for creation of adaptive super-pixels.	Euclidean	Berkeley Dataset	✗	2021 [37]
Edge aware super-pixel segmentation with unsupervised CNN	$O\left(N^{No.\ of\ layers}\right)$	Edges are detected using unsupervised convolutional neural networks then passed to super-pixel segmentation algorithms	Entropy based clustering	Berkeley Dataset	✗	2021 [38]

The gradient ascent methods for super-pixel creation seems to be a promising solution due to their simplicity of implementation, speed of processing and easy adaptation for handling the latest demands of complex visual image scenarios. However, the concerns associated with underlying proper extraction strategies is one of the challenging aspects to cater the dynamic featural requirements imposed by complex visual image scenarios such as semi-dark images.

2.2. Critical Analysis of Gradient-Ascent Super-Pixel Creation Algorithms Based on Manipulation Strategy

For the critical analysis we have considered gradient ascent based super-pixel algorithms presented over period of two decades ranging from 2001 through 2021. The studies are retrieved from Google Scholar's repository with keywords including super-pixel segmentation, pixel abstraction, content sensitive super-pixel creation, content-aware super-pixel segmentation. The search resulted in a lot of segmentation related studies in domain of image processing including the basic image transformations along with related super-pixel segmentation studies. For the critical analysis, the studies mentioning clustering based super-pixel creations were shortlisted due to their relevance with proposed algorithm. The key features of these studies are critically analyzed and comprehensively presented in Table 1, along with the critiques for respective handling concerns associated with semi-dark imagery.

Key Takeaways

The critical analysis presented in Table 1 uncovers the fact that recent research explicitly points towards the need of segmentation algorithm which considers the content relevant super-pixel segmentations. To accomplish this task several techniques are proposed with incorporation of prior transformations of image via deep learning methods, simple image processing, and probabilistic methods. Mostly the research uses and conforms the achievements of Simple Linear Iterative Clustering (SLIC). Furthermore, most of the research studies are using SLIC algorithm for super-pixel creation as base mechanisms with added features. Generally, the algorithms proposed in last decade have computational complexity of $O(N)$, whereas if neural networks are employed for automation of required parameter initialization, then the complexity becomes $O\left(N^{No.\ of\ layers}\right)$. All the proposed algorithms use two distance measures for final super-pixel creation, i.e., Euclidean, or geodesic distance measure. However, all these studies have not mentioned the occurrence of semi-dark images and their impact on the overall performance. It is estimated that huge margin of the existing image dataset already includes the problem centric image data. The Berkeley dataset that is substantially used for performance analysis of super-pixel algorithms contains up to 63% semi-dark images. The proposed study uses the semi-dark images of Berkeley dataset for benchmarking analysis of the SLIC++ algorithm.

2.3. Exclusiveness of SLIC++ w.r.t Recent Developments

The recent studies substantially focus on super-pixels with the induced key features of content sensitivity and adherence of the final segmentations; consequently several related research studies have been proposed. Generally, the desired features are good boundary adherence, compact super-pixel size and low complexity. The same features are required for super-pixel segmentation of semi-dark images. In this section we briefly review the recent developments which are closely related to our proposed method for creating super-pixel in semi-dark images.

BASS (Boundary-Aware Super-pixel Segmentation) [28], is closely related to the methodology that we have chosen, i.e., incorporation of content relevant information in the final pixel labeling which ends up with the creation of super-pixels. However, the major difference resides in the initialization of the super-pixel seeds/centers. BASS recommends the usage of forest classification method prior to the super-pixel creation. This forest classification of image space results with the creation of a binary image with highlighted boundary information over the image space. This boundary information is then utilized to aid the initialization process of the seed/ cluster centers. Theoretically, the problem with this entire configuration is additional complexity of boundary map creation which raise the complexity from $O(N)$ to $O(NlogN)$. This boundary map creation and its associated condition of addition and deletion of seeds is expected to further introduce undesired super-pixel feature under-segmentation. The under-segmentation might take place due to easy seed deletion condition and difficult addition condition which means more seeds would be deleted and less would be added. This aspect of the study is not desired for the super-pixel creation in semi-dark scenarios. On the contrary we propose regularly distributed seeds along with usage of both the recommended distance measures without prior image transformation which further reduces the overall complexity. Finally, we also propose the usage of geodesic distance for color components of the pixel rather than only using it for spatial component.

Intrinsic manifold SLIC [30], is an extension of manifold SLIC which proposes usage of manifolds to map high dimensional image data on the manifolds resembling Euclidean space near each point. IMSLIC uses Geodesic Centroidal Voronoi Tessellations (GCVT) this allows flexibility of skipping the post-processing heuristics. For computation of geodesic distance on image manifold weighted image graph is overlayed with the same graph theory of edges, nodes, and weights. For this mapping 8-connected neighbors of pixel are considered. This entire process of mapping and calculation of geodesic distances seems complex. The theoretical complexity is $O(N)$, however with the incorporation of image graph the computational complexity will increase. Moreover, the conducted study computes only geodesic distance between the pixels leaving behind the Euclidean counterpart. With substantially less complexity we propose to implement both the distance measures for all the crucial pixel components.

2.4. Summary and Critiques

The comprehensive literature survey is conducted to benefit the readers and provide a kickstart review of advancements of super-pixel segmentation over the period of two decades. Moreover, the survey resulted in critical analysis of existing segmentation techniques which steered the attention to studies conducted for adverse image scenarios such as semi-dark images. Arguably, with the increased automated solutions the incoming image data will be of dynamic nature (including lighting conditions). To deal with this dynamic image data, there is a critical need of a super-pixel segmentation technique that takes into account the aspect of semi-dark imagery and results in regular and content-aware super-pixel segmentation in semi-dark scenarios. The super-pixel segmentation techniques currently employed for the task suffer from two major issues, i.e., high complexity, and information loss. The information loss associated with the gradient-ascent methods is attributed to restrictions imposed due to usage of Euclidean image space which totally loses the context of the information present in the image by calculating straight line differences. Many attempts have been made to incorporate CNN probabilistic methods in super-pixel creation methods to optimize and aid the final segmentation results. However, to the best of our knowledge there has been no method proposed exclusively for semi-dark images scenarios keeping the simplicity and optimal performance intact.

In following sections, we describe the preliminaries which are the base for the proposed extension of SLIC namely SLIC++. We also present several distance measures incorporated in base SLIC algorithm namely SLIC+ to analyze the performance for semi-dark images.

3. Materials and Methods

3.1. The Semi-Dark Dataset

For the analysis of the content-aware super-pixel segmentation algorithm, we have used the state-of-the-art dataset which has been used in the literature for years now. The Berkeley image dataset [39] has been used for the comprehensive analysis and benchmarking of the proposed SLIC++ algorithm with the state-of-the-art algorithms. The Berkeley image dataset namely BSDS 500 has got five hundred images overall, whereas the problem under consideration is of semi-dark images. For this purpose, we have initially extracted semi-dark images using RPLC (Relative Perceived Luminance Classification) algorithm. The labels are created based on the manipulation of color model information, i.e., Hue, Saturation, Lightness (HSL) [40]. The final semi-dark images extracted from the BSDS-500 dataset turned out to be 316 images. Each image has resolution of either 321×481 or 481×321 dimensions. The BSDS-500 image dataset provides the basis for empirical analysis of segmentation algorithms. For the performance analysis and boundary detection, the BSDS-500 dataset provides ground-truth labels by at least five human annotators on average. This raises questions about the selection of annotation provided by the subjects. To deal with this problem, we have performed a simple logic over the image ground truth labels. All the image labels are iterated with 'OR' operation to generate singular ground truth image label. The 'OR' operation is performed to make sure that the final ground truth is suggested by most of the human annotators. Finally, every image is segmented and benchmarked against this single ground truth labeled image.

3.2. Desiderata of Accurate Super-Pixels

Generally, for super-pixel algorithms there are no definite features for super-pixels to be accurate. The literary studies refer accurate super-pixels in terms of boundary adherence, connectivity of super-pixels, super-pixel partitioning, compactness, regularity, efficiency, controllable number of super-pixels and so on [13,41,42]. As the proposed study is research focused on semi-dark images, we take into account features that are desired for conformation of accurate boundary extraction in semi-dark images.

1. **Boundary Adherence**

The boundary adherence is the measure to compute the accuracy to which the boundary has been extracted by super-pixels against boundary image or ground-truth images. The idea is to preserve information as much as possible by creating super-pixels over the image. The boundary adherence feature is basically a measure that results how accurately the super-pixels have followed the ground-truth boundaries. This can be easily calculated by segmentation quality metrics precision-recall.

2. **Efficiency with Less Complexity**

As super-pixel segmentation algorithms are now widely used as preprocessing step for further visually intelligent tasks. The second desired feature is efficiency with less complexity. The focus should be creation of memory-efficient and optimal usage of processing resources so that more memory and computational resources can be used by subsequent process. We take into account this feature and propose an algorithm that uses exactly same resources as of Basic SLIC with added distance measures in its discourse.

3. **Controllable Number of Super-Pixels**

The controllable number is super-pixels is a desired feature to ensure the optimal boundary is extracted using the computational resources ideally. The super-pixel algorithms are susceptible to this feature that is number of super-pixels. The number of super-pixels to be created can directly impact the overall algorithm performance. The performance is degraded in terms of under-segmentation or over-segmentation error. In the former one, the respective algorithm fails to retrieve most of the boundaries due to the smaller number of super-pixels to be created, whereas the latter one retrieves maximum boundary portions of the ground-truth images but there is surplus of computational resources.

Nevertheless, as mentioned earlier there is a huge list of accuracy measures and all those measures refer to different segmentation aspects and features. The required features and subsequent accuracy measures to be reported depend on the application of algorithm. For semi-dark image segmentation, it is mandatory to ensure that most of the optimal boundary is extracted and this requirement can be related to precision-recall metrics.

3.3. SLIC Preliminaries

Before presenting the SLIC++, we first introduce base functionality of SLIC. The overall functionality is based on creation of restricted windows in which the user defined seeds are placed, and clustering of image point is performed in this restricted window. This restricted window is called Voronoi Tessellations [43]. Voronoi Tessellations is all about partitioning the image plane into convex polygons. This polygon is square in case of SLIC initialization windows. The Voronoi tessellations are made such that each partition has one generating point and all the point in the partition are close to the generating point or the mass center of that partition. As the generating point lies in the center these partitions are also called Centroidal Voronoi Tessellation (CVT). The SLIC algorithm considers CIELAB color space for the processing, where every pixel p on image I is presented by color components and spatial components as $c(p) = (l(p), a(p), b(p))$ being colour components and $p(u, v)$ being spatial components. For any two pixels SLIC measures straight line difference or Euclidean distance between the two pixels for the entire image space $\mathbb{R}^5$.

The spatial distance between two pixels is given by d_s and color component distance d_c are given in Equations (1) and (2).

$$d_s = \sqrt{(u_1 - u_2)^2 + (v_1 - v_2)^2} \tag{1}$$

And,

$$d_c = \sqrt{(l_1 - l_2)^2 + (a_1 - a_2)^2 + (b_1 - b_2)^2} \tag{2}$$

Here d_s and d_c represent Euclidean distance between pixel p_1 and p_2. Instead of simple Euclidean, SLIC uses distance term infused with Euclidean norm given by Equation (3).

$$D_s = d_c + \frac{m}{S} d_s \tag{3}$$

The final distance term is normalized using interval S and m provides control over the super-pixel compactness which results in perceptually meaningful distance with balanced aspect of spatial and color components. Provided the number of super-pixels K seeds $(s_i)_{i=1}^{K}$ are evenly distributed in over the image I clusters are created in restricted regions of Voronoi Tessellations. The initialization seed are placed in image space within a window of $2S \times 2S$ having center s_i. After that simple K-means is performed over the pixels residing in the window to its center. SLIC computes the distance between pixels using Equation (3) and iteratively processes the pixels until convergence.

3.4. The Extension Hypothesis—Fusion Similarity Measure

The super-pixels created by the SLIC algorithm basically uses the Euclidean distance measure to create pixel clusters or the super-pixels based on the seed or cluster centers. The Euclidean distance measure takes into account the similarity among pixels using straight line differences among cluster centers and the image pixels. This property of distance measure results in distortion of extracted boundaries of image. The reason is measure remains same no matter if there is a path along the pixels. The path along the pixels will result in smoother and content relevant pixels [16,36]. The Euclidean distance overlays a segmentation map over the image without having relevance to the actual content present in the image. Moreover, large diversity in the image (light conditions/high density portions) result in unavoidable distortion. Therefore, we hypothesize to use accurate distance measure which presents content relevant information of the visual scene. For this reason, we extend the functionality of SLIC by replacing the Euclidean distance measure with four potential similarity measures including chessboard, cosine, Minkowski, and geodesic and named it as SLIC+. These distance measures have been used in the literature integrated in clustering algorithms for synthetic textual data clustering where studies mentioned to render reasonable results for focused problem solving [44]. However, we use these similarity measures to investigate the effects on visual images using SLIC approach.

Prior to implementation, a brief introductory discussion will help understand the overall integration and foundation for choosing these similarity measures. The distance measures are basically the distance transforms applied on different images, specifying the distance from each pixel to the desired pixel. For uniformity and easy understanding, let pixel p_1 and p_2 have the coordinates (x_1, y_1) and (x_2, y_2), respectively.

- **Chessboard:** This measure calculates the maximum distance between vectors. This can be referred to measuring path between the pixels based on eight connected neighborhood whose edges are one unit apart. The chessboard distance along any co-ordinate is given by identifying maximum, as presented in Equation (4).

$$D_{chess} = \max(|x_2 - x_1|, |y_2 - y_1|) \tag{4}$$

Rationale of Consideration: Since the problem with existing similarity measures is loss of information, chessboard is one of the alternate to be incorporated in super-pixel creation base. This measure is considered as it takes into account information of eight connected neighbors of pixels under consideration. However, it might add computational overhead due to the same.

- **Cosine:** This measure calculates distance based on the angle between two vectors. The cosine angular dimensions counteract the problem of high dimensionality. The inner angular product between the vectors turns out to be one if vectors were previously normalized. Cosine distance is based on cosine similarity which is then plugged-in distance equation. Equations (5) and (6) shows calculation of cosine distance between pixels.

$$cosine\ similarity = \frac{p_1 \cdot p_2}{\sqrt{p_1^2}\sqrt{p_2^2}} \tag{5}$$

$$D_{cosine} = 1 - cosine\ similarity \tag{6}$$

Rationale of Consideration: One of the aspects of content aware similarity measure is to retain the angular information thus we attempted to incorporate this measure. The resulting super-pixels are expected to retain the content relevant boundaries. However, this measure does not consider magnitude of the vectors/pixels due to which boundary performance might fall.

- **Minkowski:** This measure is a bit more intricate. It can be used for normed vector spaces, where distance is represented as vector having some length. The measure multiplies a positive weight value which changes the length whilst keeping its direction. Equation (7) presents distance formulation of Minkowski similarity measure.

$$D_{min} = \left(|p_2 - p_1|^\mu\right)^{1/\mu} \tag{7}$$

Here μ is the weight, if its value is set to 1 the resultant measure corresponds to Manhattan distance measure. $\mu = 2$, refers to euclidean and $\mu = \infty$, refers to chessboard or Chebyshev distance measure. *Rationale of Consideration*: As user-control in respective application is desired, Minkowski similarity provides the functionality by replacing merely one parameter which changes the entire operationality without changing the core equations. However, here we still have problems relating to the retainment of angular information.

- **Geodesic:** This measure considers geometric movements along the pixel path in image space. This distance presents locally shortest path in the image plane. Geodesic distance computes distance between two pixels which results in surface segmentation with minimum distortion. Efficient numerical implementation of geodesic distance is achieved using first order approximation. For approximation parametric surfaces are considered with n number of points on the surface. Given an image mask, geodesic distance for image pixels can be calculated using Equation (8).

$$D_{geo} = \min_{P_{x_i,x_j}} \int_0^1 D(P_{x_i,x_j}(t) \| \dot{P}_{x_i,x_j}(t) \| dt \tag{8}$$

where $P_{x_i,x_j}(t)$ is connected path between pixel x_i, x_j, provided $t = 0,1$. The density function $D(x)$ increments the distance and can be computed using Equation (9).

$$D(x) = e^{\frac{E(x)}{v}}, \; E(x) = \frac{\| \nabla I \|}{G_\sigma * \| \nabla I \| + \gamma'} \tag{9}$$

where v is scaling factor, $E(x)$ is edge measurement also provides normalization of gradient magnitude of image $\| \nabla I \|$. G_σ is the Gaussian function with its standard deviation being σ. γ minimizes the effect of weak intensity boundaries over density function. $D(x)$ always produces constant distance, for homogeneous appearing regions if $E(x)$ is zero $D(x)$ becomes one. *Rationale of Consideration*: For shape analysis by computing distances geodesic has been the natural choice. However, computing geodesic distance is computationally expensive and is susceptible to noise [44]. Therefore, to overcome effect of noise geodesic distance should be used in amalgamation of Euclidean properties to retain maximum possible information in terms of minimum distance among pixels and their relevant angles.

The mentioned distance measures for identification of similarity among pixels based on pixel proximity provides different functionality features including extraction of information based on the 4-connected and 8-connected pixel neighborhood, and incorporation of geometric flows to keep track of angular movements of image pixels. However, none of these similarity measures provide balanced equation with integrated features of optimal boundary extraction based on connected neighbors and their angular movements. Thus, we hypothesize boundary extraction to be more accurate and intricate in presence of a similarity measure which provides greater information of spatial component provided by neighborhood pixels along with geometric flows.

3.5. SLIC++ Proposal

3.5.1. Euclidean Geodesic—Content-Aware Similarity Measure

Considering the simplicity and fast computation as critical components for segmentation, the proposed algorithm uses fusion of Euclidean and geodesic distance measures. The depiction of Euclidean and geodesic similarity is presented in Figure 1, where straight line shows Euclidean similarity while curved line shows geodesic similarity. Since using only Euclidean similarity loses the context information due to usage of straight-line distance and geodesic similarity focuses more on the actual possible path along the pixels. We propose the fusion of both the similarities to extract accurate information of image pixels and their associations.

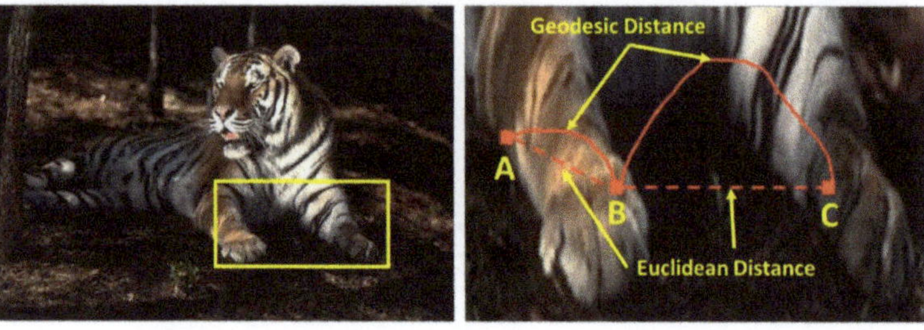

Figure 1. Irrelevance of Euclidean distance measure for super-pixel creation relating to image content.

Using the same logic as SLIC, we propose a normalized similarity measure. The normalization is based on the interval S between the pixel cluster centers. To provide the control over super-pixel same variable m is also used. Beforehand the contribution of Euclidean and geodesic distance in final similarity measure cannot be determined in terms of optimized performance. Hence, we have introduced two weight parameters for proposal of final similarity measure which based on weighted combination of Euclidean and geodesic distance. The proposed similarity measure is presented in Equation (10).

$$D_{ca} = w_1\left(d_1 + \frac{m}{S}d_2\right) + w_2\left(d_3 + \frac{m}{S}d_4\right) \tag{10}$$

where D_{ca} is content-aware distance measure, d_1 and d_2 are same as d_s and d_c (Equations (5) and (6)) calculating the Euclidean distance for spatial and color component of image pixels. Variables d_3 and d_4 presents color and spatial component distance calculation using geodesic distance equation 8. Specifically, d_3 represents geodesic calculation of color components of image pixel and d_4 represents geodesic calculation of spatial components of pixel. Here, again we introduce similar normalization as of SLIC using variable S and m to provide control super-pixel compactness using geometric flows. The weights w_1 and w_2 further provides user control to choose the contribution of Euclidean and geodesic distance in final segmentation. These weights provide user flexibility, and these values can be changed based on the application. Moreover, these weights can be further tuned in future studies.

3.5.2. Proposal of Content-Aware Feature Infusion in SLIC

The SLIC++ is proposed to extract the optimal information from a visual scene captured in semi-dark scenarios. Nevertheless, the same algorithm holds for any type of image if the objective is to retrieve maximum information from the image space. The steps involved in computing super-pixels are written in SLIC++ algorithm (refer Algorithm 1). Basically, super-pixels are perceptual cluster computed based on pixel proximity and color intensities. Some of the parameters include: Kbeing the number of super-pixels, Ntotal number of pixels, Aapproximate number of pixels also called area of super-pixel, and Slength of super-pixel.

Algorithm 1. SLIC++ Algorithm

1:   Initialize K cluster center with seed $(s_i)_{i=1}^{K}$ defined at regular intervals S
2:   Move cluster centers in $n \times n$ pixel neighborhood to lowest gradient
3:   *Repeat*
4:      *For* each cluster center s_i do
5:         Assign the pixel from $2S \times 2S$ in a square window or CVT using distance measure given by Equation (3).
6:         Assign the pixel from $2S \times 2S$ in a square window or CVT using distance measure given by Equation (10).
7:      *End for*
8:   Compute new cluster centers and residual error *err* (distance between previous centers and recomputed centers).
9:   *Until err<=* threshold
10: Enforce connectivity.

Keeping simplicity and fast computation intact we present SLIC++ algorithm, here only one of the steps mentioned on step 5 or 6 will be used. If step 5 is implemented, i.e., distance measure given by Equation (3) is used entire functionality of SLIC algorithm is implemented. Whereas, if step 6 is implemented, i.e., distance measure given by Equation (10) is used entire functionality of SLIC algorithm is implemented.

a.   Initialization and Termination

For initialization a grid of initial point is created separated by distance S in each direction as seen in Figure 2. The number of initial centers is given as parameter K. Placement of initial center in restricted squared grids can result in error if the initial center is placed on the edge of image content. This initial center is termed as confused center. To overcome this error gradient of the image is computed and the cluster center is moved in the direction of minimum gradient. The gradient is computed with 4-neighboring pixels and the centroid is moved. To solve this mathematically L2 Norm distance is computed among four connected neighbors of center pixel. The gradient calculation is given by Equation (11).

$$G(x,y) = \|(x+1,y) - (x-1,y)\|^2 + \|(x,y+1) - (x,y-1)\|^2 \tag{11}$$

$G(x,y)$ is the gradient of center pixel under consideration.

The gradient of the image pixels is calculated until stability where pixels stop changing the clusters based on the initialized clusters. Overall, the termination and optimization is controlled by parameter 'n' which represents number of iterations the overall SLIC algorithm goes through before finally resulting in super-pixel creation of the image. To keep the uniformity in presented research we have selected 'n' as 10 which has been a common practice [11,14,29,32].

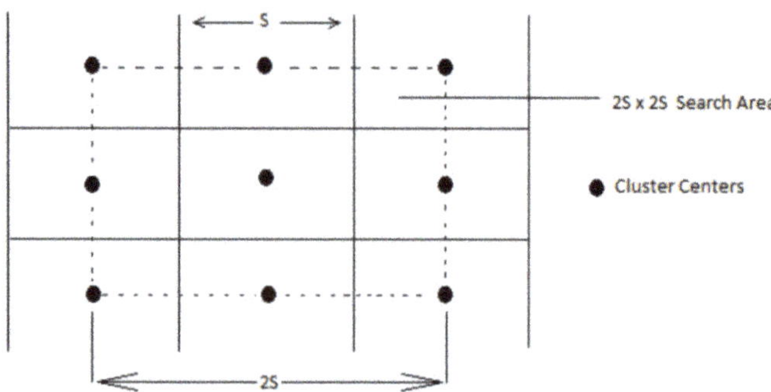

Figure 2. Restricted Image search area for super-pixel creation specified by input argument for image window under consideration [31].

How it works?

The incoming image is converted to CIELAB space. The user provides information of all the initialization parameters including 'K', 'm', 'n'. Referring to the algorithm steps presented in SLIC++ algorithm. Step 1, places K number of super-pixels provided by user on an equidistant grid. This grid is created separated by S, where S is given by $\sqrt{\frac{N}{K}}$, N is total number of image pixels. Step 2 performs reallocation of initial seed takes places subjected to the gradient condition to overcome the effect of initial centers placed over the edge pixels in image. Step 3 through step 7, are iteratively executed till the image pixels stop changing the clusters based on the cluster centers/seeds. Steps 5 or 6 are chosen based for respective implementation of SLIC or SLIC++ vice versa. Step 5 and 6 basically performs clustering over the image pixels based on different distance measures. If user opts for SLIC then Euclidean distance measure is used (base functionality). If user opts for SLIC++ then proposed hybrid distance measure is used. Step 8 checks if the new cluster center after every iteration of clustering is different than the previous center (distance between previous centers and recomputed centers). Step 9 keeps track of the threshold value for iterations as specified by the parameter 'n'. Step 10 enforces connectivity among the created super-pixel/clusters of image pixels.

The simple difference in the implementation of SLIC and SLIC++ lies in the usage of distance measure being used for the computation of image super-pixels. The presented research shows merely changing the distance measure to content-aware computational distance measure leads to better accuracy of results against the ground-truth for semi-dark images.

b.   Algorithm Complexity

The proposed algorithm follows the same steps as of Basic SLIC by introducing a new content-aware distance equation, thus the complexity of the proposed SLIC++ remains the same without any addition of new parameters, except the weights associated to the Euclidean and geodesic distance. These weights are merely scaler values to be taken into account in the core implementation of content-aware variant of SLIC, i.e., SLIC++. Hence, the complexity for the pixel manipulation is up to $O(N)$ where N is the total number of image pixels. With the minimum possible imposed requirements of computation SLIC++ manages to find accurate balance of implementation with infused functionality of Euclidean and geodesic distance. This fusion results in optimal boundary detection verified in terms of precision-recall in Section 4.

4. Validation of the Proposed Algorithm

4.1. Experimental Setup and Implementation Details

Following the proposed algorithm and details of implementation scheme, SLIC++ is implemented in MATLAB. The benchmarking analysis and experiments are conducted in MATLAB workspace version R2020a using the core computer vision and machine learning toolboxes. For experiments, the semi-dark images of Berkeley dataset have been used. The reported experiments are conducted on processor with specs core i7 10750H CPU, 16 GB RAM and 64-bit operating system.

The images are extracted form a folder using Fullfile method then incoming RGB images are converted in CIELAB space. After that parameter initialization takes place to get the algorithm started. Based on the number of K seed are initialized on the CIELAB image space and the condition relating to the gradient is checked using several different built-in methods. After that each pixel is processed using the proposed similarity measure and super-pixels are created until the threshold specified by user is reached. Similarly, the performance of reported state-of-the-arts is checked using the same environmental setup using the relevant parameters. Finally, the reported boundary performance is reported in form of precision recall measure to check the boundary adherence of super-pixel methods including Meanshift, SLIC and SLIC++. For analysis in terms of precision recall bfscore method is used which takes in the segmented image, ground-truth image and compares the extracted boundary with the ground-truth boundary by returning parameters precision, recall and score.

4.2. Parameter Selection

In this section we introduce the parameter associated with Meanshift, SLIC and SLIC++. Starting off with the proposed algorithm, SLIC++ uses several parameters as of Basic SLIC. Scaling factor m is set to 10, threshold on the iteration is set to value 10 represented by variable n and parameter S is computed based on N number of image pixels divided by user defined number of super-pixels in terms of variable K. The variable K provides user control for the number of super-pixels. Compact super-pixels are created as the value of K is increased but it increases the computational overhead. We have reported the performance using four different set of values of K, i.e., 500, 1000, 1500 and 2000. All these parameters including m, n and K are kept same as for the basic SLIC experiments. However, there are some additional parameters associated with SLIC++ which are w_1 and w_2 and their values are set to 0.3175 and 0.6825, respectively. The weights are cautiously picked based on trial-and-error experimentation procedure. The images were tested for a range of different weights. The weight values were varied to have weight ratios including 10:90, 30:70, 50:50, 70:30, and 90:10 for Euclidean and Geodesic distance, respectively. The ratio of 30:70 retains empirically maximum and perceptually meaningful super-pixels resulting in the optimal performance against the ground-truth. For Meanshift implementation the bandwidth parameter is set to 16 and 32, keeping rest of the implementation parameter default. Table 2 shows the averaged performance of the proposed SLIC++ algorithm acquired by varying different values of weights for random test cases.

Table 2. Summary statistics of average performance of SLIC++ for varying weights.

Row	Ratio	w_1 (Euclidean)	w_2 (Geodesic)	Precision	Recall	Score
			Test Case 1 (Image ID = 14037):			
1	10:90	0.1123	0.8877	0.47882	0.88930	0.62248
2	70:30	0.6825	0.3175	0.38850	0.92210	0.54660
3	50:50	0.4863	0.5137	0.37780	0.93040	0.53740
4	*30:70*	*0.3175*	*0.6825*	*0.48854*	*0.89124*	*0.63113*
5	90:10	0.8877	0.1123	0.38840	0.87340	0.53770
			Test Case 2 (Image ID = 26031):			
6	10:90	0.1123	0.8877	0.21623	0.78808	0.33935
7	70:30	0.6825	0.3175	0.18370	0.82790	0.30070
8	50:50	0.4863	0.5137	0.18910	0.85000	0.31000
9	*30:70*	*0.3175*	*0.6825*	*0.22661*	*0.86520*	*0.35920*
10	90:10	0.8877	0.1123	0.18650	0.79000	0.30220
			Test Case 3 (Image ID = 108082):			
11	10:90	0.1123	0.8877	0.27023	0.89832	0.41548
12	70:30	0.6825	0.3175	0.21840	0.82160	0.34510
13	50:50	0.4863	0.5137	0.22640	0.86800	0.35920
14	*30:70*	*0.3175*	*0.6825*	*0.28547*	*0.91629*	*0.43532*
15	90:10	0.8877	0.1123	0.22360	0.79470	0.34900

Empirically optimized performance of SLIC++ over 30:70 weight ratio for Euclidean and Geodesic distance hybridization has been tabulated in Table 2 row number 4, 9, 14 (formatted bold and italics). Moreover, the parameter values have been set as $K = 500$, $m = 10$ and $n = 10$ for the conducted experiments.

4.3. Performance Analysis

For performance analysis we considered two different experimental setups including qualitative analysis and quantitative analysis. Initially, we extended and analyzed the performance of SLIC with different distance measures to propose the most relevant distance measure for optimal boundary extension in semi-dark images. Then we compare the proposed algorithm with state-of-the-art super-pixel segmentation algorithms. The detail of the analysis is presented in following sub-sections.

4.3.1. Numeric Analysis of SLIC Extension with Different Distance Measures

For the detailed analysis of the proposed algorithm, we first compare the performance of basic SLIC with the variants of SLIC+ proposed in this study. The evaluation is presented in form of precision recall. For the optimal boundary detection greater values of precision are required. High precision rates relate to low number of false positives eventually resulting in high chance of accurate boundary retrieval, whereas high recall rates are relevant to matching of ground-truth boundaries to segmented boundary. Mathematically, precision is probability of valid results and recall is probability of detected ground-truths data [42]. For analysis of image segmentation modules, both high precision and recall are required to ensure maximum information retrieval [45].

Table 3 shows performance analysis of basic SLIC and its variants over randomly picked semi-dark images.

Table 3. Performance analysis of SLIC extensions.

Row	K	m	n	Parameters		Score	Precision	Recall	Distance Measure
				Test Case 1 (Image ID = 14037):					
1	500	10	10			0.54430	0.39120	0.89390	Euclidean—SLIC
2	500	10	10			0.61234	0.46563	0.89406	Chessboard—SLIC+
3	500	10	10			0.59713	0.44407	0.91118	Cosine—SLIC+
4	500	10	10	$\mu = 4$		0.62792	0.47345	0.93199	Min4—SLIC+
5	500	10	10			0.56128	0.43777	0.78186	Geodesic—SLIC+
6	500	10	10	$w_1 = 0.3175$	$w_2 = 0.6825$	0.63113	0.48854	0.89124	Euclidean Geodesic—SLIC++
				Test Case 2 (Image ID = 26031):					
7	500	10	10			0.30420	0.18690	0.81740	Euclidean—SLIC
8	500	10	10			0.35454	0.22098	0.89623	Chessboard—SLIC+
9	500	10	10			0.35698	0.22329	0.88957	Cosine—SLIC+
10	500	10	10	$\mu = 4$		0.34057	0.20959	0.90798	Min4—SLIC+
11	500	10	10			0.33715	0.21369	0.79842	Geodesic—SLIC+
12	500	10	10	$w_1 = 0.3175$	$w_2 = 0.6825$	0.3592	0.22661	0.86525	Euclidean Geodesic—SLIC++
				Test Case 3 (Image ID = 108082):					
13	500	10	10			0.35410	0.22720	0.80260	Euclidean—SLIC
14	500	10	10			0.42099	0.27720	0.87476	Chessboard—SLIC+
15	500	10	10			0.38368	0.24251	0.91811	Cosine—SLIC+
16	500	10	10	$\mu = 4$		0.42465	0.27694	0.91004	Min4—SLIC+
17	500	10	10			0.40382	0.26764	0.82216	Geodesic—SLIC+
18	500	10	10	$w_1 = 0.3175$	$w_2 = 0.6825$	0.43532	0.28547	0.91629	Euclidean Geodesic—SLIC++

Table 2 depicts all the extension of SLIC perform better in terms of precision-recall. The parameters are kept uniform for all the experiments specifically parameter m and n as in SLIC [11]. Moreover, there is up to 3–9% gain in precision percentage using SLIC++ as compared to the basic SLIC algorithm. The relevant scores based on precision and recall also shoot up by margin of 5–9% using SLIC++ (row 1 vs. 6 and row 7 vs. 12). However, the performance of other variants of SLIC is subjective to dimensions of incoming data, magnitudes, and memory overload. There usually is no defined consensus regarding best generalized performer in terms of similarity measure so far [44]. Thus, we propose an integration of two similarity measures which takes into account minimal processing resources and still provides optimal boundary detection.

For further detailed qualitative analysis using the same test cases by changing the number of super-pixels we extend the analysis of SLIC versus SLIC++. The precision recall and score graphs are shown in Figure 3.

In Figure 3, solid lines represents performance of SLIC and SLIC++ for Test case 1 and dashed lines represents performance for Test case 2. Figure 3a shows precision curves of SLIC++ are substantially better than the SLIC presented by brown (dark and light) lines for test case 1 and 2, respectively. Figure 3b shows the SLIC++ recall is less the resulting recall of SLIC for the same images. Subsequently, based on the precision, recall and the final scores SLIC++ outperforms basic SLIC on semi-dark images. For number of pixels set to 1000 there is a drop observed in precision and recall of SLIC++, this behavior can be attributed to accuracy measure's intolerance, i.e., even mutual refinements may result in low precision and recall values [45]. Nevertheless, performance for retrieval increases with increasing number of super-pixels and SLIC++ outperforms SLIC up to margin of 10%.

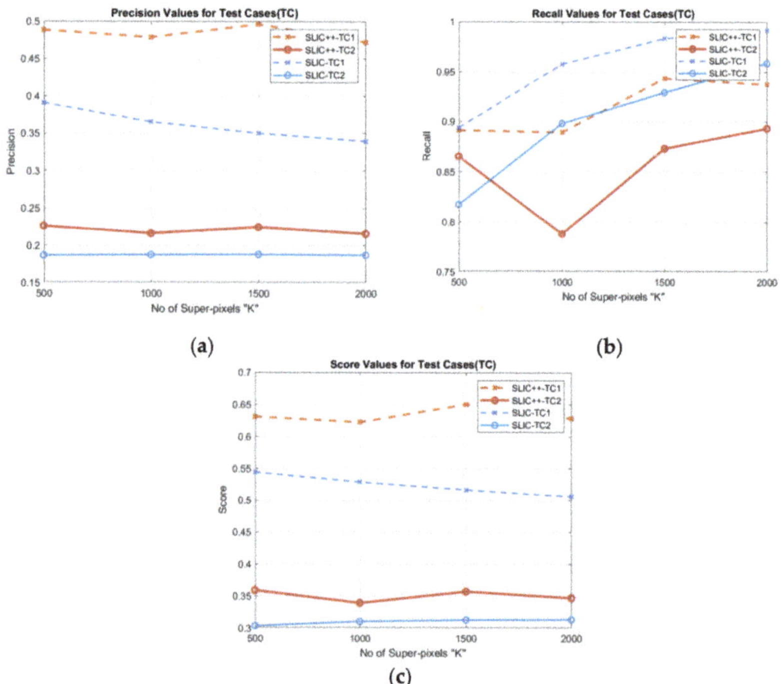

Figure 3. SLIC v/s SLIC++ performance over different number of pixels: (**a**) precision values; (**b**) recall value; (**c**) score values.

4.3.2. Comparative Analysis with State-of-the-Art

For the benchmarking of SLIC++ two different algorithms, i.e., SLIC and Meanshift are considered. To investigate the performance of SLIC and SLIC++ for the analysis over entire Berkeley dataset (semi-dark images), we set the number of super-pixels to 1500. The number of super-pixels is chosen 1500 because the peak performance of both the algorithms in experiment for test case 1 and 2 (refer Figure 3) is achieved by setting this parameter to value 1500. For the comparative analysis we also used Meanshift algorithm with input parameter, i.e., bandwidth set to 32. The bandwidth of Meanshift decides the complexity of the algorithm as this value is decreased the segmentation becomes more intricate with the overhead of computational complexity. To maintain computational resources throughout the experiment and keeping it uniform the parameters are chosen. The summary statistics of the obtained super-pixel segmentation results are shown in Table 4. The numerals presented in table are averaged values of precision, recall, and scores obtained for 316 images separately. The average precision, recall, and scores are presented in Table 4.

Table 4. Summary statistics of average performance for Berkeley dataset.

Algorithm	Score	Precision	Recall
SLIC	0.47020	0.31604	0.97719
SLIC++	**0.54799**	**0.39776**	**0.93470**
Meanshift-32	0.55705	0.57573	0.68416

Table 4 shows SLIC++ achieves average percentage score up to 54%, whereas SLIC maintains a score of 47%. Finally, Meanshift achieves a score of 55%, which is greater than SLIC++ but as stated earlier for segmentation application greater values of precision and recall are required. So, comparing the recall of SLIC++ versus Meanshift a huge difference is observed. This difference is in terms of low recall of Meanshift which means algorithm fails to capture salient image structure [45] which is not desired for semi-dark image segmentation.

4.3.3. Boundary Precision Visualization against Ground-Truth

To validate the point-of-view relating to high precision and high recall we present perceptual results of Meanshift, SLIC and SLIC++. Notice that, the high precision means the algorithm has retrieved most of the boundary as presented by the ground-truth, whereas high recall means most of the salient structural information is retrieved from the visual scene. Meanshift resulted in a minimum recall, which hypothetically means the structural information was lost. Table 5 presents how Meanshift, SLIC, and SLIC++ performed in terms of perceptual results for visual information retrieval. The reported results are for parameters $K = 1500$ for SLIC and SLIC++ and bandwidth = 32 for Meanshift.

Table 5. Semi-dark perceptual results conforming boundary retrieval.

Row ID	Image	Groundtruth Image	Prediction	Prediction Map Compared with Groundtruth
		Test Case 1:		
1	Input Image	Ground Truth by Human	Prediction by SLIC	Compared with Ground Truth
2	Input Image	Ground Truth by Human	Prediction by SLIC++	Compared with Ground Truth
3	Input Image	Ground Truth by Human	Prediction by Meanshift	Compared with Ground Truth
		Test Case 2:		
4	Input Image	Ground Truth by Human	Prediction by SLIC	Compared with Ground Truth

Table 5. *Cont.*

Row ID	Image	Groundtruth Image	Prediction	Prediction Map Compared with Groundtruth
5	Input Image	Ground Truth by Human	Prediction by SLIC++	Compared with Ground Truth
6	Input Image	Ground Truth by Human	Prediction by Meanshift	Compared with Ground Truth
		Test Case 3:		
7	Input Image	Ground Truth by Human	Prediction by SLIC	Compared with Ground Truth
8	Input Image	Ground Truth by Human	Prediction by SLIC++	Compared with Ground Truth
9	Input Image	Ground Truth by Human	Prediction by Meanshift	Compared with Ground Truth

As super-pixels are not just about the boundary detection, resulting applications also expect the structural information present in the visual scene. Consequently, we are not just interested in the object boundaries but also the small structural information present in the visual image specifically semi-dark images. Table 5 shows SLIC and SLIC++ not only retrieves boundaries correctly with minimal computational power consumed but also retrieves the structural information. Column 4 shows the fact by mapping prediction over ground-truth image. For test case 1, in column 4 row id 3 Meanshift fails to extract the structural information as few green lines are observed. Whereas, for the same image SLIC and SLIC++ perform better as a lot of green textured lines are observed (refer column 4 row id 1 and 2). Meanwhile for test case 2, all three algorithms perform equally likely. Similar performance is observed with test case 3, SLIC and SLIC++ retains structural information better than Meanshift. Since Meanshift resulted in minimum recall over the entire semi-dark Berkeley dataset (refer Table 4) it does not qualify to be a good fit for super-pixel segmentation. The reason is less reliability of structural information fetching and its performance is highly subjective to the incoming input images.

4.3.4. Visualizing Super-Pixels on Images

For one more layer of subjective analysis of super-pixel performance we present super-pixel masks in this section. Initially, we present the input image in Figure 4 with the highlighted boxes to look closely for retrieval of structural information from the image. Here, the red box shows the texture information present on the hill whereas the green box shows water flowing in a very dark region of the semi-dark image.

Figure 4. Input image with highlighted regions for detailed analysis.

Using the input image presented in Figure 4, we conducted experiments by changing the initialization parameters of all three algorithms. Table 5 shows the perceptual analysis visualizing the retrieval of salient structural information.

Table 6 shows that Meanshift extracts the boundaries correctly, whereas it loses all the contextual information when the bandwidth parameter is set to 32. This loss of information is attributed to low recall scores, whereas decreasing the value of the bandwidth increases the computational complexity and at the cost of additional complexity Meanshift now retrieves contextual information. SLIC and SLIC++ with minimal computational power retains structural information as seen in the red and green boxes in rows 1 and 2 of Table 5. Moreover, as the number of super-pixels 'K' increases, better and greater structural information retrieval is observed.

Figure 5 shows a zoomed in view of the super-pixels created by SLIC and SLIC++, residing in the red box. Here, we can see that SLIC++ retrieves content-aware information and SLIC ends up creating circular super-pixels (Figure 5a) due to the content irrelevant distance measure being used in its operational discourse.

Key Takeaways

The benchmarking analysis shows that the proposed algorithm SLIC++ achieves robust performance over different cases. The results of SLIC++ are more predictable as compared to the state-of-the-art methods Meanshift and SLIC. The performance of Meanshift is highly subjective as the recall keeps changing. Less recall values eventually result in less scores at the cost of information loss. Whereas, SLIC achieves 7% less scores and 8% less precision values in terms of boundary retrieval. The results of SLIC++ indicate that the proposed content-aware distance measure integrated in base SLIC demonstrates superior results. The significant improvement to the existing knowledge of super-pixel creation research is hybridization of proximity measures. Based on the comprehensive research it is seen that the hybrid measure performs better than the singular proximity measure counterparts of the same algorithm. These measures substantially control the end results of super-pixel segmentation in terms of accuracy. The proposed hybrid proximity measure carefully finds a balance between the two existing distance measure and performs clustering over image pixels making sure to retain content-aware information.

Table 6. Detailed perceptual analysis with increasing parameters.

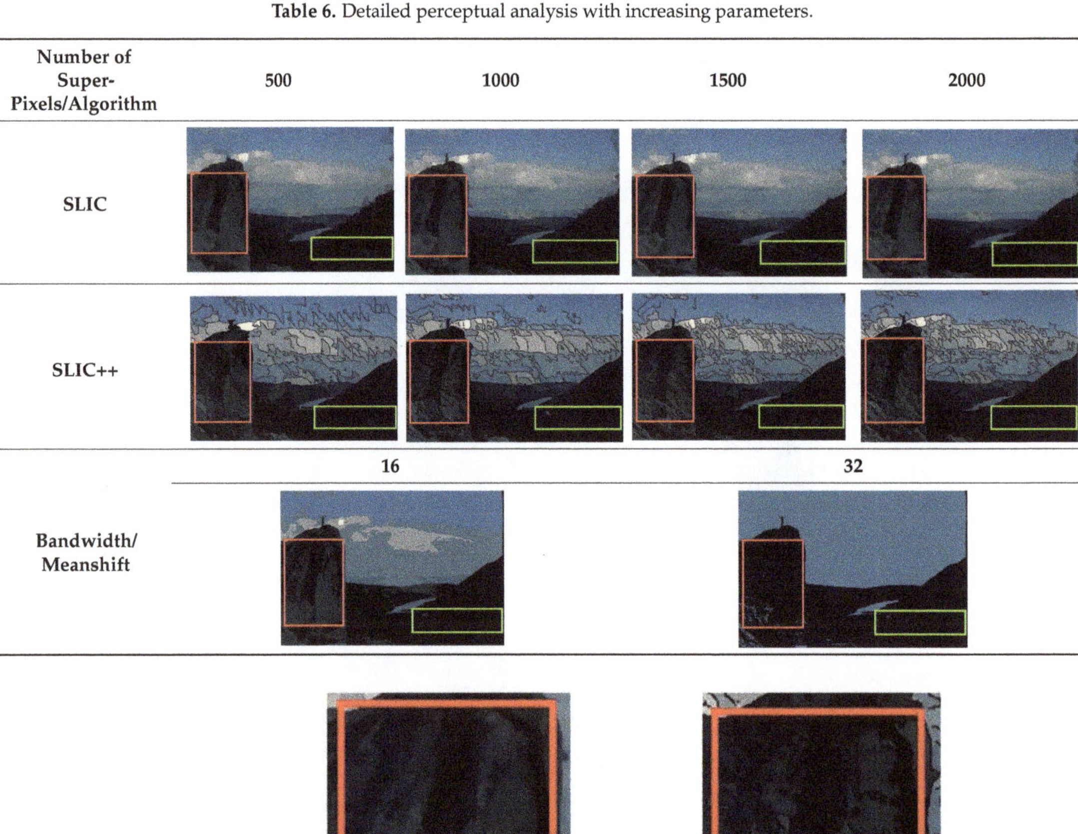

Number of Super-Pixels/Algorithm	500	1000	1500	2000
SLIC				
SLIC++				
	16		32	
Bandwidth/ Meanshift				

(a) SLIC Super-pixels, K = 1500   **(b)** SLIC++ Super-pixels, K = 1500

Figure 5. Zoomed in view of test case image for content-aware super-pixel analysis created by SLIC++ (**b**) against SLIC (**a**).

5. Limitations of Content-Aware SLIC++ Super-Pixels

The super-pixel segmentation algorithms are considered pre-processing step for wide range of computer vision applications. To obtain the optimal performance of sophisticated applications, the base super-pixel algorithm SLIC uses set of input parameters. These parameters allow the user control over different aspects of image segmentation. The idea is to extract uniform super-pixels throughout the image grid to maintain reliable learning statistics throughout the process. To make this possible the SLIC initially allows user to choose number of pixels 'K' (values ranging from 500–2000), parameter 'm' (where $m = 10$) which decides the extent of enforcement of color similarity over spatial similarity, number of iterations 'N' (where N = 10) which decides the convergence of the algorithm, neighborhood window 'w' (where w = 3) for gradient calculation to relocate cluster centers (if placed on edge pixel). This makes four input parameters for the base SLIC, whereas the proposed extension SLIC++ introduces two more weight parameters, w_1 and w_2 (0.3175 and 0.6825, respectively), to decide the impact of each distance measure in the hybrid distance measure. All these parameters significantly control the accuracy of segmentation results. Incorrect selection of these parameters leads to overall poor performance. Hence, for diverse applications, initial parameter search is necessary, which in turn requires several runs. For the reported research, using the state-of-the-art segmentation dataset, i.e., Berkeley dataset we chose the parameters as selected by the base SLIC. These parameters offer good performance over the image size of 321 × 481 or 481 × 321, whereas, as we increased the resolution of images during the extended research we observed that a higher value of 'K' is required for better segmentation accuracy.

For the existing research, we conducted experiments focused to identify the gains associated with usage of the proposed content-aware distance measure over the straight line distance measure. For the extended research, the input parameters shall be considered for optimization.

6. Emerging Successes and Practical Implications

Several decades of research in computer vision for boosted implementations resulting in fast accurate decisions, super-pixels have been a topic of research for long time now. The super-pixel segmentation is taken as entry stage providing pre-processing functionality for sophisticated intelligent workflows such as semantic image segmentation. To speed up the overall process of training and testing of these intelligent systems super-pixels are probable to provide remedies. As the intelligent automated vision systems have critical applications in medicine [46,47], manufacturing [48], surveillance [49], tracking [2] and so on. For this reason, fast and accurate visual decision are required. As the environmental conditions in form of visual dynamicity is challenging task to tackle by pre-processing modules. These modules are required to provide reliable visual results. Many super-pixel creation algorithms have been proposed over time to solve focused problems of image content sparsity [30], initialization optimization [28], and accurate edge detection [38]. However, the topic of the lightning condition in this domain remains untouched and needs attention. The dynamic lightning condition is a key component in autonomous vehicles, autonomous robots, surgical robots. The Berkeley dataset is comprised of images of different objects, ranging from humans, flowers, mountains, animals and so on. The conducted research holds for applications of autonomous robots and autonomous vehicles. However, the proposed algorithm is backed by the core concepts of image segmentation. For this reason, the presented work can be extended for variety of applications. Depending on the nature of application, the ranges of input parameters would be changed based on the required sensitivity of the end results, such as for the segmentation application in the medical domain compact where content-aware information is required. Consequently, the input values including the number of super-pixels to be created will be carefully selected. To handle the pre-processing problems associated with dynamic lightning conditions focusing autonomous robotics, the proposed extension of SLIC is a good fit. SLIC++ imposing minimum prerequisite conditions provides direct control over the working functionality and still manages to provide optimal information retrieval from the visual scenes for not only normal images but rather inclusive of semi-dark images.

7. Conclusions and Future Work

7.1. Conclusions

In this paper, we introduced a content-aware similarity measure which not only solved the problem of boundary retrieval in semi-dark images but is also applicable to other image types such as bright and dark. The content-aware measure is integrated in SLIC to create content-aware super-pixels which can then be used by other automated applications for fast implementations of high-level vision task. We also report results of integration of SLIC with existing similarity measures and describe their limitations of applicability for visual image data. To validate out proposed algorithm along with the proposed similarity measure, we conduct qualitative and quantitative evaluations on semi-dark images extracted from Berkeley dataset. We also compare SLIC++ with state-of-the-art super-pixel algorithms. Our comparisons show that the SLIC++ outperforms the existing super-pixel algorithms in terms of precision and score values by a margin of 8% and 7%, respectively. Perceptual experimental results also confirm that the proposed extension of SLIC, i.e., SLIC++ backed by content-aware distance measure outperforms the existing super-pixel creation methods. Moreover, SLIC++ results in consistent and reliable performance for different test image cases characterizing a generic workflow for super-pixel creation.

7.2. Future Work

For the extended research, density-based similarity measures integrated with content-aware nature may lead the future analysis. The density-based feature is expected to aid the overall all working functionality with noise resistant properties against the noisy incoming image data. Moreover, another aspect is the creation of accurate super-pixels in the presence of non-linearly separable data properties. Finally, the input parameter selection for the initialization of SLIC variants depending on the application domain and incoming image size remains an open area of research.

Author Contributions: Project administration, supervision, conceptualization, formal analysis, investigation, review, M.A.H.; conceptualization, methodology, analysis, visualization, investigation, writing—original draft preparation, data preparation and curation, M.M.M.; project administration, resources, review, K.R.; funding acquisition, conceptualization, review, S.H.A.; methodology, software, analysis, visualization, S.S.R.; validation, writing—review and editing, M.U. All authors have read and agreed to the published version of the manuscript.

Funding: This research study is a part of the funded project under a matching grant scheme supported by Iqra University, Pakistan, and Universiti Teknologi PETRONAS (UTP), Malaysia. Grant number: 015MEO-227.

Institutional Review Board Statement: Not applicable.

Informed Consent Statement: Not applicable.

Data Availability Statement: Following dataset is used. Berkeley: https://www2.eecs.berkeley.edu/Research/Projects/CS/vision/bsds/ (accessed on 27 September 2021).

Conflicts of Interest: The authors declare no conflict of interest.

References

1. Al-Huda, Z.; Peng, B.; Yang, Y.; Algburi, R.N. Object Scale Selection of Hierarchical Image Segmentation with Deep Seeds. *Int. J. Pattern Recogn. Artif. Intell.* **2021**, *35*, 2154026. [CrossRef]
2. Riera, L.; Ozcan, K.; Merickel, J.; Rizzo, M.; Sarkar, S.; Sharma, A. Detecting and Tracking Unsafe Lane Departure Events for Predicting Driver Safety in Challenging Naturalistic Driving Data. In Proceedings of the 2020 IEEE Intelligent Vehicles Symposium (IV), Las Vegas, NV, USA, 19 October–13 November 2020; pp. 238–245.
3. Nadeem, A.; Jalal, A.; Kim, K. Automatic human posture estimation for sport activity recognition with robust body parts detection and entropy markov model. *Multimed. Tools Appl.* **2021**, *80*, 21465–21498. [CrossRef]
4. Rashid, M.; Khan, M.A.; Alhaisoni, M.; Wang, S.-H.; Naqvi, S.R.; Rehman, A.; Saba, T. A sustainable deep learning framework for object recognition using multi-layers deep features fusion and selection. *Sustainability* **2020**, *12*, 5037. [CrossRef]
5. Jaffari, R.; Hashmani, M.A.; Reyes-Aldasoro, C.C. A Novel Focal Phi Loss for Power Line Segmentation with Auxiliary Classifier U-Net. *Sensors* **2021**, *21*, 2803. [CrossRef] [PubMed]
6. Memon, M.M.; Hashmani, M.A.; Rizvi, S.S.H. Novel Content Aware Pixel Abstraction for Image Semantic Segmentation. *J. Hunan Univ. Nat. Sci.* **2020**, *47*, 9.
7. Tian, Z.; Shen, C.; Chen, H. Conditional convolutions for instance segmentation. In Proceedings of the Computer Vision–ECCV 2020 16th European Conference, Glasgow, UK, 23–28 August 2020; Part I 16; pp. 282–298.

8. Junejo, A.Z.; Memon, S.A.; Memon, I.Z.; Talpur, S. Brain Tumor Segmentation Using 3D Magnetic Resonance Imaging Scans. In Proceedings of the 2018 1st International Conference on Advanced Research in Engineering Sciences (ARES), Dubai, United Arab Emirates, 15 June 2018; pp. 1–6.

9. Hamadi, A.B. Interactive Automation of COVID-19 Classification through X-Ray Images using Machine Learning. *J. Indep. Stud. Res. Comput.* **2021**, *18*. [CrossRef]

10. Zhao, L.; Li, Z.; Men, C.; Liu, Y. Superpixels extracted via region fusion with boundary constraint. *J. Vis. Commun. Image Represent.* **2020**, *66*, 102743. [CrossRef]

11. Achanta, R.; Shaji, A.; Smith, K.; Lucchi, A.; Fua, P.; Süsstrunk, S. *Slic Superpixels*; EPFL Technical Report 149300; School of Computer and Communication Sciences, Ecole Polytechnique Fedrale de Lausanne: Lausanne, Switzerland, 2010; pp. 1–15.

12. Kim, K.-S.; Zhang, D.; Kang, M.-C.; Ko, S.-J. Improved simple linear iterative clustering superpixels. In Proceedings of the 2013 IEEE International Symposium on Consumer Electronics (ISCE), Hsinchu, Taiwan, 3–6 June 2013; pp. 259–260.

13. Wang, M.; Liu, X.; Gao, Y.; Ma, X.; Soomro, N.Q. Superpixel segmentation: A benchmark. *Signal Process. Image Commun.* **2017**, *56*, 28–39. [CrossRef]

14. Achanta, R.; Shaji, A.; Smith, K.; Lucchi, A.; Fua, P.; Süsstrunk, S. SLIC superpixels compared to state-of-the-art superpixel methods. *IEEE Trans. Pattern Anal. Mach. Intell.* **2012**, *34*, 2274–2282. [CrossRef]

15. Liu, Y.-J.; Yu, C.-C.; Yu, M.-J.; He, Y. Manifold SLIC: A fast method to compute content-sensitive superpixels. In Proceedings of the IEEE Conference on Computer Vision and Pattern Recognition, Las Vegas, NV, USA, 26 June–1 July 2016; pp. 651–659.

16. Wang, P.; Zeng, G.; Gan, R.; Wang, J.; Zha, H. Structure-sensitive superpixels via geodesic distance. *Int. J. Comput. Vis.* **2013**, *103*, 1–21. [CrossRef]

17. Comaniciu, D.; Meer, P. Mean shift: A robust approach toward feature space analysis. *IEEE Trans. Pattern Anal. Mach. Intell.* **2002**, *24*, 603–619. [CrossRef]

18. Sheikh, Y.A.; Khan, E.A.; Kanade, T. Mode-seeking by medoidshifts. In Proceedings of the 2007 IEEE 11th International Conference on Computer Vision, Rio de Janeiro, Brazil, 14–21 October 2017; pp. 1–8.

19. Vedaldi, A.; Soatto, S. Quick shift and kernel methods for mode seeking. In *Proceedings of the European Conference on Computer Vision; ECCV 2008*; Lecture Notes in Computer Science; Springer: Berlin/Heidelberg, Germany, 2008; Volume 5305. [CrossRef]

20. Levinshtein, A.; Stere, A.; Kutulakos, K.N.; Fleet, D.J.; Dickinson, S.J.; Siddiqi, K. Turbopixels: Fast superpixels using geometric flows. *IEEE Trans. Pattern Anal. Mach. Intell.* **2009**, *31*, 2290–2297. [CrossRef]

21. Moore, A.P.; Prince, S.J.; Warrell, J.; Mohammed, U.; Jones, G. Scene shape priors for superpixel segmentation. In Proceedings of the 2009 IEEE 12th International Conference on Computer Vision, Kyoto, Japan, 29 September–2 October 2009; pp. 771–778.

22. Veksler, O.; Boykov, Y.; Mehrani, P. Superpixels and supervoxels in an energy optimization framework. In *Proceedings of the European Conference on Computer Vision; ECCV 2010*; Lecture Notes in Computer Science; Springer: Berlin/Heidelberg, Germany, 2010; Volume 6315. [CrossRef]

23. Van den Bergh, M.; Boix, X.; Roig, G.; de Capitani, B.; Van Gool, L. Seeds: Superpixels extracted via energy-driven sampling. *Int. J. Comput. Vis.* **2012**, *111*, 298–314. [CrossRef]

24. Weikersdorfer, D.; Schick, A.; Cremers, D. Depth-adaptive supervoxels for RGB-D video segmentation. In Proceedings of the 2013 IEEE International Conference on Image Processing, Melbourne, Australia, 15–18 September 2013; pp. 2708–2712.

25. Conrad, C.; Mertz, M.; Mester, R. Contour-relaxed superpixels. In *Proceedings of the International Workshop on Energy Minimization Methods in Computer Vision and Pattern Recognition; EMMCVPR 2013*; Lecture Notes in Computer Science; Springer: Berlin/Heidelberg, Germany; Germany, 2013; Volume 8081. [CrossRef]

26. Xu, L.; Zeng, L.; Wang, Z. Saliency-based superpixels. *Signal Image Video Process.* **2014**, *8*, 181–190. [CrossRef]

27. Li, Z.; Chen, J. Superpixel segmentation using linear spectral clustering. In Proceedings of the IEEE Conference on Computer Vision and Pattern Recognition, Boston, MA, USA, 12 June 2015; pp. 1356–1363.

28. Rubio, A.; Yu, L.; Simo-Serra, E.; Moreno-Noguer, F. BASS: Boundary-aware superpixel segmentation. In Proceedings of the 2016 23rd International Conference on Pattern Recognition (ICPR), Cancun, Mexico, 4–8 December 2016; pp. 2824–2829.

29. Wang, H.; Peng, X.; Xiao, X.; Liu, Y. BSLIC: Slic superpixels based on boundary term. *Symmetry* **2017**, *9*, 31. [CrossRef]

30. Liu, Y.-J.; Yu, M.; Li, B.-J.; He, Y. Intrinsic manifold SLIC: A simple and efficient method for computing content-sensitive superpixels. *IEEE Trans. Pattern Anal. Mach. Intell.* **2017**, *40*, 653–666. [CrossRef]

31. Akyilmaz, E.; Leloglu, U.M. Similarity ratio based adaptive Mahalanobis distance algorithm to generate SAR superpixels. *Can. J. Remote Sens.* **2017**, *43*, 569–581. [CrossRef]

32. Lowekamp, B.C.; Chen, D.T.; Yaniv, Z.; Yoo, T.S. Scalable Simple Linear Iterative Clustering (SSLIC) Using a Generic and Parallel Approach. *arXiv* **2018**, arXiv:1806.08741.

33. Chuchvara, A.; Gotchev, A. Content-Adaptive Superpixel Segmentation Via Image Transformation. In Proceedings of the 2019 IEEE International Conference on Image Processing (ICIP), Taipei, Taiwan, 22–25 September 2019; pp. 1505–1509.

34. Uziel, R.; Ronen, M.; Freifeld, O. Bayesian adaptive superpixel segmentation. In Proceedings of the IEEE/CVF International Conference on Computer Vision, Seoul, Korea, 27 October–2 November 2019; pp. 8470–8479. [CrossRef]

35. Yang, F.; Sun, Q.; Jin, H.; Zhou, Z. Superpixel Segmentation With Fully Convolutional Networks. In Proceedings of the 2020 IEEE/CVF Conference on Computer Vision and Pattern Recognition (CVPR), Seattle, WA, USA, 13–19 June 2020; pp. 13961–13970.

36. Wu, J.; Liu, C.; Li, B. Texture-aware and structure-preserving superpixel segmentation. *Comput. Graph.* **2021**, *94*, 152–163. [CrossRef]

37. Chuchvara, A.; Gotchev, A. Efficient Image-Warping Framework for Content-Adaptive Superpixels Generation. *IEEE Signal Process. Lett.* **2021**, *28*, 1948–1952. [CrossRef]

38. Yu, Y.; Yang, Y.; Liu, K. Edge-Aware Superpixel Segmentation with Unsupervised Convolutional Neural Networks. In Proceedings of the 2021 IEEE International Conference on Image Processing (ICIP), Anchorage, AK, USA, 19–22 September 2021; pp. 1504–1508.

39. Arbelaez, P.; Maire, M.; Fowlkes, C.; Malik, J. Contour detection and hierarchical image segmentation. *IEEE Trans. Pattern Anal. Mach. Intell.* **2010**, *33*, 898–916. [CrossRef]

40. Memon, M.M.; Hashmani, M.A.; Junejo, A.Z.; Rizvi, S.S.; Arain, A.A. A Novel Luminance-Based Algorithm for Classification of Semi-Dark Images. *Appl. Sci.* **2021**, *11*, 8694. [CrossRef]

41. Stutz, D.; Hermans, A.; Leibe, B. Superpixels: An evaluation of the state-of-the-art. *Comput. Vis. Image Underst.* **2018**, *166*, 1–27. [CrossRef]

42. Neubert, P.; Protzel, P. Superpixel benchmark and comparison. *J. Comput. Phys.* **1983**, *51*, 191–207.

43. Tanemura, M.; Ogawa, T.; Ogita, N. A new algorithm for three-dimensional Voronoi tessellation. *J. Comput. Phys.* **1983**, *51*, 191–207. [CrossRef]

44. Yu, J.; Kim, S.B. Density-based geodesic distance for identifying the noisy and nonlinear clusters. *Inf. Sci.* **2016**, *360*, 231–243. [CrossRef]

45. Monteiro, F.C.; Campilho, A.C. Performance evaluation of image segmentation. In *Image Analysis and Recognition. ICIAR 2006*; Lecture Notes in Computer Science; Campilho, A., Kamel, M.S., Eds.; Springer: Berlin/Heidelberg, Germany, 2006; Volume 4141. [CrossRef]

46. Gauriau, R. *Shape-Based Approaches for Fast Multi-Organ Localization and Segmentation in 3D Medical Images*; Telecom ParisTech: Paris, France, 2015.

47. Dai, W.; Dong, N.; Wang, Z.; Liang, X.; Zhang, H.; Xing, E.P. Scan: Structure correcting adversarial network for organ segmentation in chest x-rays. In *Deep Learning in Medical Image Analysis and Multimodal Learning for Clinical Decision Support*; Springer: Berlin/Heidelberg, Germany, 2018; pp. 263–273.

48. Khan, A.; Mineo, C.; Dobie, G.; Macleod, C.; Pierce, G. Vision guided robotic inspection for parts in manufacturing and remanufacturing industry. *J. Remanuf.* **2021**, *11*, 49–70. [CrossRef]

49. Fedorov, A.; Nikolskaia, K.; Ivanov, S.; Shepelev, V.; Minbaleev, A. Traffic flow estimation with data from a video surveillance camera. *J. Big Data* **2019**, *6*, 1–15. [CrossRef]

Article

Human Segmentation and Tracking Survey on Masks for MADS Dataset

Van-Hung Le [1,*] and Rafal Scherer [2]

[1] Department of Information Technology, Tan Trao University, Tuyen Quang 22000, Vietnam
[2] Department of Intelligent Computer Systems, Czestochowa University of Technology, 42-218 Czestochowa, Poland; rafal.scherer@pcz.pl
* Correspondence: Van-hung.le@mica.edu.vn; Tel.:+84-973-512-275

Abstract: Human segmentation and tracking often use the outcome of person detection in the video. Thus, the results of segmentation and tracking depend heavily on human detection results in the video. With the advent of Convolutional Neural Networks (CNNs), there are excellent results in this field. Segmentation and tracking of the person in the video have significant applications in monitoring and estimating human pose in 2D images and 3D space. In this paper, we performed a survey of many studies, methods, datasets, and results for human segmentation and tracking in video. We also touch upon detecting persons as it affects the results of human segmentation and human tracking. The survey is performed in great detail up to source code paths. The MADS (Martial Arts, Dancing and Sports) dataset comprises fast and complex activities. It has been published for the task of estimating human posture. However, before determining the human pose, the person needs to be detected as a segment in the video. Moreover, in the paper, we publish a mask dataset to evaluate the segmentation and tracking of people in the video. In our MASK MADS dataset, we have prepared 28 k mask images. We also evaluated the MADS dataset for segmenting and tracking people in the video with many recently published CNNs methods.

Keywords: MADS dataset; human segmentation; human tracking; convolutional neural networks

Citation: Le, V.-H.; Scherer, R. Human Segmentation and Tracking Survey on Masks for MADS Dataset. *Sensors* **2021**, *21*, 8397. https://doi.org/10.3390/s21248397

Academic Editor: Jing Tian

Received: 10 November 2021
Accepted: 8 December 2021
Published: 16 December 2021

Publisher's Note: MDPI stays neutral with regard to jurisdictional claims in published maps and institutional affiliations.

1. Introduction

Human segmentation and tracking in the video are two crucial problems in computer vision. Segmentation is the process of separating human data from other data in a complex scene of an image. This problem is widely applied in recognizing the activities of humans in the video. Human tracking extracts the person's position during the video and is applied in many tasks such as monitoring and surveillance.

The MADS dataset is a benchmark dataset for evaluating human pose estimation. This dataset includes activities in traditional martial arts (tai-chi and karate), dancing (hip-hop and jazz), and sports (basketball, volleyball, football, rugby, tennis, badminton). The fast execution speed of actions poses many challenges for human segmentation and tracking methods. In [1], the authors report on the results of human tracking in the video; however, the results are evaluated based on baseline tracking methods. Human tracking and segmentation evaluation are based on the segmented person data based on the constraint of the context. In the MADS dataset, only two sets of human mask data are provided (tai-chi and karate).

The human data needs to be segmented and tracked in the video to reduce the estimated space and computations in human pose estimation in 2D images or 3D space. Specifically, to reduce the space to estimate 3D human pose on the 3D data/point cloud in future studies, we have prepared manually masked data of humans in the single view (depth video) of various actions (jazz, hip-hop, sports). Nowadays, with the strong development of CNNs, many studies have applied CNN methods in the task of estimating, detecting, and tracking humans. Human tracking in the video can be done based on two methods. The first method is based on human detection in each frame, the results of human being detected and marked with a bounding box. The CNNs (Faster R-CNN [2], SSD [3], YOLO [4–6], etc.) used to detect humans were surveyed in the study of Xu et al. [7] and Tsai et al. [8]. The second method to track humans is motion-based [9]. In this paper, we survey studies that use CNNs to segment and track humans in the video. We manually prepared the masked data for the MADS dataset (MASK MADS). We also fine-tuned a set of parameters on the depth videos of the MADS dataset. Moreover, we trained and evaluated human segmentation and tracking by various state-of-the-art CNNs.

The main contribution of the paper is as follows:

- We manually prepared human masks for nearly 28 k images captured from a single-view. The marking process was performed using the interactive-segmentation tool (http://web.archive.org/web/20110827170646/, http://kspace.cdvp.dcu.ie/public/interactive-segmentation/index.html (accessed on 18 April 2021)).
- We summarised significant studies on human segmentation and human tracking in RGB images, in which we focus on survey studies that use CNNs for human segmentation and tracking in video. Our survey process is based on methods, datasets, evaluations, and metrics for human segmentation and human tracking. We ultimately analyze the challenges during the process. We also refer to the implementation or the source code path of each study. In particular, we investigate the challenges and results in human segmentation and human tracking in images and video.
- We fine-tuned a set of parameters of the recently published CNNs (Mask R-CNN, PointRend, TridentNet, TensorMask, CenterMask, etc.) to retrain the model for human segmentation and tracking on the video that was captured from single-views of the MADS dataset, as illustred in Figure 1. The data of the MADS dataset was divided into different ratios for training and evaluation.
- We evaluated the results of human segmentation in images based on the retrained CNN models (Mask R-CNN, PointRend, TridentNet, TensorMask, CenterMask) according to the data rates of the MADS dataset. We used the most significant CNNs in recent years for object segmentation.

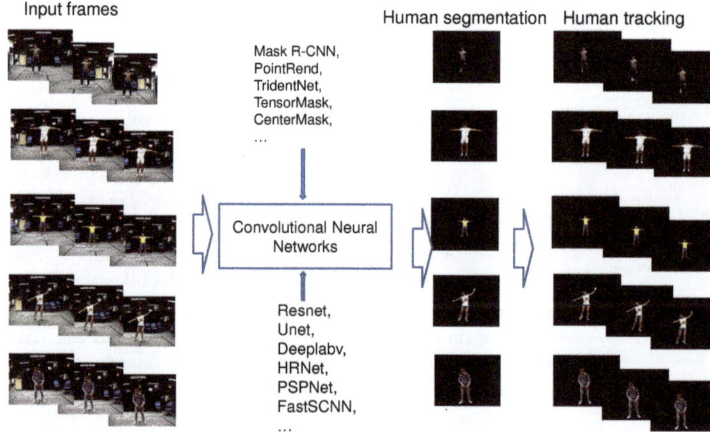

Figure 1. Illustrating of the process of segmenting and tracking humans in image sequences and video.

The paper is organized as follows. Section 2 discusses related works by the methods, results of the human segmentation, and tracking. Section 3 presents the survey of human segmentation and tracking methods based on CNNs, and Section 4 our MASK MADS dataset. Section 5 shows and discusses the experimental results of human segmentation and tracking on state-of-the-art CNNs, and Section 6 concludes the paper.

2. Related Works

Human segmentation and tracking in video are highly applicable in activity recognition and surveillance. Therefore, these two issues have spurred research interest for many years. Especially in the past five years, with the advent of CNNs, their performance improved significantly in terms of processing time and accuracy. Since then, many studies have been investigated these two issues based on CNNs.

In the research of Xu et al. [10], the authors performed a survey of segments of human data in images obtained from still cameras using CNN. In their work, a person is detected and marked with a bounding box in their work. In this study, the authors also presented the method to segment the human data based on detecting humans in images. Among these, there are some awe-inspiring results of CNNs: Fast R-CNN [11], Faster R-CNN and YOLO, etc. The precision of SSD300, SSD512, YOLO on the Pascal VOC 2007 is reported 79.4%, 82.5%, 63.5% [12], respectively. The processing speed of Faster R-CNN and YOLO is 10 fps and 45 fps, respectively. In this study, the authors are also presented pixel-level segmentation of human data that uses CNNs for fine-tuning. The CNNs are introduced as Fully Convolutional Network [13], AlexNet [14], VGGNet [15].

Yao et al. [9] conducted an overall survey of methods and challenges in object segmentation and tracking in video. They divided the methods for segmenting and tracking objects into five groups as follows: unsupervised methods for object segmentation, semi-supervised methods for object segmentation, interactive methods for object segmentation, weakly supervised methods for object segmentation, and segmentation-based methods for object tracking. The authors also presented challenges in the process of object segmentation and tracking in video, namely: complex background of the scene, low resolution, occlusion, deformation, motion blur, and scale variation. The authors attempt to present approaches to answering questions such as: What is the application context of segmentation and audience tracking? What form is the object represented (point, superpixel, pixel)? What are the features that can be extracted from the image for object segmentation and tracking? How to build an object's motion model? Which datasets are suitable for object segmentation and tracking? Although the authors have introduced datasets and metrics for the evaluating of subject tracking in the video, the results on these datasets have not been presented.

Ciaparrone et al. [16] performed a surveyed of deep learning-based multi-object tracking in the video captured from a single camera. The authors have presented the methods, measurements, and datasets of multi-object tracking. Therein, the used methods for multi-object tracking are presented in the following direction: object detection for tracking. According to the objects, the stages of the process are listed as follows: object detection with output marked with the bounding box, appearance feature extraction to predict the motion; computing a similarity between pairs of detections; assigning an ID for each object. In this study, the authors also presented metrics that evaluate the tracking of objects in the image: an evaluation based on trajectories, ground-truth trajectories are marked on the frames in the video; evaluation based on Jaccard similarity coefficient based on accuracy, precision, and recall of the detected, labeled bounding box of each object [17]. In [1], the authors were followed by the Bayesian tracker algorithm [18] and twin Gaussian processes algorithm [19] for multi-view tracking; Personalized Depth Tracker [20] and Gaussian Mixture Model tracker [21]. Although these methods are the start of human tracking, many CNNs have recently been introduced to solve this problem. We will introduce them in the next section.

The process of detecting, segmenting, and tracking people in the video is sequential in computer vision. To track people in videos, people need to be detected and segmented in each frame; to segment people in videos at the pixel level, people's data must be detected and marked with a bounding box. As described above, each step of detecting, segmenting, and tracking people in videos is surveyed in our study. To provide an overview of the whole process, we conducted a survey covering all three stages: detecting, segmenting, and tracking people in the video.

In our future studies, we will use human mask data to segment human point cloud (3D point) data with the scene, supporting the estimation and evaluation of 3D human pose estimation. The point cloud data of a human is generated based on depth data and color data a human segmented from a human mask. The process of converting to point cloud data is done [22]. Each 3D point (P) is created from a pixel with coordinates (x, y) on the depth image and a corresponding pixel on the color image that has a color value $C(r, g, b)$. P includes the following information: coordinates (P_x, P_y, P_z) in 3D space, the color value of that point (P_r, P_g, P_b), where the depth value (D) of point $P(x, y)$ must be greater than 0. P is computed according to the Formula (1).

$$
\begin{aligned}
P_x &= \frac{(x - c_x) * D}{f_x} \\
P_y &= \frac{(y - c_y) * D}{f_y} \\
P_z &= D \\
P_r &= C_r \\
P_g &= C_g \\
P_b &= C_b
\end{aligned}
\tag{1}
$$

where (f_x, f_y—focal length), (c_x, c_y—center of the images) are intrinsics of the depth camera.

There are now many CNNs for estimating 3D human pose from human point cloud data such as HandpointNet [23], V2V [24], Point-to-Point [25]. Especially, there are many studies on 3D human pose estimation with amazing results on depth image data and point cloud data [26–28]. These studies have been examined in detail in our previous study on 3D human pose estimation [29]. In the future, we will study more deeply about using point cloud data in the MASK MADS dataset.

3. Human Segmentation and Tracking by CNNs—Survey

3.1. CNN-Based Human Segmentation

3.1.1. Methods

Human segmentation is applicable in many practical and real-world scenarios, for example, surveillance person activities, virtual reality, action localization, 3D human modeling, etc. Human segmentation is the process of separating human data and scene data [10]. The methods for segmenting human data can be divided into three directions: top-down (semantic segmentation), bottom-up (instance segmentation), and combined. The top-down methods are based on training human-extracted features (shapes, appearance characteristics, and texture) to generate a classification model to classify the human pixels and scene pixels. This family of methods only segments persons into one class; despite there being many people and cars in the image, people are classified into one class, cars into another class. The bottom-up methods are based on generating the candidate regions that include a human and then identifying these regions following texture and bounding contours. Thus they segment the details of each person in the image. Figure 2 shows the differences between the approaches for human segmentation in images. The combined methods synergistically promote the advantages of both top-down and bottom-up methods to obtain the best effect. The human segmentation process is usually based on three steps described later on, as illustrated in Figure 3.

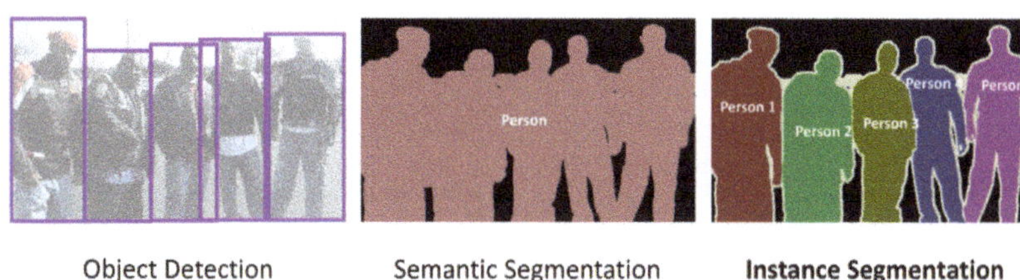

Figure 2. An explanation of object detection and semantic and instance segmentation [30].

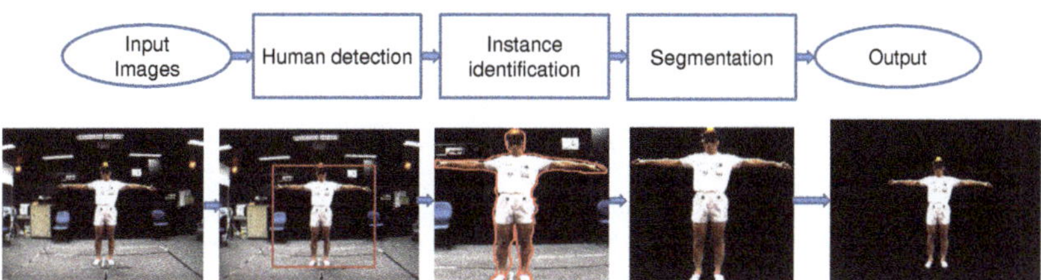

Figure 3. Illustration of the model of human segmentation in images.

a. Human detection

Object detection, and specifically human detection in images or videos, is one of the most important problems in computer vision. During appearance-based methods, traditional machine learning often uses hand-crafted features and a classification algorithm (e.g., SVM, AdaBoost, random forest, etc.) to train the human detection model. In recent years, most of the studies and applications have used CNNs to detect persons and objects in general and demonstrated many impressive results.

Girshick et al. [31] proposed a Region-based Convolutional Neural Network (R-CNN) for object detection. This network can be applied as a bottom-up method for localizing and segmenting objects of region proposals and improved classification efficiency by using supervised pre-training for labeled training data. He et al. [32] proposed SPPnet (Spatial Pyramid Pooling network) to train the object detection model. Traditional CNNs include two main components: convolutional layers and fully connected layers. To overcome the fixed-size constraint of the network, SPPnet adds an SPP layer to the last convolutional layer. The fixed-length output features are generated from the SPP layer pools. SPP is robust with object deformations. The extracted features of variable scales are pooled by SPP. Karen et al. [15] based their assumptions on the characteristics of CNNs that the depth of the CNNs affects the accuracy. The greater the depth, the greater the identification detection accuracy. Therefore, the authors have proposed the VGG16 network with the input size of the convolutional layer of 224 × 224 RGB image. After that, the input image passed a stack of convolutional layers. The final output size of the convolutional layer is 3 × 3. Recently, Xiangyu et al. [33] improved the VGG model in Fast R-CNN for object classification and detection; Haque et al. [34] also applied the VGG model to ResNet to detect objects. Implementation details of VGG for object detection are shown under the links (https://www.robots.ox.ac.uk/~vgg/research/very_deep/, https://neurohive.io/en/popular-networks/vgg16/ (accessed on 20 May 2021)). To improve the results of R-CNN and SPPnet, Girshick et al. [11] proposed Fast R-CNN, which input is the entire image and a set of region proposals. Fast R-CNN performs two main computational steps: processes several convolutional and max-pooling layers on the whole image to generate a feature map. Each proposal interest region of the pooling layer then extracts a fixed-length feature vector from the generated feature map, and the input of a sequence of fully connected layers is the extracted feature vector. Implementation details of Fast R-CNN for object detection are shown under link (https://github.com/rbgirshick/fast-rcnn (accessed on 25 May 2021)). The SPPnet [32] and Fast R-CNN [11] models work on region proposals that could be the object, which reduces the computation burden of these CNNs. However, the accuracy of these networks has not been greatly improved. Ren et al. [2] proposed an RPN (Region Proposal Network) that shares the full-image convolutional features with the detection network, which makes a nearly cost-free region proposal. The architecture of Faster R-CNN consists of two parts: a deep, fully convolutional network (RPN) and a Fast R-CNN detector that uses the proposed regions. Implementation details of Faster R-CNN for object detection are available under link (https://towardsdatascience.com/faster-r-cnn-object-detection-imple mented-by-keras\-for-custom-data-from-googles-open-images-125f62b9141a (accessed on 10 July 2021)). Especially recently, Goon et al. [35] used Faster R-CNN for detecting pedestrians from drone images. The CNNs presented (R-CNN, SPPnet, VGG, Fast R-CNN, Faster R-CNN) are mainly concerned with the high accuracy, but the computation time for object detection is high. Therefore, Redmon et al. [4] proposed the YOLO network with a computation speed of about 67 fps of YOLO version 2 on the VOC 2007 dataset. The bounding boxes are predicted directly using the fully connected layers on top of the convolutional feature extractor. Currently, the YOLO network has four versions (YOLO version 1 to 4). Implementation details of YOLO versions for object detection are available under the link (https://pjreddie.com/darknet/yolov1/, https://pjreddie.com/darknet/yolov2/, https://pjreddie.com/darknet/yolo/ and https://github.com/AlexeyAB/darknet (accessed on June 2021)), respectively. Lui et al. [3] proposed the Single Shot Detector (SSD) network for object detection. It uses the following mechanism: the base network is used for high-quality image classification, fixed-size bounding boxes and scores are generated from a feed-forward convolutional network, and the final detections are generated by a non-maximum suppression step (https://github.com/weiliu89/caffe/tree/ssd (accessed on 12 June 2021)). Jonathan et al. [36] have performed a comparative study for objects detection, which focuses on comparing object detection results-based on typical CNNs: Faster R-CNN [2], R-FCN [37], and SSD [3]. The CNNs used the feature extractors as VGG or ResNet, calling them "meta-architectures". The authors evaluated many configurations of each CNN and analyzed the effect of configurations, image size on the detection results.

b. Human segmentation

The next step in the model shown in Figure 3 is human segmentation, which is the process of labeling each pixel as either human or non-human data. It uses the result of object detection in the form of bounding boxes or identifies the human region, and then classifies the pixels in the bounding box or the area as human or non-human giving the most accurate results [10]. Previously there were studies by Meghna et al. [38,39] that suggested human pose-based segmentation. Lately, much research was proposed based on deep learning, for example, He et al. [40]. In [40], the authors proposed Mask R-CNN using Faster R-CNN for object detection and predicting an object mask on each Region of Interest (RoI) conducted in parallel, in which predicting a segmentation mask in a pixel-to-pixel basis. Implementation details of Mask R-CNN are available under link (https://github.com/matterport/Mask_RCNN (accessed on 14 June 2021)), and as Detectron2 (https://github.com/facebookresearch/detectron2 (accessed on 14 June 2021)) [41]. In the Detectron2 toolkit from Facebook AI Research [41], the authors also developed source code to train and test the segmentation of the image on some CNNs models: DeepLabv3 [42,43], details are shown in link (https://github.com/facebookresearch/detect ron2/tree/master/projects/DeepLab (accessed on 12 June 2021)). DeepLabv3 is the human semantic segmentation group, this CNN is an improvement of the DeepLab2 [44] method. This method has been applied in parallel to the Atrous Spatial Pyramid Pooling (ASPP) method for multi-scale context. In [42], the authors improved the DeepLabv3 network with a combination of the spatial pyramid pooling module and the encoder-decoder structure. Therein, the rich semantic features are obtained from the encoder module; the detected objects by the bounding boxes are recovered by the decoder module. This network architecture of DeepLabv3+ made a trade-off between precision and processing time based on the extracted encoder features by atrous convolution. DensePose [45,46], details are shown in link (https://github.com/facebookresearch/detectron2/tree/master/projects/DensePose (accessed on 12 June 2021)). Riza et al. [46] proposed DensePose-RCNN for estimating human pose. DensePose-RCNN is a combination of the DenseReg and the Mask-RCNN to improve the accuracy, with the cascaded extensions. Cheng et al. [47,48] proposed the Panoptic-DeepLab (details are shown in the link (https://github.com/facebookresearch/detectron2/tree/master/projects/Panoptic-DeepLab (accessed on 14 June 2021)), which predicts the semantic segmentation and instance segmentation based on the dual-context and dual-decoder modules. The ASPP is employed in the decoder module. Kirillov et al. [49] proposed the PointRend method, details are shown in link (https://github.com/facebookresearch/detectron2/tree/master/projects/PointRend (accessed on 14 June 2021)). The PointRend network is applied in both semantic segmentation and instance segmentation. It was applied to each region in the coarse-to-fine method (from large to small size). Chen et al. [50] proposed a TensorMask network applied to the instance segmentation. Before the sliding window method used for object detection, the results are displayed on bounding boxes. After that, they use Mask R-CNN for object segmentation on the data inside in bounding box. Details are shown in link (https://github.com/facebookresearch/detectron2/tree/master/projects/TensorMask (accessed on 20 June 2021)); Li et al. [51] proposed a parallel multi-branch architecture called the TridentNet with ResNet-101 backbone retrained, details are shown in link (https://github.com/facebookresearch/detectron2/tree/master/projects/TridentNet (accessed on 15 June 2021)). This CNN used the image with multi-scales as the input. After that the image pyramid methods are used for feature extraction and object detection for each scale. Especially recently, the Centermask network is proposed and published by Lee et al. [52], this CNN implemented the human instance segmentation, details are shown in link (https://github.com/youngwanLEE/CenterMask (accessed on 16 June 2021)). This network used the anchor-free object detector (FCOS) [53,54] to predict the per-pixel object detection. After that, the SAG-Mask branch was added to predict a segmentation mask on each detected box. The feature extractions and feature map used the pyramid method of VoVNetV2 [55] backbone network.

George et al. [56] proposed the PersonLab model to estimate human pose and segment human instance from the images. This model used CNNs to predict all key points of each person in the image, after that the author predicted instance-agnostic semantic person segmentation maps by a greedy decoding process to group them into instances. This means that the determination of an *i*th human pixel is based on the probable distance from that pixel to the nearest detected keypoint. Implementation details are shown in link (https://github.com/scnuhealthy/Tensorflow_PersonLab (accessed on 16 June 2021)). Zhang et al. [57] have proposed a model based on the human instance segmentation method where the object detection step is based on the results of human pose estimation. The human pose is estimated based on the combination of scale, translation, rotation, and left-right flip, and it is called Affine-Align. The Affine-Align operation uses human pose templates to align the people which does not use a bounding box as in Faster R-CNN or Mask R-CNN, in which the human pose templates are divided into clusters and center of clusters used to compute the error function with detected poses by the affine transformation. The human segmentation module is concatenated from the Skeleton features to the instance feature map after Affine-Align. Implementation details are available under link (https://github.com/liruilong940607/Pose2Seg (accessed on 18 June 2021)).

3.1.2. Datasets, Metrics and Results

a. Human detection

Object detection in images and videos is the first operation applied in computer vision pipelines such as object segmentation, object identification, or object localization. Object detection methods have been evaluated on many benchmark datasets. In this section, we present some typical benchmark datasets.

Everingham et al. [58] introduced the Pascal VOC (PV) 2007 with 20 object classes with 9963 images which are collected from both indoor and outdoor environments. The interesting objects are divided into the following groups: person; animal (bird, cat, cow, dog, horse, sheep); vehicle (airplane, bicycle, boat, bus, car, motorbike, train); indoor (bottle, chair, dining table, potted plant, sofa, tv/monitor). This data set is divided into 50% for training/validation and 50% for testing. The dataset can be downloaded from link (http://host.robots.ox.ac.uk/pascal/VOC/voc2007/ (accessed on 19 June 2021)). In [59], the authors updated the Pascal VOC dataset (PV 2010) with 10,103 images containing 23,374 annotated objects and 4203 segmentations. In this dataset, the authors changed the way of computing the average precision and used all data points rather than TREC style sampling. In 2012, the authors updated the PV 2012 dataset [60] with the training/validation data of 11,530 images containing 27,450 ROI annotated objects and 6929 segmentations. These versions of the PV dataset are presented, compared in link (http://host.robots.ox.ac.uk/pascal/VOC/voc2012/ (accessed on 18 June 2021)).

In [61], Lin et al. published the benchmark MS COCO (Microsoft Common Objects in COntext) dataset. It includes 328,000 images with 80 common object categories and 2,500,000 labeled instances. The objects in the images are person, car, elephant and the background is grass, wall, or sky. After six years, this dataset is now available with more than 200,000 images and 80 object categories, and over 500,000 object instances segmented.

The ImageNet Large Scale Visual Recognition Challenge 2014 detection [62] task involves 200 categories. There are 450 k/20 k/40 k images in the training/validation/testing sets. The authors focus on the task of the provided data-only track (the 1000-category CLS training data is not allowed to use).

Most of the research on object detection presented above use the mean Average Precision (mAP) measurement for evaluation. It is calculated according to Equation (2).

$$mAP = \frac{\sum_{Q}^{q=1} AverageP(q)}{Q} \tag{2}$$

where Q is the number of frames and $AverageP(q)$ is the Average Precision (AP) of object detection for each frame. *Precision* is calculated as in [17]. Some results of human detection are shown in Tables 1 and 2.

Table 1. Human detection results (mAP) on the several benchmark datasets.

Measurement/ Dataset/Methods	mAP (Mean Average Precision) (%)						
	PV 2007	PV 2010	PV 2012	COCO (Year)	IC 2013	IC 2014	IC 2015
R-CNN [31]	58.7	58.1	-	-	31.4	-	-
SPPnet [32]	58.9	-	-	-	-	35.11	-
VGG vs. Fast R-CNN [15]	66.1	-	-	-	-	-	-
VGG vs. ResNet [34]	-	-	93.7	-	-	-	-
VGG 16 [15]	89.3	-	89	-	-	-	-
Fast R-CNN [11]	69.9	72.7	72.0	35.9 (2015)	-	-	42.9
Faster R-CNN [2]	78.8	-	75.9	42.7 (2015)	-	-	32.6
YOLO v1 [63]	63.4	-	63.5	-	-	-	-
YOLO v2 [4]	78.6	-	81.3	21.6 (2015)			
YOLO v3 [5]	-	-	-	60.6 (2015)	-	-	-
YOLO v4 [6]	-	-	-	65.7 (2017)	-	-	-
SSD300 [3]	74.3	-	79.4	23.2 (2015)	-	-	-
SSD500 [3]	76.8	-	83.3	26.8 (2015)	-	-	-

Table 2. The frame rate of human detection (fps—frames per second) on the PV 2007 benchmark dataset.

Measurement/ Dataset/Methods	Procesing Time (fps)
	PV 2007
R-CNN [31]	0.076
SPPnet [32]	0.142
VGG vs. Fast R-CNN [15]	7
VGG vs. ResNet [34]	5
Fast R-CNN [11]	0.5
Faster R-CNN [2]	7
YOLO v1 [63]	45
YOLO v2 [4]	40
YOLO v3 [5]	45
YOLO v4 [6]	54
SSD300 [3]	46
SSD500 [3]	19

b. Human segmentation

Human data segmentation is the process of classifying whether each pixel belongs to a human object or not. To evaluate the human segmentation, the studies of Kaiming et al. [40] and Zhang et al. [57] were evaluated on the COCO [61] dataset and introduced the above. The metric used to evaluate is AP (Average Precision), precision is calculated as shown in [17].

The results of the human segmentation on the PV 2007 benchmark are shown in Table 3. The processing time of Mask R-CNN, Pose2Seg for human segmentation is 5 fps, 20 fps, respectively. We also listed the results of object segmentation on the PV 2017 benchmark dataset with the validation/testing sets from [61], and they are presented in Table 4.

Table 3. The frame rate of human detection (fps—frames per second) on the PV 2007 benchmark dataset. AP_M, AP_L are the *Median, Large AP* categories, respectively.

Measurement/ Method/	Backbone	AP_M	AP_L
Mask R-CNN [40]	Resnet50-fpn	43.3	64.8
PersonLab [56]	Resnet101	47.6	59.2
PersonLab [56]	Resnet101(ms scale)	49.2	62.1
PersonLab [56]	Resnet152	48.3	59.5
PersonLab [56]	Resnet152(ms scale)	49.7	62.1
Pose2Seg [57]	Resnet50-fpn	49.8	67.0
Pose2Seg(GTKpt) [57]	Resnet50-fpn	53.9	67.9

Table 4. The object segmentation results (*m*—mask, *b*—box) on the COCO 2017 Val/Test dataset [61] (%—percent).

Measurement/ CNNs	Backbone Network	AP^m	AP^b	AP^b_S	AP^b_M	AP^b_L	IS	SS
CenterMask	VoVNetV2-99	38.3	43.5	25.8	47.8	57.3	√	-
TridentNet	ResNet-101	-	42.0	24.9	47.0	56.9	-	√
TensorMask	ResNet-101-FPN	-	37.1	17.4	39.1	51.6	√	-
PointRend (IS)	X101-FPN	40.9	-	-	-	-	√	√
Panoptic-DeepLab	Xception-71	39.0	-	-	-	-	√	√

3.1.3. Discussions

Table 1 shows the human detection accuracy based on some benchmark datasets such as PV 2007, PV 2010, PV 2012, COCO, or IC datasets. Table 2 shows the processing time of human detection on the PV 2007 dataset. The results of human detection can show us that accuracy increases the following time and that the processing time of human detection has reached real-time. We only surveyed on the PV 2007 dataset to agree on the dataset. Each different version of the PV dataset has a different number of images and complexity. So the processing time will be different. When compared to the same dataset, we can see the processing speed of CNNs (e.g, R-CNN, Fast R-CNN, YOLO, etc) for detecting humans. Our survey is based on an evaluation of CNNs for human detection using the TensorFlow framework on the PV 2007 dataset. A part of which we refer to in the survey of Sanchez et al. [64].

As presented above, the human detection results significantly affect the process of human segmentation. As shown in first table and second table of [56], the keypoint detection results of human pose on the COCO dataset are between 61% and 75% on the AP_M, AP_L measurements, in the human segmentation step is between 47% and 62%. Therefore, the error from detecting humans is from 25% to 39%, and the human segment error is about 14%. These results also show that the segmentation of the human data on the COCO dataset still poses many challenges.

3.2. CNN-Based Human Tracking

3.2.1. Methods

Human tracking is the process of estimating the movement trajectory of people in the scene based on the video captured by the scene. A tracker is a set of object labels and classifications between that object and other objects and the background [65]. Human tracking is the process of reusing the results of human detection or human segmentation on each frame of the video. From there, the person's trajectory is drawn on the video. Watada et al. [65] performed a human tracking survey, which uses the results of human detection (which is the process of detecting points of interest in the human body). This detector is called a *"point detector"* and the second method is to use the results of the human segment in the image. In this paper, we are interested in studies using CNNs to track people in the video. Dina et al. [66] experimented with a method for tracking multiple persons in images using Faster R-CNN to detect people. The results of people detection in the image are shown as bounding boxes. The author used VGG16 with thirteen convolutional layers of various sizes (64, 128, 256, 512), two fully connected layers, and a softmax layer to build the Faster R-CNN. Javier [67] has generalized the object tracking problem in video, in which the author presented Siamese Network Algorithm (SNA) for object tracking. SNA assumes that the object to track always be unknown, and there is no need to learn it. SNA uses the CNNs to parallel object detection in the images, and then it computes the differences of the pairs of images. Implementation details are shown in link (https://github.com/JaviLaplaza/Pytorch-Siamese (accessed on 20 June 2021)). Gulraiz et al. [68] proposed a model of human detection and tracking that used Faster R-CNN to detect the human with five implementation steps. Then, they use more Deep Appearance Features (DAF) to track humans. Especially, the method provided the motion information and appearance information. The authors also presented the challenges of object tracking systems. The first one concerns real-time human tracking. There has been much research trying to achieve the goal of real-time tracking of people. The second is the identity switch in all frames or specific time duration. The third is the fragmentation problem when the person is not detected at some frames, which causes the person's moving trajectory to be interrupted. Particularly, we proposed solutions to improve the detection results in some frames by using CNNs for human detection or just detecting parts of the person, such as the head or shoulder. To address identity switches, the authors suggested using the appearance, localization features, and the size of a human in frames or perhaps using facial recognition. The proposed system is better than both SORT and Deep SORT [69,70] in real-time scenarios for pedestrian tracking. Ejaz et al. [71] proposed a method for improving human detection and tracking accuracy in noisy and occluded environments. To improve the accuracy of human detection and classification, a softmax layer is used in the CNN model. Special tactics of enhancing learning complex data (data augmentation) are applied.

3.2.2. Datasets, Metrics and Results

To evaluate the results of human tracking, Laura et al. [72] proposed a huge MOTChellange dataset, which combined of 22 different subsets. Some results of human detection in images of MOTChellange dataset before human tracking based on the CNNs are shown in Table 5.

Table 5. The results (accuaracy, precision (%)) of human segmentation on the MOTChallenge dataset. SNN is a human tracking evaluation based on the Siamese Neural Network. Ecdist is a human tracking evaluation based on the simple minimum Euclidean distance.

Measurement/Model	MOTA (MOT Accuracy) [%]	MOTP (MOT Precision) [%]
SORT [73]	59.8	79.6
Deep SORT [69]	61.4	79.1
Faster RCNN + DAF [68]	75.2	81.3
Faster RCNN [66] (Ecdist)	61.26	-
Faster RCNN [66] (SNN)	47.38	-

However, Gulraiz et al. [68] were only interested in 10-minute videos of individuals with 61,440 rectangles of human detection. It is composed of 14 different sequences with proper annotations by expert annotators. This dataset collection camera is set up in multiple states (dynamic or static). The camera position can be positioned horizontally with people or lower. The lighting conditions are also quite varied: lighting, shadows, and blurring of the pedestrians are inconsistent. The processing times of human detection for various ResNets are shown in Table 6.

Table 6. The processing time (ms—millisecond) of Human detection for various ResNets. The calculation process is performed on a computer with the following configuration: GeForce GTX 1080 Ti GPU with Ubuntu OS installed in the system. The experiments are performed using the TensorFlow framework [68].

Measurement/Model	Processing Time [ms]
ResNet-34	52.09
ResNet-50	104.13
ResNet-101	158.35
ResNet-152	219.06
ResNet-30	48.93

In [34], the authors evaluated the INRIA human dataset and Pedestrian parsing on surveillance scenes (PPSS) dataset. The authors used the INRIA dataset that includes 2416 images for training and 1126 images for testing. The persons in the INRIA dataset were captured from many different positions: pose and occluded background, crowd scenes. The PPSS dataset included a total of 3673 images captured from 171 videos of different scenes and 2064 images in this dataset that the people are occluded. Haque et al. [34] used 100 videos for training and 71 videos for testing. The results of human tracking on the INRIA dataset and PPSS dataset are shown in Table 7.

Table 7. The results of Human segmentation on the INRIA dataset and PPSS dataset [34].

Dataset	Model	Accuracy [%]
INRIA dataset	VGG-16	96.4
	CNNs + DA [34]	98.8
PPSS dataset	VGG-16	94.3
	CNNs + DA [34]	98.3

4. Human Mask of MADS Dataset

The MADS dataset includes martial arts (tai-chi and karate), dancing (hip-hop and jazz), and sports (basketball, volleyball, football, rugby, tennis, and badminton) actions. The activities in this dataset are fast and dynamic, and many body parts are active, especially arms and legs. The MADS dataset consists of two sets of data: The RGB image data collected from multi-view settings; RGB image data and depth image data collected from a single viewpoint (captured from the depth sensor). Figure 4 shows the mask image of the human (left) when segmented to the pixel level, and the point cloud data of a human (right) generated from the person data segmented on the depth image with the camera's intrinsic parameter based on Equation (1). We also illustrate a result of 3D human pose estimation based on point cloud data (right). Therefore in this paper, we are only interested in the dataset collected from the depth sensor (the data is collected from a single viewpoint). We will use human point cloud data in further studies. An example of the RGB and depth data from the dataset is illustrated in Figure 5. To evaluate the results of human segmentation and tracking humans in videos, we implemented pixel marking of the human area in the RGB images that were captured from a single viewpoint (depth sensor).

Figure 4. Illustrating of 3D human annotation data correction results according to point cloud data based on the human mask from the image. The red human skeleton is a result of 3D human pose estimation in the point cloud data.

Figure 5. Illustration of image data of an unmarked human and masked image. Human depth data is delimited by a yellow border. The depth value of the human pixels is greater than 0 (which is the distance from the camera to the surface of the body) and is a gray color, the other pixels are the background and is black color. The depth image is the result of mapping from the human mask to the depth image obtained from the environment.

To mask people in the image, we have manually prepared the mask data of the human using the Interactive Segmentation tool (http://web.archive.org/web/20110827170646/http://kspace.cdvp.dcu.ie/public/interactive-segmentation/index.html (accessed on 18 April 2021)). We have prepared about 28,000 frames and make available at the link (https://drive.go

5. Human Segmentation and Tracking of MADS Dataset

5.1. Methods

In this paper, we evaluate in detail the human instance segmentation on the MASK MADS dataset with the Mask R-CNN benchmark [40] method and some of its improvements in the Detectron2 toolkit [41].

- Mask R-CNN [40] is an improvement of Faster R-CNN [2] for image segmentation at the pixel level. The operation of Mask R-CNN for human instance segmentation does the following several steps.

  Backbone Model: Using ConvNet like Resnet to extract human features from the input image.

  Region Proposal Network (RPN): The model uses the extracted feature applied to the RPN network to predict whether the object is in that area or not. After this step, bounding boxes at the possible areas of objects from the prediction model will be obtained.

  Region of Interest (RoI): The bounding boxes from the human detection areas will have different sizes, so through this step, all those bounding boxes will be merged to a certain size at 1 person. These regions are then passed through a fully connected layer to predict the layer labels and bounding boxes. The gradual elimination of bounding boxes through the calculation of the IOU. If the IOU is greater than or equal to 0.5 then be taken into account else be discarded.

  Segmentation Mask: Mask R-CNN adds the third branch to predict the person's mask parallel to the current branches. Mask detection is a Fully-Connected Network (FCN) applied to each RoI. The architecture of the Mask-RCNN is illustrated in Figure 6.

  In this paper, we use Mask-RCNN's code developed in [41]. The backbone model used is ResNet-50 and the pre-trained weights is "COCO-InstanceSegmentation/mask_rcnn_R_50_FPN_1x.yaml".

  It is trained with ResNet-50-FPN on COCO *trainval35k* takes 32 h in our synchronized 8-GPU implementation (0.72 s per 16-image mini-batch) [40]. The code that we used for training, validation, testing is shared under the link (https://github.com/duonglo ng289/detectron2 (accessed on 10 June 2021)).

- PointRend [49]: PointRend is an enhancement of the Mask R-CNN for human instance and human semantic segmentation. This network only differs from Mask R-CNN in the prediction step on bounding-boxes (FCN), Mask R-CNN [40] performs the coarse prediction on a low-resolution (28×28) grid for instance segmentation, the grid is not irrespective of object size. However, it is not suitable for large objects, it generates undesirable "blobby" output that over smooths the fine-level details of large objects. PointRend predicts on the high-resolution output grid (224×224), to avoid computations over the entire high-resolution grid. PointRend suggests 3 strategies: choose a small number of real-value points to make predictions, extract features of selected points, a small neural network trained to predict a label from this point-wise feature representation of a point head. In this paper, the pre-trained weights that we use is "InstanceSegmentation/pointrend_rcnn_R_50_FPN_1x_coco.yaml".

  That means the backbone we use is the ResNet-50. It is trained on the COCO [61] dataset with *train2017* ($\sim$118 k images). The code that we used for training, validation, testing is shared in the link (https://github.com/duonglong289/detectron2/tree/ma ster/projects/PointRend (accessed on 15 June 2021)).

- TridentNet [51]: TridentNet is proposed for human detection by bounding-box on images that are based on the start-of-the-art Faster R-CNN. TridentNet can improve the limitations of two groups of networks for object detection (one-stage methods: YOLO, SSD, and two-stage methods: Faster R-CNN, R-FCN). TridentNet generates scale-specific feature maps with a uniform representational power for training with multiple branches; trident blocks share the same parameters with different dilation rates. TridentNet training with ResNet-50 backbone on 8 GPUs, the pre-trained weights initialized in file "tridentnet_fast_R_50_C4_1x.yaml". The code that we used for training, validation, testing is shared at the link (https://github.com/duonglong 289/detectron2/tree/master/projects/TridentNet (accessed on 16 June 2021)).

- TensorMask [50]: TensorMask is an improvement of Mask R-CNN to use structured 4D tensors ((V, U) represent relative mask position; (H, W) represent the object position) to represent mask image content in a set of densely sliding windows. The dense mask predictor of TensorMask extends the original dense bounding box predictor of Mask R-CNN. TensorMask performs multiclass classification in parallel to mask prediction. The code we use has the pre-trained weights initialized in the file "tensormask_R_50_FPN_1x.yaml" and the ResNet-50 backbone on 8 GPUs are used. The code that we used for training, validation, testing is shared in the link (https://github .com/duonglong289/detectron2/tree/master/projects/TensorMask (accessed on 16 June 2021)).

- CenterMask [52]: CenterMask is an improvement of Mask R-CNN. During the implementation of Mask R-CNN, Centermask added a novel spatial attention-guided mask (SAG-Mask) branch to anchor-free one stage object detector (FCOS), the SAG-Mask branch predicts a segmentation mask on each box with the spatial attention the map that helps to focus on informative pixels and suppress noise. Although, in the present paper [52] used the backbone network VoVNetV2 based on VoVNet [55] to ease optimization and boosts the performance, that shows better performance and faster speed than ResNet [74] and DenseNet [75]. In this paper, we still use the pre-trained weights initialized in file "centermask_R_50_FPN_1x.yaml" The code that we used for training, validation, testing is shared in the link (https: //github.com/duonglong289/centermask2 (accessed on 16 June 2021)).

Start-of-the-art Backbone: As discussed above, the backbone used to train the pre-train weights is ResNet-50.

ResNet is a very efficient deep learning network designed to work with hundreds or thousands of convolutional layers. When building a CNN with many convolutional layers, Vanishing Gradient will occur, leading to a suboptimal learning process. To solve the problem, ResNet proposes the idea of going from the output layer to the input layer and computing the gradient of the corresponding cost function for each parameter (weight) of the network. Gradient descent is then used to update those parameters. It is proposed to use a uniform "identity shortcut connection" connection to traverse one or more layers, illustrated in Figure 7. Such a block is called a Residual Block. The input of this layer is not only the output of the layer above but also the input of the layers that shorten to it.

In residual bock, the input x is added directly and the output of the network is $F(x) + x$, and this path is called an identity shortcut connection. The output of Residual Block is called $H(x) = F(x) + x$. So, when $F(x) = 0$, then $H(x) = x$ is said to be a homogeneous mapping when the input of the network equals its output. To add $F(x) + x$, the shape of both must be the same. If the shapes are not the same then we multiply a matrix W_s by the input x. $H(x) = F(x) + W_s * (x)$, where W_s can be trained. When the input of the network and the output of the network are the same, ResNet uses an identity block, otherwise uses a convolutional block, as presented in Figure 8.

Using residual blocks, in the worst case, the deeper layers learn nothing, and performance is not affected, thanks to skipping connections. But in most cases, these classes will also learn something that can help improve performance.

Recently, Resnet version 1 (v1) has been improved to ameliorate classification performance, and was called ResNet version 2 (v2) [76], where Resnet v2 has two changes in the residual block [77]: The use of a stack of $(1 \times 1) - (3 \times 3) - (1 \times 1)$ at the steps BN, ReLU, Conv2D, respectively; the Batch normalization, and ReLU activation that comes before 2D convolution. The difference between ResNet v1 and ResNet v2 is shown in Figure 9.

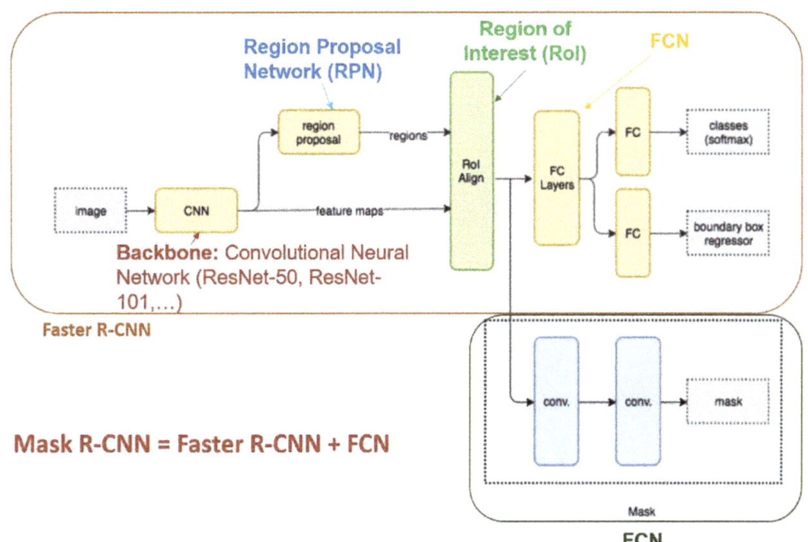

Figure 6. Mask R-CNN architecture human instance segmentation in images.

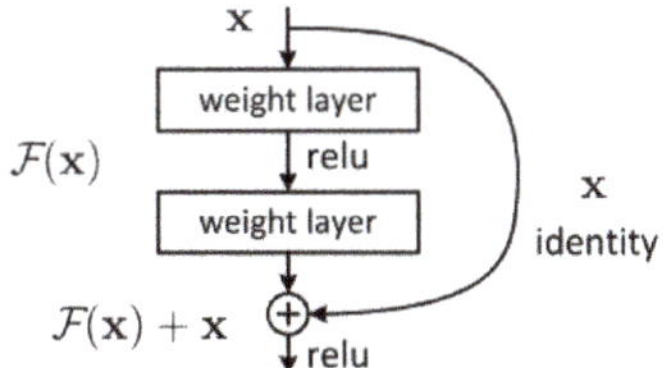

Figure 7. A Residual Block across two layers of ResNet.

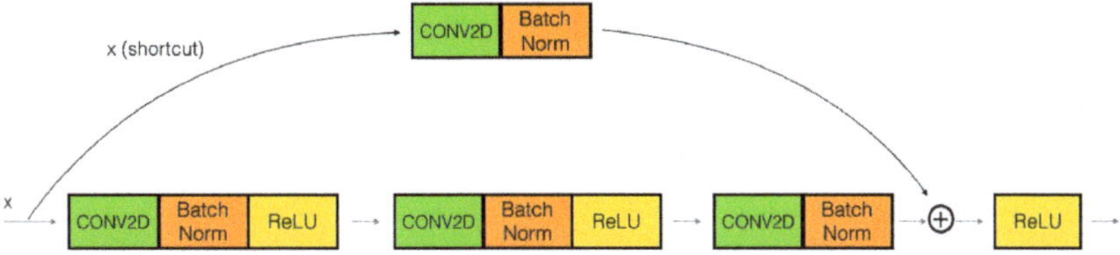

Figure 8. Illustrating convolutional block of ResNet.

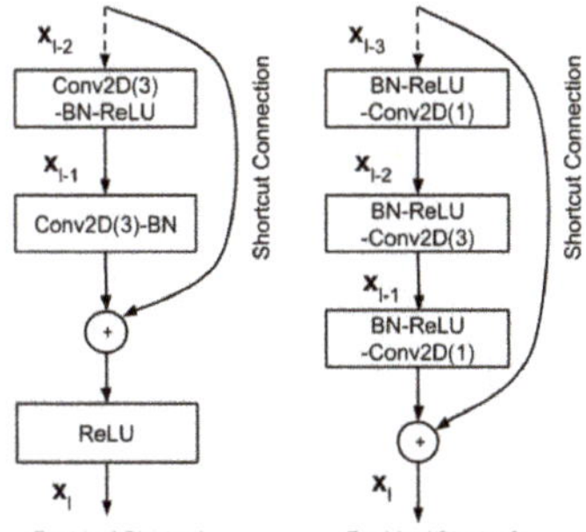

Figure 9. A comparison of residual blocks between ResNet v1 and ResNet v2 [77].

5.2. Experiments

To evaluate the results of human detection and human segmentation on the MADS dataset, we divide the MADS database into training and testing sets according to the following ratios: 50% for training and 50% for testing (rate_50_50), 60% for training, and 40% for testing (rate_60_40), 70% for training and 30% for testing (rate_70_30), 80% for training and 20% for testing (rate_80_20). The images are randomly assigned. The number of frames in the ratios is shown in Table 8.

Table 8. The number of frames in the ratios (%—percent) for training and testing of MASK MADS dataset.

Ratios (%)	The Number of Frames for Training	The Number of Frames for Testing
rate_50_50	14,414	14,414
rate_60_40	17,047	11,492
rate_70_30	19,141	8616
rate_80_20	21,473	5742

In this paper, we used Colab Notebook with GPU Tesla P100, 16 GB for fine-tuning, training, testing the CNNs on the MASK MADS dataset. The processing steps, code fine-tuning, training, testing, and development process were performed in Python language ($\geq$3.6 version) with the support of the OpenCV, Pytorch ($\geq$3.6 version), CUDA/cuDNN libraries, gcc/& g++ ($\geq$5.4 version), In addition, there are a number of other libraries such as Numpy, scipy, Pillow, cython, matplotlib, scikit-image, tensorflow $\geq$ 1.3.0, keras $\geq$ 2.0.8, opencv-python, h5py, imgaug, IPython. The parameters that we use are as follows: the batch size is 2, trained on 90 thousand iterations, the learning rate of 0.02; the weight decay is 0.0001, the momentum is 0.9, and other parameters are the same as when the default training of Mask R-CNN [40] and Detectron2 [41].

In this paper, we also use metrics of the standard COCO metrics including AP (Average Precision) of over IoU thresholds, AP_{50}, AP_{75}, and AP_S, AP_M, AP_L (AP at different scales) [40].

5.3. Results

We compare the results of human segments based on Mask R-CNN, PointRend, TridentNet, TensorMask, CenterMask on the MADS dataset, which have been divided by the ratios. The results based on box (b) are shown in Table 9. In Tables 9 and 10, the human segmentation results on the box (b) and mask (m) of the CenterMask are the highest ($AP^b = 69.47\%$, $AP^m = 61.28\%$). Human segmentation results on the MADS dataset are also shown in Figure 10, and the wrong human segmentation results are also shown in Figure 11.

Table 9. The results (%—percent) of human segmentation (box-m) on the MADS dataset is evaluated on the CNNs.

CNN Model	Training/Testing Ratios (%)	AP^b (%)	AP^b_{50} (%)	AP^b_{75} (%)	AP^b_S (%)	AP^b_M (%)	AP^b_L (%)
Mask R-CNN [40]	rate_50_50	59.25	71.45	65.82	7.22	86.85	86.80
	rate_60_40	59.12	72.02	65.29	7.18	86.45	86.22
	rate_70_30	59.44	71.85	65.84	6.72	87.50	86.55
	rate_80_20	59.96	71.98	66.35	7.16	88.04	86.58
PointRend [49]	rate_50_50	63.25	78.98	72.19	8.03	67.89	84.71
	rate_60_40	64.58	79.81	72.87	9.64	68.84	85.78
	rate_70_30	67.90	81.40	74.30	13.82	72.11	88.69
	rate_80_20	66.67	80.91	73.53	11.29	71.74	87.52

Table 9. *Cont.*

CNN Model	Training/Testing Ratios (%)	AP^b (%)	AP^b_{50} (%)	AP^b_{75} (%)	AP^b_S (%)	AP^b_M (%)	AP^b_L (%)
TridentNet [51]	rate_50_50	61.17	70.84	65.89	6.42	91.39	88.77
	rate_60_40	55.50	71.80	66.25	0.75	50.81	79.66
	rate_70_30	61.28	70.87	65.85	5.71	91.53	88.86
	rate_80_20	61.50	70.96	66.67	6.18	92.01	88.23
TensorMask [50]	rate_50_50	67.11	80.95	74.15	11.78	69.58	88.49
	rate_60_40	67.41	81.71	74.01	12.36	73.79	87.67
	rate_70_30	64.37	79.81	72.38	7.93	67.09	85.74
	rate_80_20	64.78	80.20	73.13	8.95	69.69	85.63
CenterMask [52]	rate_50_50	65.94	79.78	73.09	8.39	71.59	87.75
	rate_60_40	64.75	79.89	72.08	9.21	68.03	86.24
	rate_70_30	65.40	79.36	71.57	6.77	68.72	87.95
	rate_80_20	**69.47**	**81.63**	**75.04**	**12.95**	**74.44**	**91.10**

The results based on the mask (*m*) are shown in Table 10.

Table 10. The results (%—percent) of human segmentation (mask-*m*) on the MADS dataset evaluated on the CNNs.

CNN Model	Training/Testing Ratios (%)	AP^m (%)	AP^m_{50} (%)	AP^m_{75} (%)	AP^m_S (%)	AP^m_M (%)	AP^m_L (%)
PointRend [49]	rate_50_50	58.73	78.86	69.17	9.35	59.70	80.63
	rate_60_40	59.39	79.26	68.48	10.81	55.64	81.07
	rate_70_30	62.66	81.23	70.70	14.11	60.39	83.55
	rate_80_20	61.93	80.58	70.48	11.48	63.29	83.17
TensorMask [50]	rate_50_50	54.10	79.38	65.71	8.48	52.30	74.35
	rate_60_40	57.08	80.01	67.97	10.23	55.36	77.26
	rate_70_30	47.71	77.97	55.07	5.88	38.55	69.53
	rate_80_20	50.87	78.39	60.42	6.70	44.74	72.12
CenterMask [52]	rate_50_50	53.43	79.18	65.85	8.47	56.48	71.71
	rate_60_40	52.24	78.69	64.28	8.80	53.35	70.93
	rate_70_30	52.67	78.96	64.60	7.19	54.36	71.27
	rate_80_20	**61.28**	**81.19**	**72.10**	**13.61**	**66.89**	**79.72**

Figure 11 shows that there are a lot of wrongly segmented pixels (segmented background pixels of human data), and there are also some segmented areas of human data. The problem is the result of the wrong person detection step in the image. In this paper, we also have shared the complete revised source code of CNNs on links (https://github.com/duonglong289/detectron2.git), (https://github.com/duonglong289/centermask2.git), and the retrained model of CNNs on link (https://drive.google.com/drive/folders/16YHR8MxOn4l8fMdNCJZv56AcLKfP_K4-?usp=sharing (accessed on 16 June 2021)). Although there is only one person in the image of the MADS dataset (the data captured by the stereo sensor), it still poses many challenges. Due to the low quality of the images obtained from the stereo sensor, the images are blurred, the lighting is not perfect, and the activities of the people in the image are fast (martial arts, dancing, and sports), so the gestures of the legs and arms are blurred.

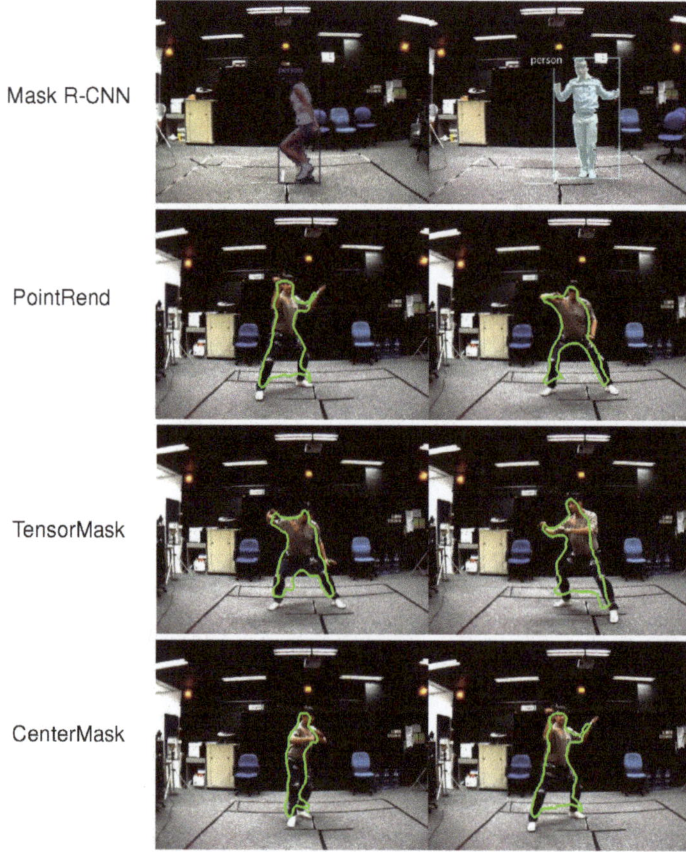

Mask R-CNN

PointRend

TensorMask

CenterMask

Figure 10. Examples of human segmentation results on the MADS dataset by CNNs.

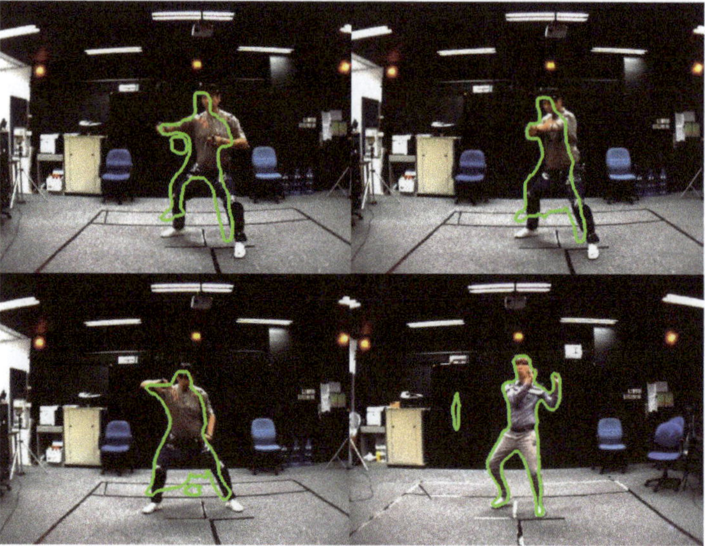

Figure 11. Examples of false human segmentation results on the MADS dataset.

6. Conclusions

In this paper, we performed manual annotation of human masks in videos of data captured from a single view of the MADS dataset on 28 thousand images. This is called the Mask MADS dataset; it is shared for the community to use. We have conducted complete and detailed surveys on using CNNs to detect, segment, and track the people in the video. This survey goes from the mode of methods (CNNs), datasets, metrics, results, analysis, and some discussion. In particular, links to the source code of the CNNs are provided in this survey. Finally, we fine-tuned a set of parameters from the masked human data. We have represented the architecture of start-of-the-art methods and backbone model to fine-tune the human detection, segmentation model. We performed detailed evaluations with many recently published CNNs and published the results on the mask MADS dataset (Tables 8–10).

Author Contributions: Conceptualization, V.-H.L. and R.S.; methodology, V.-H.L.; validation, V.-H.L. and R.S.; formal analysis, V.-H.L. and R.S.; resources, V.-H.L.; writing—original draft preparation, V.-H.L.; writing—review and editing, V.-H.L. and R.S.; funding acquisition, R.S. All authors have read and agreed to the published version of the manuscript.

Funding: This research is funded by Vietnam National Foundation for Science and Technology Development (NAFOSTED) under grant number 102.01-2019.315.

Institutional Review Board Statement: Not Applicable.

Informed Consent Statement: Not Applicable.

Data Availability Statement: The data is available upon the request.

Conflicts of Interest: The authors declare no conflict of interest.

References

1. Zhang, W.; Liu, Z.; Zhou, L.; Leung, H.; Chan, A.B. Martial Arts, Dancing and Sports dataset: A Challenging Stereo and Multi-View Dataset for 3D Human Pose Estimation. *Image Vis. Comput.* **2017**, *61*, 22–39. [CrossRef]
2. Ren, S.; He, K.; Girshick, R.; Sun, J. Faster R-CNN: Towards Real-Time Object Detection with Region Proposal Networks. *Adv. Neural Inf. Process. Syst.* **2015**, *28*, 91–99. [CrossRef]
3. Liu, W.; Anguelov, D.; Erhan, D.; Szegedy, C.; Reed, S.; Fu, C.Y.; Berg, A.C. SSD: Single shot multibox detector. In Proceedings of the European Conference on Computer Vision, Amsterdam, The Netherlands, 11–14 October 2016; Volume 9905, pp. 21–37. [CrossRef]
4. Redmon, J.; Farhadi, A. YOLO9000: Better, Faster, Stronger. *arXiv* **2016**, arXiv:1612.08242.
5. Redmon, J.; Farhadi, A. YOLOv3: An Incremental Improvement. *arXiv* **2018**, arXiv:1804.02767.
6. Bochkovskiy, A.; Wang, C.Y.; Liao, H.Y.M. YOLOv4: Optimal Speed and Accuracy of Object Detection. *arXiv* **2020**, arXiv:2004.10934.
7. Xu, Y.; Zhou, X.; Chen, S.; Li, F. Deep learning for multiple object tracking: A survey. *IET Comput. Vis.* **2019**, *13*, 411–419. [CrossRef]
8. Tsai, J.-K.; Hsu, C.-C.; Wang, W.-Y.; Huang, S.-K. Deep Learning-Based Real-Time Multiple-Person Action Recognition System. *Sensors* **2020**, *20*, 4758. [CrossRef]
9. Yao, R.; Lin, G.; Xia, S.; Zhao, J.; Zhou, Y. Video object segmentation and tracking: A survey. *arXiv* **2019**, arXiv:1904.09172.
10. Xu, J.; Wang, R.; Rakheja, V. Literature Review: Human Segmentation with Static Camera. *arXiv* **2019**, arXiv:1910.12945v1.
11. Girshick, R. Fast R-CNN. In Proceedings of the IEEE International Conference on Computer Vision (ICCV), Santiago, Chile, 7–13 December 2015; Volume 2015, pp. 1440–1448. [CrossRef]
12. Wang, H. Detection of Humans in Video Streams Using Convolutional Neural Networks. In *Degree Project Computer Science and Engineering*; KTH, School of Computer Science and Communication (CSC): Stockholm, Sweden, 2017.
13. Jonathan, L.; Evan, S.; Trevor, D. Fully convolutional networks for semantic segmentation. In Proceedings of the IEEE Conference on Computer Vision and Pattern Recognition (CVPR), Boston, MA, USA, 7–12 June 2015; pp. 3431–3440.
14. Krizhevsky, A.; Sutskever, I.; Hinton, G.E. Handbook of approximation algorithms and metaheuristics. In Proceedings of the 25th International Conference on Neural Information Processing Systems, Red HooK, NY, USA, 3–6 December 2012; pp. 1–1432. [CrossRef]
15. Simonyan, K.; Zisserman, A. Very deep convolutional networks for large-scale image recognition. In Proceedings of the 3rd International Conference on Learning Representations, ICLR 2015—Conference Track Proceedings, San Diego, CA, USA, 7–9 May 2015; pp. 1–14.

16. Ciaparrone, G.; Luque Sánchez, F.; Tabik, S.; Troiano, L.; Tagliaferri, R.; Herrera, F. Deep learning in video multi-object tracking: A survey. *Neurocomputing* **2020**, *381*, 61–88. [CrossRef]

17. Renuka, J. Accuracy, Precision, Recall and F1 Score: Interpretation of Performance Measures. 2016. Available online: https://blog.exsilio.com/all/accuracy-precision-recall-f1-score-interpretation-of-performance-measures/ (accessed on 4 January 2021).

18. Zhang, W.; Shang, L.; Chan, A.B. A robust likelihood function for 3D human pose tracking. *IEEE Trans. Image Process.* **2014**, *23*, 5374–5389. [CrossRef]

19. Liefeng, B.; Cristian, S. Twin Gaussian Processes for Structured Prediction. *Int. J. Comput. Vis.* **2010**, *87*, 28–52.

20. Helten, T.; Baak, A.; Bharaj, G.; Muller, M.; Seidel, H.P.; Theobalt, C. Personalization and evaluation of a real-time depth-based full body tracker. In Proceedings of the 2013 International Conference on 3D Vision, Seatle, BC, Canada, 1–29 July 2013; pp. 279–286. [CrossRef]

21. Ye, M.; Shen, Y.; Du, C.; Pan, Z.; Yang, R. Real-time simultaneous pose and shape estimation for articulated objects using a single depth camera. *IEEE Trans. Pattern Anal. Mach. Intell.* **2016**, *38*, 1517–1532. [CrossRef] [PubMed]

22. Nicolas, B. Calibrating the Depth and Color Camera. 2018. Available online: http://nicolas.burrus.name/index.php/Research/KinectCalibration (accessed on 10 January 2018).

23. Ge, L.; Cai, Y.; Weng, J.; Yuan, J. Hand PointNet: 3D Hand Pose Estimation Using Point Sets. In Proceedings of the 2018 IEEE Conference on Computer Vision and Pattern Recognition, Salt Lake City, UT, USA, 18–22 June 2018; pp. 8417–8426. [CrossRef]

24. Moon, G.; Chang, J.; Lee, K.M. V2V-PoseNet: Voxel-to-Voxel Prediction Network for Accurate 3D Hand and Human Pose Estimation from a Single Depth Map. In Proceedings of the IEEE Conference on Computer Vision and Pattern Recognition (CVPR), Salt Lake City, UT, USA, 18–23 June 2018.

25. Ge, L.; Ren, Z.; Yuan, J. Point-to-Point Regression PointNet for 3D Hand Pose Estimation. In Proceedings of the 15th European Conference on Computer Vision (ECCV), Munich, Germany, 8–14 September 2018.

26. Haque, A.; Peng, B.; Luo, Z.; Alahi, A.; Yeung, S.; Li, F.-F. Towards viewpoint invariant 3D human pose estimation. In *Proceedings of the Lecture Notes in Computer Science (Including Subseries Lecture Notes in Artificial Intelligence and Lecture Notes in Bioinformatics)*; Springer: Amsterdam, The Netherlands, 2016; Volume 9905, pp. 160–177. [CrossRef]

27. Zhang, Z.; Hu, L.; Deng, X.; Xia, S. Weakly Supervised Adversarial Learning for 3D Human Pose Estimation from Point Clouds. *IEEE Trans. Vis. Comput. Graph.* **2020**, *26*, 1851–1859. [CrossRef] [PubMed]

28. D'Eusanio, A.; Pini, S.; Borghi, G.; Vezzani, R.; Cucchiara, R. RefiNet: 3D Human Pose Refinement with Depth Maps. In Proceedings of the International Conference on Pattern Recognition (ICPR), Milan, Italy, 10–15 January 2020.

29. Zhang, Z.; Hu, L.; Deng, X.; Xia, S. A Survey on 3D Hand Skeleton and Pose Estimation by Convolutional Neural Network. *Adv. Sci. Technol. Eng. Syst. J.* **2020**, *5*, 144–159.

30. Harshall, L. Understanding Semantic Segmentation with UNET. 2019. Available online: https://towardsdatascience.com/understanding-semantic-segmentation-with/-unet-6be4f42d4b47 (accessed on 4 January 2021).

31. Girshick, R.; Donahue, J.; Darrell, T.; Malik, J. Rich feature hierarchies for accurate object detection and semantic segmentation. In Proceedings of the IEEE Computer Society Conference on Computer Vision and Pattern Recognition, Columbus, OH, USA, 23–28 June 2014; pp. 580–587. [CrossRef]

32. He, K.; Zhang, X.; Ren, S.; Sun, J. Spatial Pyramid Pooling in Deep Convolutional Networks for Visual Recognition. *IEEE Trans. Pattern Anal. Mach. Intell.* **2015**, *37*, 1904–1916. [CrossRef]

33. Zhang, X.; Zou, J.; He, K.; Sun, J. Accelerating Very Deep Convolutional Networks for Classification and Detection. *IEEE Trans. Pattern Anal. Mach. Intell.* **2016**, *38*, 1943–1955. [CrossRef] [PubMed]

34. Haque, M.F.; Lim, H.; Kang, D.S. Object Detection Based on VGG with ResNet Network. In Proceedings of the 2019 International Conference on Electronics, Information, and Communication (ICEIC), Auckland, New Zealand, 22–25 January 2019; pp. 1–3.

35. Hung, G.L.; Sahimi, M.S.B.; Samma, H.; Almohamad, T.A.; Lahasan, B. Faster R-CNN Deep Learning Model for Pedestrian Detection from Drone Images. In *SN Computer Science*; Springer: Singapore, 2020; Volume 1, pp. 1–9. [CrossRef]

36. Huang, J.; Rathod, V.; Sun, C.; Zhu, M.; Korattikara, A.; Fathi, A.; Fischer, I.; Wojna, Z.; Song, Y.; Guadarrama, S.; et al. Speed/accuracy trade-offs for modern convolutional object detectors. In Proceedings of the 2017 IEEE Conference on Computer Vision and Pattern Recognition, Honolulu, HI, USA, 21–26 July 2017; pp. 3296–3305. [CrossRef]

37. Dai, J.; Li, Y.; He, K.; Sun, J. R-FCN: Object detection via region-based fully convolutional networks. In *Advances in Neural Information Processing Systems*; Curran Associates Inc.: Red Hook, NY, USA, 2016; pp. 379–387.

38. Singh, M.; Basu, A.; Mandal, M.K. Human activity recognition based on silhouette directionality. *IEEE Trans. Circuits Syst. Video Technol.* **2008**, *18*, 1280–1292. [CrossRef]

39. Singh, M.; Mandai, M.; Basu, A. Pose recognition using the radon transform. *Midwest Symp. Circuits Syst.* **2005**, *2005*, 1091–1094. [CrossRef]

40. He, K.; Gkioxari, G.; Dollar, P.; Girshick, R. Mask R-CNN. *arXiv* **2017**, arXiv:1703.06870v3.

41. Wu, Y.; Kirillov, A.; Massa, F.; Lo, W.Y.; Girshick, R. Detectron2. 2019. Available online: https://github.com/facebookresearch/detectron2 (accessed on 14 June 2021).

42. Chen, L.C.; Zhu, Y.; Papandreou, G.; Schroff, F.; Adam, H. Encoder-Decoder with Atrous Separable Convolution for Semantic Image Segmentation. In Proceedings of the European Conference on Computer Vision (ECCV), Munich, Germany, 8–14 September 2018.

43. Chen, L.C.; Papandreou, G.; Schroff, F.; Adam, H. Rethinking atrous convolution for semantic image segmentation. *arXiv* **2017**, arXiv:1706.05587.

44. Sandler, M.; Howard, A.; Zhu, M.; Zhmoginov, A.; Chen, L.C. MobileNetV2: Inverted Residuals and Linear Bottlenecks. In Proceedings of the IEEE Conference on Computer Vision and Pattern Recognition, Salt Lake City, UT, USA, 18–23 June 2018.

45. Neverova, N.; Novotny, D.; Vedaldi, A. Correlated Uncertainty for Learning Dense Correspondences from Noisy Labels. In Proceedings of the Advances in Neural Information Processing Systems 32 (NeurIPS 2019), Vancouver, BC, Canada, 8–14 December 2019; Volume 32.

46. Guler, R.A.; Natalia Neverova, I.K. DensePose: Dense Human Pose Estimation In The Wild. In Proceedings of the IEEE Conference on Computer Vision and Pattern Recognition, Salt Lake City, UT, USA, 18–23 June 2018

47. Cheng, B.; Collins, M.D.; Zhu, Y.; Liu, T.; Huang, T.S.; Adam, H.; Chen, L.C. Panoptic-DeepLab: A Simple, Strong, and Fast Baseline for Bottom-Up Panoptic Segmentation. In Proceedings of the IEEE/CVF Conference on Computer Vision and Pattern Recognition, Seattle, WA, USA, 13–19 June 2020.

48. Cheng, B.; Collins, M.D.; Zhu, Y.; Liu, T.; Huang, T.S.; Adam, H.; Chen, L.C. Panoptic-DeepLab. *arXiv* **2019**, arXiv:1910.04751.

49. Kirillov, A.; Wu, Y.; He, K.; Girshick, R. PointRend: Image Segmentation as Rendering. In Proceedings of the IEEE/CVF Conference on Computer Vision and Pattern Recognition, Long Beach, CA, USA, 15–20 June 2019.

50. Chen, X.; Girshick, R.; He, K.; Dollár, P. Tensormask: A Foundation for Dense Object Segmentation. In Proceedings of the IEEE/CVF International Conference on Computer Vision, Seoul, Korea, 27–28 October 2019.

51. Li, Y.; Chen, Y.; Wang, N.; Zhang, Z. Scale-Aware Trident Networks for Object Detection. In Proceedings of the IEEE/CVF International Conference on Computer Vision, Seoul, Korea, 27 October–2 November 2019.

52. Lee, Y.; Park, J. CenterMask: Real-Time Anchor-Free Instance Segmentation. In Proceedings of the IEEE/CVF Conference on Computer Vision and Pattern Recognition, Seattle, WA, USA, 13–19 June 2020.

53. Tian, Z.; Shen, C.; Chen, H.; He, T. FCOS: Fully Convolutional One-Stage Object Detection. In Proceedings of the IEEE/CVF International Conference on Computer Vision, Seoul, Korea, 27–28 October 2019.

54. Tian, Z.; Shen, C.; Chen, H.; He, T. FCOS: A Simple and Strong Anchor-free Object Detector. *IEEE Trans. Pattern Anal. Mach. Intell.* **2021**. [CrossRef] [PubMed]

55. Lee, Y.; Hwang, J.W.; Lee, S.; Bae, Y.; Park, J. An Energy and GPU-Computation Efficient Backbone Network for Real-Time Object Detection. In Proceedings of the IEEE Conference on Computer Vision and Pattern Recognition Workshops, Long Beach, CA, USA, 16–20 June 2019.

56. Papandreou, G.; Zhu, T.; Chen, L.C.; Gidaris, S.; Tompson, J.; Murphy, K. PersonLab: Person pose estimation and instance segmentation with a bottom-up, part-based, geometric embedding model. In Proceedings of the European Conference on Computer Vision (ECCV), Munich, Germany, 8–14 September 2018.

57. Zhang, S.H.; Li, R.; Dong, X.; Rosin, P.; Cai, Z.; Han, X.; Yang, D.; Huang, H.; Hu, S.M. Pose2Seg: Detection free human instance segmentation. In Proceedings of the IEEE Computer Society Conference on Computer Vision and Pattern Recognition, Long Beach, CA, USA, 16–20 June 2019; pp. 889–898. [CrossRef]

58. Everingham, M.; Van Gool, L.; Williams, C.K.I.; Winn, J.; Zisserman, A. The PASCAL Visual Object Classes Challenge 2007 (VOC2007) Results. Available online: http://www.pascal-network.org/challenges/VOC/voc2007/workshop/index.html (accessed on 14 June 2021).

59. Everingham, M.; Van Gool, L.; Williams, C.K.I.; Winn, J.; Zisserman, A. The PASCAL Visual Object Classes Challenge 2010 (VOC2010) Results. Available online: http://www.pascal-network.org/challenges/VOC/voc2010/workshop/index.html (accessed on 15 June 2021).

60. Everingham, M.; Van Gool, L.; Williams, C.K.I.; Winn, J.; Zisserman, A. The PASCAL Visual Object Classes Challenge 2012 (VOC2012) Results. Available online: http://www.pascal-network.org/challenges/VOC/voc2012/workshop/index.html (accessed on 16 June 2021).

61. Lin, T.Y.; Maire, M.; Belongie, S.; Hays, J.; Perona, P.; Ramanan, D.; Dollár, P.; Zitnick, C.L. Microsoft COCO: Common objects in context. In *Lecture Notes in Computer Science (Including Subseries Lecture Notes in Artificial Intelligence and Lecture Notes in Bioinformatics)*; Springer: Zurich, Switzerland, 2014; Volume 8693, pp. 740–755. [CrossRef]

62. Russakovsky, O.; Deng, J.; Su, H.; Krause, J.; Satheesh, S.; Ma, S.; Huang, Z.; Karpathy, A.; Khosla, A.; Bernstein, M.; et al. ImageNet Large Scale Visual Recognition Challenge. *Int. J. Comput. Vis. (IJCV)* **2015**, *115*, 211–252. [CrossRef]

63. Redmon, J.; Divvala, S.; Girshick, R.; Farhadi, A. You Only Look Once: Unified, Real-Time Object Detection. In Proceedings of the IEEE Conference on Computer Vision and Pattern Recognition, Las Vegas, NV, USA, 27–30 June 2016.

64. Sanchez, S.; Romero, H.; Morales, A. A review: Comparison of performance metrics of pretrained models for object detection using the TensorFlow framework. In Proceedings of the IOP Conference Series Materials Science and Engineering, Chennai, India, 30–31 October 2020.

65. Watada, J.; Musa, Z.; Jain, L.C.; Fulcher, J. Human tracking: A state-of-art survey. In *Lecture Notes in Computer Science (Including Subseries Lecture Notes in Artificial Intelligence and Lecture Notes in Bioinformatics)*; Springer,: Berlin/Heidelberg, Germany, 2010; Volume 6277, pp. 454–463. [CrossRef]

66. Chahyati, D.; Fanany, M.I.; Arymurthy, A.M. Tracking People by Detection Using CNN Features. In *Procedia Computer Science*; Elsevier: Amsterdam, The Netherlands, 2017; Volume 124, pp. 167–172. [CrossRef]

67. Laplaza Galindo, J. Tracking and Approaching People Using Deep Learning Techniques. Master's Thesis, Universitat Politècnica de Catalunya, Barcelona, Spain, 2018.
68. Khan, G.; Tariq, Z.; Usman Ghani Khan, M. Multi-Person Tracking Based on Faster R-CNN and Deep Appearance Features. In *Visual Object Tracking with Deep Neural Networks*; IntechOpen: London, UK, 2019; pp. 1–23. [CrossRef]
69. Wojke, N.; Bewley, A.; Paulus, D. Simple Online and Realtime Tracking with a Deep Association Metric. In Proceedings of the 2017 IEEE International Conference on Image Processing (ICIP), Beijing, China, 17–20 September 2017; pp. 3645–3649. [CrossRef]
70. Wojke, N.; Bewley, A. Deep Cosine Metric Learning for Person Re-identification. In Proceedings of the 2018 IEEE Winter Conference on Applications of Computer Vision (WACV), Lake Tahoe, NV, USA, 12–15 March 2018; pp. 748–756. [CrossRef]
71. Haq, E.U.; Huang, J.; Li, K.; Haq, H.U. Human detection and tracking with deep convolutional neural networks under the constrained of noise and occluded scenes. *Multimed. Tools Appl.* **2020**, *79*, 30685–30708. [CrossRef]
72. Leal-Taixe, L.; Milan, A.; Reid, I.; Roth, S.; Schindler, K. MOTChallenge 2015: Towards a Benchmark for Multi-Target Tracking. *arXiv* **2015**, arXiv:1504.01942 .
73. Bewley, A.; Ge, Z.; Ott, L.; Ramos, F.; Upcroft, B. Simple online and realtime tracking. In Proceedings of the 2016 IEEE International Conference on Image Processing (ICIP), Phoenix, AZ, USA, 25–28 September 2016; pp. 3464–3468. [CrossRef]
74. He, K.; Zhang, X.; Ren, S.; Sun, J. Deep Residual Learning for Image Recognition. In Proceedings of the 2016 IEEE Conference on Computer Vision and Pattern Recognition, Las Vegas, NV, USA, 27–30 June 2016; pp. 770–778. [CrossRef]
75. Huang, G.; Liu, Z.; van der Maaten, L.; Weinberger, K.Q. Densely Connected Convolutional Networks. In Proceedings of the IEEE Conference on Computer Vision and Pattern Recognition, Honolulu, HI, USA, 21–26 July 2017.
76. He, K.; Zhang, X.; Ren, S.; Sun, J. Identity mappings in deep residual networks. In *Lecture Notes in Computer Science (Including Subseries Lecture Notes in Artificial Intelligence and Lecture Notes in Bioinformatics)*; Springer: Amsterdam, The Netherlands, 2016; Volume 9908, pp. 630–645. [CrossRef]
77. Hieu, N.V.; Hien, N.L.H. Recognition of Plant Species using Deep Convolutional Feature Extraction. *Int. J. Emerg. Technol.* **2020**, *11*, 904–910.

Article

Heuristic Attention Representation Learning for Self-Supervised Pretraining

Van Nhiem Tran [1,2], Shen-Hsuan Liu [1,2], Yung-Hui Li [2,*] and Jia-Ching Wang [1]

[1] Department of Computer Science and Information Engineering, National Central University, Taoyuan 3200, Taiwan; tvnhiemhmus@g.ncu.edu.tw (V.N.T.); 109522071@cc.ncu.edu.tw (S.-H.L.); jcw@csie.ncu.edu.tw (J.-C.W.)

[2] AI Research Center, Hon Hai Research Institute, Taipei 114699, Taiwan

[*] Correspondence: yunghui.li@foxconn.com; Tel.: +886-2-2268-3466

Abstract: Recently, self-supervised learning methods have been shown to be very powerful and efficient for yielding robust representation learning by maximizing the similarity across different augmented views in embedding vector space. However, the main challenge is generating different views with random cropping; the semantic feature might exist differently across different views leading to inappropriately maximizing similarity objective. We tackle this problem by introducing **H**euristic **A**ttention **R**epresentation **L**earning (HARL). This self-supervised framework relies on the joint embedding architecture in which the two neural networks are trained to produce similar embedding for different augmented views of the same image. HARL framework adopts prior visual object-level attention by generating a heuristic mask proposal for each training image and maximizes the abstract object-level embedding on vector space instead of whole image representation from previous works. As a result, HARL extracts the quality semantic representation from each training sample and outperforms **existing** self-supervised baselines on several downstream tasks. In addition, we provide efficient techniques based on conventional computer vision and deep learning methods for generating heuristic mask proposals on natural image datasets. Our HARL achieves +1.3% advancement in the ImageNet semi-supervised learning benchmark and +0.9% improvement in AP_{50} of the COCO object detection task over the previous state-of-the-art method BYOL. Our code implementation is available for both TensorFlow and PyTorch frameworks.

Keywords: heuristic attention; perceptual grouping; self-supervised learning; visual representation learning; deep learning; computer vision

Citation: Tran, V.N.; Liu, S.-H.; Li, Y.-H.; Wang, J.-C. Heuristic Attention Representation Learning for Self-Supervised Pretraining. *Sensors* **2022**, *22*, 5169. https://doi.org/10.3390/s22145169

Academic Editor: Jing Tian

Received: 4 June 2022
Accepted: 7 July 2022
Published: 10 July 2022

Publisher's Note: MDPI stays neutral with regard to jurisdictional claims in published maps and institutional affiliations.

1. Introduction

Visual representation learning has been an extended research area on supervised and unsupervised methods. Most supervised learning models learn visual representations by training with many labeled datasets, then transferring the knowledge to other tasks [1–5]. Most supervised learning frameworks try to tune their parameters such that they maximally compress mapping the particular input variables that preserve the information on the output variables [6–8]. As a result, most deep neural networks fail to generalize and maintain robustness if the test samples are different from the training samples on variant distribution and domains.

The new approaches are self-supervised representation learning to overcome the existing drawbacks of supervised learning [9–15]. These techniques have attracted significant attention for efficient, generalization, and robustness representation learning when transferring learned representation on multiple downstream tasks achieving on-par or even outperforming supervised baselines. Furthermore, self-supervised learning methods overcome the human supervision capability of leveraging the enormous availability of unlabeled data. Despite various self-supervised frameworks, these methods involve certain forms of the joint embedding architectures of the two branches neural network such as the Siamese network [16]. The neural networks of two branches are usually weights-sharing or different. In the joint embedding self-supervised framework, the common objective is to maximize the agreement between embedding vectors from different views of the same image. However, the biggest challenge is avoiding collapsing to a trivial constant solution, which is that all output embedding vectors are the same. Several strategies to prevent the collapsing phenomenon can be categorized into two main approaches: contrastive learning and non-contrastive learning. Self-supervised contrastive learning [9,17] prevents collapse via negative sample pairs. However, contrastive learning requires a large number of negative samples leading to the requirement of high computational resources. The efficient alternative approach is non-contrastive learning [13,14,18]. These frameworks rely only on positive pairs with a momentum encoder [13] or using an extra neural network on one branch with the block gradient flow [14,18].

Most existing contrastive and non-contrastive objectives are optimized based on the whole image semantic features across different augmented views. However, under this assumption, several challenges exist. First, popular contrastive methods such as SimCLR [9] and MoCo [17] require more computation and training samples than supervised methods. Second, more importantly, there is no guarantee that semantic representation of different objects will differentiate between different cropping views of the same image. For instance, several meaningful objects (vehicles, humans, animals, etc.) may exist in the same image. The semantic representation of vehicles and humans is different, so contrasting the similarity between different views based on the whole-image semantic feature may be misleading. Research in cognitive psychology and neural science [19–22] showed that early visual attention helps humans focus on the main group of important objects. In computer vision, the perceptual grouping principle is used to group visual features into meaningful parts that allow a much more effective learning representation of the input context information [21].

Motivated by perceptual grouping, we proposed the **Heuristic Attention Representation Learning (HARL)** framework that comprises two main components. First, the early attention mechanism uses unsupervised techniques to generate the heuristic mask to extract object-level semantic features. Second, we construct a framework to abstract and maximize similarity object-level agreement (foreground and background) across different views beyond augmentations of the same image [13,18,23]. This approach helps enrich the quantity and quality of semantic representation by leveraging foreground and background features extracted from the training dataset.

We can summarize our main findings and contributions as follows:

1. We introduce a new self-supervised learning framework (HARL) that maximizes the similarity agreement of object-level latent embedding on vector space across different augmented views. The framework implementation is available in the Supplementary Material section.
2. We utilized two heuristic mask proposal techniques from conventional computer vision and unsupervised deep learning methods to generate a binary mask for the natural image dataset.
3. We construct the two novel heuristic binary segmentation mask datasets for the ImageNet ILSVRC-2012 [24] to facilitate the research in the perceptual grouping for self-supervised visual representation learning. The datasets are available to download in the Data Availability Statement section.
4. Finally, we demonstrate that adopting early visual attention provides a diverse set of high-quality semantic features that increase more effective learning representation for self-supervised pretraining. We report promising results when transferring HARL's learned representation on a wide range of downstream vision tasks.

The remainder of this paper is organized as follows. In Section 2, we discussed related works. Section 3 introduces the HARL framework in detail. Section 4.1 briefly describes the implementation of the HARL framework in self-supervised pretraining. Section 4.2

evaluates and benchmarks HARL performance on the ImageNet evaluation, transfers learning to other downstream tasks and compares it to previous state-of-the-art methods. In Section 5, we provide the analysis of the components impacting the performance and understanding of the behavior of our proposed method. Finally, this paper is concluded in Section 6.

2. Related Works

Our method is mostly related to unsupervised visual representation learning methods, aiming to exploit input signals' internal distributions and semantic information without human supervision. The early works focused on several design-solving pretext tasks, and image generation approaches. Pretext tasks focus on the aspects of image restoration such as denoising [25], predicting noise [26], colorization [27,28], inpainting [29], predicting image rotation [30], solving jigsaw puzzles [31] and more [32,33]. However, these methods, the learned representation of neural networks pre-trained on pretext tasks, still failed in generalization and robustness when performed on different downstream tasks. The generative adversarial learning [34–36] and variational auto-encoding [25,37,38] operate directly on pixel space and high-level details for image generations, which require costly computation that may not be essential and efficient for visual representation learning.

Self-supervised contrastive learning. The popular self-supervised contrastive learning frameworks [9,39,40] aim to pull semantic features from different cropping views of the same image while pushing other features away from other images. However, the downside of contrastive methods is that they require a considerable number of negative pairs, leading to significant computation resources and memory footprint. The efficient alternative approach is non-contrastive learning [13,18], which only maximizes the similarity of two views from the same image without contrast to other views from different images.

Self-supervised non-contrastive learning. Distillation learning-based framework [13,18] inspired by knowledge distillation [41] is applied to joint embedding architecture. One branch is defined as a student network, and another is described as a teacher network. The student network is trained to predict the representation of the teacher network; the teacher network's weights are optimized from the student network by a running average of the student network's weights [13] or by sharing with the student's weights and blocking the gradient flow through the teacher network [18]. Non-contrastive frameworks are effective and computationally efficient compared to self-supervised contrastive frameworks [9,17,39].

However, most contrastive or non-contrastive self-supervised techniques maximize similarity agreements of the whole-image context representation of different augmented views. While developing localization attention to separate the semantically features [42,43] by the perceptual grouping of semantic information proved that adopting prior mid-level visible in pretraining gains efficiency for representation learning. The most recent study related to our [39] leveraging visual attention with segmentation obtained impressive results when transferring the learned representation to downstream tasks on object detection and segmentation in multiple datasets. In contrast to our work, previous work employs pixel-level models for contrastive learning, which uses backbones specialized for semantic segmentation and uses different loss functions. It is important to note that the primary work objective is difficult to transfer to other self-supervised frameworks. It also did not investigate the masking feature method or the impact of the dimension and size of the output spatial feature maps on the latent embedding representation, which we will examine next.

3. Methods

In contrastive or non-contrastive learning-based frameworks, HARL object-level objectives are applicable. For example, our study implements a non-contrastive learning framework using an exponential moving average weight parameter of one encoder to another and an extra predictor inspired by BYOL [13]. HARL's objective maximizes the agreement of the object-level (foreground and background) latent embedding vector across different cropping views beyond augmentations shown in Figure 1.

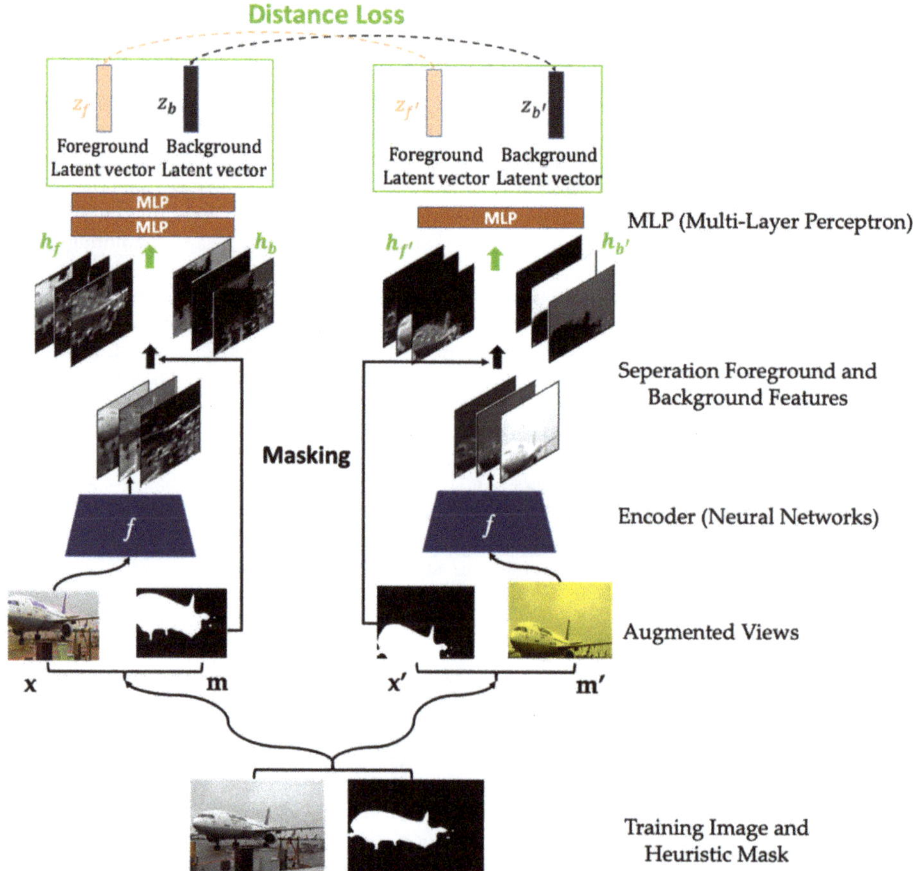

Figure 1. The HARL's architecture. The heuristic binary mask can be estimated by using either conventional computer vision or deep learning approaches. After that, data augmentation transformation is applied to both the image and its mask (**bottom**). Then, the image pairs flow to a convolutional feature extraction module. The heuristic mask is used to mask the feature maps (which are the outputs of the feature extraction module) in order to separate the foreground from the background features (**middle**). These features are further processed by non-linear multi-layer perceptron modules (MLP). Finally, the similarity objective maximizes foreground and background embedding vectors across different augmented views from the same image (**top**).

3.1. HARL Framework

The HARL framework consists of three essential steps. In step 1, we estimate the heuristic binary mask for the input image, which segments an image into foreground and background (see described detail in Section 3.2). Next, these masks can be computed using either conventional computer vision methods such as DRFI [44] or unsupervised deep learning saliency prediction [42]. After the mask is estimated, we perform the same image transformation (cropping, flipping, resizing, etc.) to both the image and its mask. Finally, if it is the RGB image, transformations such as color distortion can be applied to the image, such as the image augmentation pipeline of SimCLR [9]. The detailed augmentation pipeline is described in Appendix A.1. After data augmentation, each image and mask pair generated two augmented images x, x′ aligned with two augmented masks m and m′ as illustrated in Figure 1.

In step 2, we implement standard ResNet-50 [45] convolution residual neural network for feature extractor denotation as f. Each image through the feature extractor encodes the output to obtain the spatial feature maps of size $7 \times 7 \times 2048$, and this feature extraction process can be formulated as $h = f(x)$, where $h \in \mathbb{R}^{H \times W \times D}$. Then, the feature maps can be separated into the foreground and background feature maps by performing element-wise multiplication with the heuristic binary mask. In addition, we provide ablation studies to analyze the impact of the spatial feature map in various sizes and dimensions, as described in Section 5.1. The foreground and background features are denoted as, h_f h_b (Appendix A.2 provides detail of the masking feature method). The foreground and background spatial features are down-sampled using global average pooling to project to a smaller dimension with non-linear multi-layer perceptron (MLP) architecture g.

HARL framework structure adapts from BYOL [13], in which one augmented image (x) is processed with the encoder f_θ, and projection network g_θ, where θ is the learned parameters. Another augmented image (x′) is processed with f_ζ and g_ζ, where ζ is an exponential moving average of θ. The first augmented image is further processed with the predictor network q_θ. The projection and predictor network architectures are the same using the non-linear multi-layer perceptron (MLP), as detailed in Section 4. The definition of encoder, projection, and prediction network is adapted from the BYOL. Finally, the latent representation embedding vectors corresponding to the augmented image's foreground and background features are denoted as z_f, z_b, $z_{f'}$ and $z_{b'} \in \mathbb{R}^d$.

$$\text{where :} \qquad z_f, z_b \triangleq g_\theta \circ q_\theta \left(h_f, h_b \right),$$
$$z_{f'}, z_{b'} \triangleq g_\zeta \left(h_{f'}, h_{b'} \right).$$

In step 3, we compute the HARL's loss function of the given foreground and background latent representations (z_f, z_b, $z_{f'}$ and $z_{b'}$ are extracted from two augmented images x, x′) which is defined as mask loss, as illustrated in Equation (1). We apply ℓ_2-normalization to these latent vectors, then minimize their negative cosine similarity agreement with the weighting coefficient α. We study the impact of α value and the combination of the whole image and object-level latent embedding vector in the loss objective provided in Section 5.2.

$$\mathcal{L}_\theta^{Maskloss} = -\left(\alpha \cdot \frac{z_f}{\|z_f\|_2} \cdot \frac{z_{f'}}{\|z_{f'}\|_2} + (1 - \alpha) \cdot \frac{z_b}{\|z_b\|_2} \cdot \frac{z_{b'}}{\|z_{b'}\|_2} \right), \tag{1}$$

where $\|.\|_2$ is ℓ_2-norm, and it is equivalent to the mean squared error of ℓ_2-normalized vectors. The weighting coefficient α is in the range [0–1].

We symmetrized loss $\mathcal{L}$ by separately feeding augmented image and mask of view one to the online network and augmented image and mask of view two to the target network and vice versa to compute the loss at each training step. We perform a stochastic optimization step to minimize the symmetrized loss $\mathcal{L}_{symmetrized} = \mathcal{L} + \mathcal{L}^{\sim}$.

$$\mathcal{L}_{symmetrized} = \mathcal{L}_\theta^{Maskloss} + \mathcal{L}_\theta^{\sim Maskloss}. \tag{2}$$

After pretraining processing is complete, we only keep the encoder $_\theta$ and discard all other parts of the networks. The whole training procedure summary is in the python pseudo-code Algorithm 1.

Algorithm 1: HARL: Heuristic Attention Representation Learning

---

Input:

    D, M, T, and T': set of images, mask and distributions of transformations

    θ, f_θ, g_θ, and Q_θ : initial online parameters, encoder, projector, and predictor

    ξ, f_ξ, g_ξ; // initial target parameters, target encoder, and target projector

    Optimizer; //optimizer, updates online parameters using the loss gradient

    K and N; //total number of optimization steps and batch size

    $\{T_K\}_{k=1}^{K}$ and $\{\eta_k\}_{k=1}^{K}$; //target network update schedule and learning rate

schedule

1.   For k = 1 to K do

2. $B \leftarrow \{x_i \sim D\}_{i=1}^{N}$; //sample a batch of N images

3. $C \leftarrow \{m_i \sim M\}_{i=1}^{N}$; //sample a batch of N mask

4.   For $x_i \in B, m_i \in C$

5. $h \leftarrow f_\theta(t(x_i))$; //compute the encoder feature map

6. $h' \leftarrow f_\theta(t'(x_i))$; //compute the target encoder feature map

7. $h_f, h_b \leftarrow m_i * h$; //separate the feature map

8. $h_{f\prime}, h_{b\prime} \leftarrow m_i * h'$; //separate the target feature map

9. $z_f, z_b \leftarrow q_\theta\left(g_\theta\left(h_f, h_b\right)\right)$; //compute projections

10. $z_{f\prime}, z_{b\prime} \leftarrow g_\xi\left(h_{f\prime}, h_{b\prime}\right)$; compute target projections

11. $l_i \leftarrow -2 \cdot \left(\alpha \cdot \frac{z_f}{\|z_f\|_2} \cdot \frac{z_f}{\|z_f\|_2} + (1 - \alpha) \cdot \frac{z_{b\prime}}{\|z_{b\prime}\|_2} \cdot \frac{z_{b\prime}}{\|z_{b\prime}\|_2}\right)$; //compute loss

12.   End for

13. $\delta\theta \leftarrow \frac{1}{N}\Sigma_N^{i=1} \partial l_i$ //compute the total loss gradient w.r.t. θ

14. $\theta \leftarrow optimizer(\theta, \delta\theta, \eta_k)$; //update online parameters

15. $\xi \leftarrow \tau_k \xi + (1 - \tau_k)\theta$; //update target parameters encoder f_θ

3.2. Heuristic Binary Mask

Our heuristic binary mask estimation technique does not rely on external supervision, nor is it trained with the limited annotated dataset. We proposed two approaches using conventional computer vision and unsupervised deep learning to carry it out, and these methods appear to be well generalized for various image datasets. First, we use the traditional computer vision method DRFI [44] to generate a diverse set of binary masks by varying the two hyperparameters (the Gaussian filter variance σ and the minimum cluster size s). In our implementation, we defined $\sigma = 0.8$ and $s = 1000$ for generating binary masks in the ImageNet [24] dataset. In the second approach, we leverage the self-supervised encoder feature extractor of the pre-trained ResNet-50 backbone from [9,42], then pass the output feature maps into a 1×1 convolutional classification layer for saliency prediction. The classification layer predicts the saliency or "foregroundness" of a pixel. Therefore, we take the output values of the classification layer and set a threshold of 0.5 to decide which pixels belong to the foreground. Pixel saliency values greater than the threshold are determined as foreground objects. Figure 2 shows the example heuristic mask estimated by these two methods. The detailed implementation of the two methods, DRFI and deep learning feature extractor combined with 1×1 convolutional layer is described in Appendix C. In most of our experiments, we used the mask generated by the deep learning method because it is faster than DRFI by running with GPU instead of only with CPU.

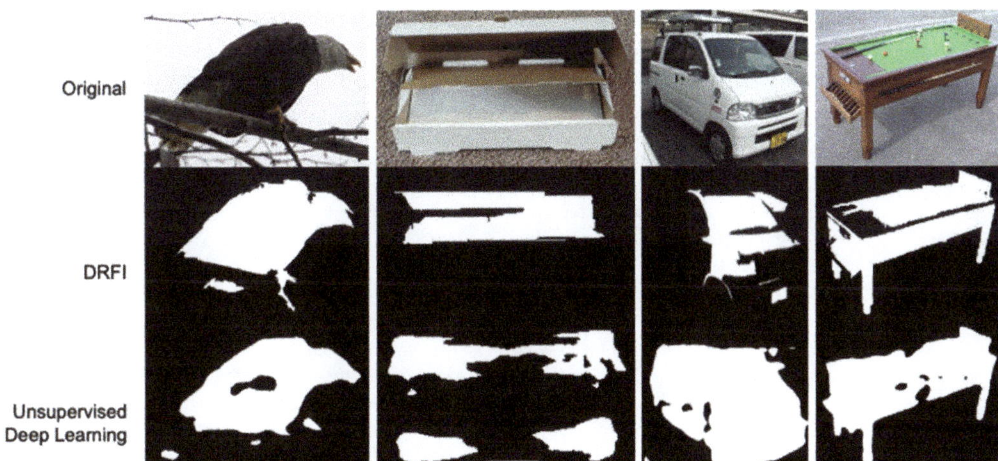

Figure 2. Example of heuristic binary masks used for mask contrastive learning framework. First row: random images from the ImageNet [24] training set. Second row: mask generated based on DRFI algorithm with a predefined sigma σ value of 0.8 and component size values of 1000. The third row is the mask obtained from the self-supervised pre-trained feature extractor ResNet-50 backbone directly followed by a 1 × 1 convolutional classification foreground and background prediction.

4. Experiments

4.1. Self-Supervised Pretraining Implementation

HARL is trained on RGB images and the corresponding heuristic mask of the ImageNet ILSVRC-2012 [24] training set without labels. We implement standard encoder ResNet [45]. According to previous works by SimCLR and BYOL [9,13], the encoder representation output is projected into a smaller dimension using a multi-layer perceptron (MLP). In our implementation, the MLP comprised a linear layer with an output size of 4960 followed by batch normalization [46], rectified linear units (ReLU) [47] and the final linear layer with 512 output units. We apply the LARS optimizer [48] with the cosine decay learning rate schedule without restarts [49], over 1000 epochs on the base learning rate of 0.2, scaled linearly [50] with the batch size (LearningRate = 0.2 × BatchSize/256) and the warmup epochs of 10. Furthermore, we apply a global weight decay parameter of 5×10^{-7} while excluding the biases and normalization parameters from the LARS adaptation and weight decay. The optimization of the online network and target network follow the protocol of BYOL [13]. We use a batch size of 4096 splits over 8 Nvidia A100GPUs. This setup takes approximately 149 h to train a ResNet-50 (×1).

The computational self-supervised pretraining stage requirements are largely due to forward and backward passes through the convolutional backbone. For the typical ResNet-50 architecture applied to 224 × 224 resolution images, a single forward pass requires approximately 4B FLOPS. The projection head MLP (2048 × 4096 + 4096 × 512) requires roughly 10M FLOPS. In our implementation, the convolution network backbone and MLP network are similar compared to baselines BYOL. Since we forward to the foreground and background representation through the projection head two times instead of one, it results in an additional 10M FLOPS in our framework, less than 0.25% of the total. Finally, the cost of computing the heuristic mask images is negligible because they can be computed once and reused throughout training. Therefore, the complexity of each iteration between our method and the baseline BYOL is almost the same for "computational cost" and "training time".

4.2. Evaluation Protocol

We evaluate the learned representation from the self-supervised pretraining stage on various natural image datasets and tasks, including image classification, segmentation and object detection. First, we assess the obtained representation on the linear classification and semi-supervised learning on the ImageNet following the protocols of [9,51]. Second, we evaluate the generalization and robustness of the learned representation by conducting transfer learning to other natural image datasets and other vision tasks across image classification, object detection and segmentation. Finally, in Appendix B, we provide a detailed configuration and hyperparameters setting of the linear and fine-tuning protocol in our transfer learning implementation.

4.2.1. Linear Evaluation and Semi-Supervised Learning on the ImageNet Dataset

The evaluation for linear and semi-supervised learning follows the procedure in [9,52,53]. For the linear evaluation, we train a linear classifier on top of the frozen encoder representation and report Top-1 and Top-5 accuracies in percentage for the test set, as shown in Table 1. We then evaluate semi-supervised learning, which is fine-tuning the pre-trained encoder on a small subset with 1% and 10% of the labeled ILSVRC-2012 ImageNet [24] training set. We also report the Top-1 and Top-5 accuracies for the test set in Table 1. HARL obtains 54.5% and 69.5% in Top-1 accuracy for semi-supervised learning using the standard ResNet-50 ($\times 1$). It represents a +1.3% and +0.7% advancement over the baseline framework BYOL [13] and significant improvement compared to the strong supervised baseline in the accuracy metric.

Table 1. Evaluation on the ImageNet. The linear evaluation and semi-supervised learning with a fraction (1% and 10%) on ImageNet labels report Top-1 and Top-5 accuracies (in%) using the pre-trained ResNet-50 backbone. The best result is bolded.

Method	Linear Evaluation		Semi-Supervised Learning			
	Top-1	Top-5	Top-1		Top-5	
			1%	10%	1%	10%
Supervised	76.5	-	25.4	56.4	48.4	80.4
PIRL [11]	63.6	-	-	-	57.2	83.8
SimCLR [9]	69.3	89.0	48.3	65.6	75.5	87.8
MoCo [17]	60.6	-	-	-	-	-
MoCo v2 [54]	71.1	-	-	-	-	-
SimSiam [18]	71.3	-	-	-	-	-
BYOL [13]	74.3	91.6	53.2	68.8	78.4	89.0
HARL (ours)	74.0	91.3	**54.5**	**69.5**	**79.2**	**89.3**

4.2.2. Transfer Learning to Other Downstream Tasks

We evaluated the HARL's quality of representation learning on linear classification and fine-tuned model following the evaluation setup protocol [9,13,39,55] as detailed in Appendix B.2. HARL's learned representation can perform well for all six different natural distribution image datasets. It has competitive performance in various distribution datasets compared to baseline BYOL [13] and improves significantly compared to the SimCLR [9] approach over six datasets, as shown in Table 2.

We further evaluated HARL's generalization ability and robustness with different computer vision tasks, including object detection of VOC07 + 12 [56] using Faster R-CNN [57] architecture with R50-C4 backbone and instance segmentation task of COCO [58] using Mask R-CNN [59] with R50-FPN backbone. The fine-tuning setup procedure and setting hyperparameter are detailed in Appendix B.3. We report the performance of the standard AP, AP_{50} and AP_{75} metrics in Table 3. HARL outperforms the baselines BYOL [13] and also has a significantly better performance than other self-supervised frameworks such as SimCLR [9], MoCo_v2 [17] and supervised baseline on object detection and segmentation.

Table 2. Transfer via fine-tuning on the image classification task. The transfer learning performance between HARL framework and other self-supervised baseline benchmarks across six natural image classification datasets with the self-supervised pre-trained representation on the ImageNet 1000 classes using the standard ResNet-50 backbone. The best result is bolded.

Method	Food101	CIFAR10	CIFAR100	SUN397	Cars	DTD
Linear evaluation:						
HARL (ours)	75.0	**92.6**	77.6	61.4	67.3	**77.3**
BYOL [13]	**75.3**	91.3	**78.4**	**62.2**	**67.8**	75.5
MoCo v2 (repo)	69.2	91.4	73.7	58.6	47.3	71.1
SimCLR [9]	68.4	90.6	71.6	58.8	50.3	74.5
Fine-tuned:						
HARL (ours)	88.0	97.6	**85.6**	**64.1**	91.1	**78.0**
BYOL [13]	**88.5**	97.4	85.3	63.7	**91.6**	76.2
MoCo v2 (repo)	86.1	97.0	83.7	59.1	90.0	74.1
SimCLR [9]	88.2	**97.7**	85.9	63.5	91.3	73.2

Table 3. Transfer learning to other downstream vision tasks. Benchmark the transfer learning performance between HARL framework and other self-supervised baselines on object detection and instance segmentation task. We use Faster R-CNN with C4 backbone for object detection and Mask-RCNN with FPN backbone for instance segmentation. Object detection and instance segmentation backbone initialize with the pre-trained ResNet-50 backbone on ImageNet 1000 classes. The best result is bolded.

Method	Object Detection						Instance Segmentation		
	VOC07 + 12 Detection			COCO Detection			COCO Segmentation		
	AP_{50}	AP	AP_{75}	AP_{50}	AP	AP_{75}	AP_{50}^{mask}	AP^{mask}	AP_{75}^{mask}
Supervised	81.3	53.5	58.8	58.2	38.2	41.2	54.7	33.3	35.2
SimCLR-IN [18]	81.8	55.5	61.4	57.7	37.9	40.9	54.6	33.3	35.3
MoCo [17]	82.2	57.2	63.7	58.9	38.5	42.0	55.9	35.1	37.7
MoCo v2 [54]	82.5	57.4	64.0	-	39.8	-	-	36.1	-
SimSiam [18]	82.4	57.0	63.7	59.3	39.2	42.1	56.0	34.4	36.7
BYOL [13]	-	-	-		40.4	-	-	37.0	-
BYOL (repo)	82.6	55.5	61.9	61.2	40.2	43.9	58.2	36.7	39.5
HARL (ours)	**82.7**	56.3	62.4	**62.1**	**40.9**	**44.5**	**59.0**	37.3	40.0

5. Ablation and Analysis

We study the HARL's components to give the intuition of its behavior and impact on performance. We reproduce the HARL framework with multiple running experiments. For this reason, we hold the same set of hyperparameter configurations and change the configuration of the corresponding component, which we try to investigate. We perform our ablation experiments on the ResNet-50 and ResNet-18 architecture on the ImageNet training set without labels. We evaluate the learned representation on the ImageNet linear evaluation during the self-supervised pretraining stage. To do so, we attach the linear classifier on top of the base encoder with the block gradient flow on the linear classifier's input, which stops influencing and updating the encoder with the label information (a similar approach to SimCLR [9]). We run ablations over 100 epochs and evaluate the performance of the public validation set of the original ILSVRC2012 ImageNet [24] in the Top-1 accuracy metric at every 100 or 200 steps per epoch following the protocol as described in Appendix B.1.

5.1. The Output of Spatial Feature Map (Size and Dimension)

In our HARL framework, separating foreground and background features from the output spatial feature map is essential to maximize the similarity objective across different augmented views. To verify this hypothesis, we analyze several spatial outputs in various sizes and dimensions by modifying the ResNet kernel's stride to generate the different feature map sizes with the same dimension. For illustration, the standard ResNet is the sequence of four convolution building blocks (conv2_x, conv3_x, conv4_x, conv5_x). For ResNet-50 architecture, the dimension of conv_5x block output feature map is $7 \times 7 \times 2048$. After changing the kernel stride of the conv_4x block from two to one, its new dimension will be $14 \times 14 \times 2048$. In this modified ResNet-50 architecture, the conv5_x block's spatial feature map size is the same as the conv4_x block output.

We conduct the experiment for three different sizes including a deep ResNet-50 ($7 \times 7 \times 2048$, $14 \times 14 \times 2048$, $28 \times 28 \times 2048$) and a shallow ResNet-18 ($7 \times 7 \times 512$, $14 \times 14 \times 512$, $28 \times 28 \times 512$). Figure 3 shows the experimental results of various output sizes and dimensions in the pretraining stage that impact the learned representation when evaluating transfer representation on the ImageNet with linear evaluation protocol. Both shallow and deep ResNet architecture yields better learning ability on the larger output spatial feature map size 14×14 than 7×7. In our experiments, the performance decreases as we continue to go to a larger output size, 28×28 or 56×56.

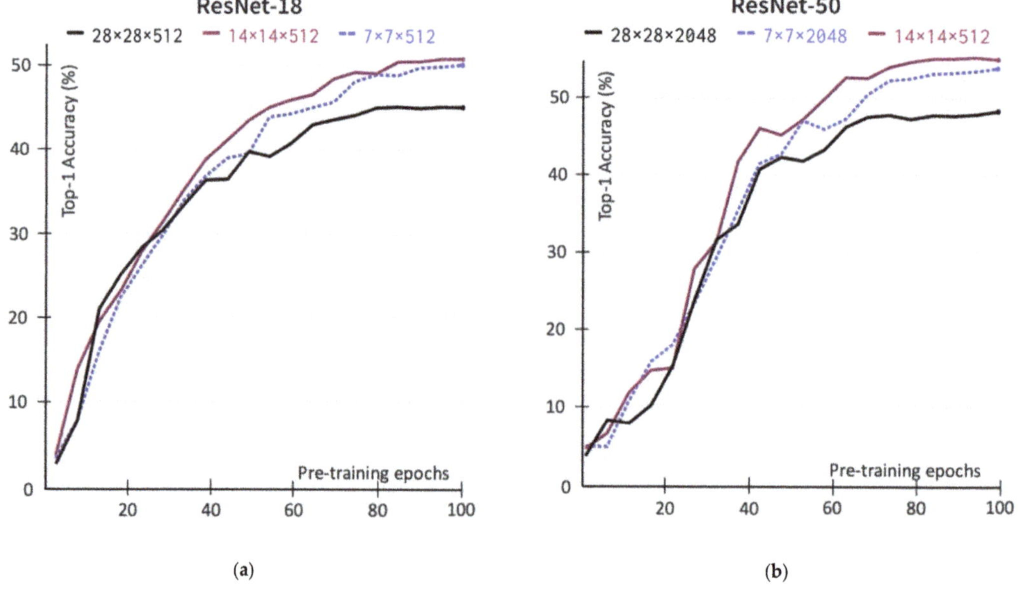

Figure 3. The ImageNet linear evaluation Top-1 accuracy (in %) of spatial output feature maps in various sizes and dimensions during the self-supervised pretraining stage. (**a**) The self-supervised pre-trained encoder uses the ResNet-18 backbone; (**b**) The self-supervised pre-trained encoder uses the ResNet-50 backbone.

5.2. Objective Loss Functions

HARL framework structure reuses elements of BYOL [13]. We use two neural networks denoted as *online network* and *target network*. Each network is defined by a set of parameters θ and ξ. The optimization objective minimizes the loss $\mathcal{L}_{\theta, \xi}$ with respect to learnable parameters θ, while the set ξ is parameterized by using an exponential moving average of the θ, as shown in Equation (3):

$$\xi \leftarrow \tau \xi + (1 - \tau)\theta. \tag{3}$$

Unlike previous approaches that minimize loss function only based on the whole image latent embedding vector between two augmented views, HARL minimized the similarity of object-level latent representation, which associated the same spatial regions abstracting from segmentation mask and thus same semantic meanings. As shown in Figure 1, we use the mask information to separate the spatial semantic object-level feature (foreground and background) of the two augmented views. Then, we minimize their negative cosine similarity, denoting mask loss in Equation (1). In addition to our mask loss objective, we combine the distance loss of the whole image representation and object-level, resulting in hybrid loss as described in Equation (4). We study these two loss objectives in the self-supervised pretraining stage and then evaluate the obtained representation on the ImageNet with a linear evaluation protocol.

5.2.1. Mask Loss

The mask loss objective converges to minimizing the distance loss objective between foreground and background latent embedding on vector space $\mathcal{L}_{\text{foreground}}(\theta, \xi)$ and $\mathcal{L}_{\text{background}}(\theta, \xi)$ with the weighting coefficient α as described in Equation (1). We study the impact of α when it is set to a few predefined values and when it varies according to the cosine scheduling rule. In the first approach, we perform self-supervised pretraining sweeping over three different values {0.3, 0.5, 0.7}. In the second approach, we schedule the α based on a cosine schedule, $\alpha \triangleq (1 - (1 - \alpha_{base})) \cdot (\cos \pi k / K) + 1)/2$, to gradually increase from the starting α_{base} value to 1 corresponding current training step k over total training step K. We tried three α_{base} values, including 0.3, 0.5 and 0.7. We report the Top-1 accuracy on the ImageNet linear evaluation set during the self-supervised pretraining stage, as shown in Figure 4. The weighting coefficient α value of 0.7 yields the consistent learned representation of both approaches. Furthermore, the experimental results demonstrate that the foreground is more important than the background latent representation. For example, in the ImageNet training set, many images exist in which the background information is more than 50% of the image.

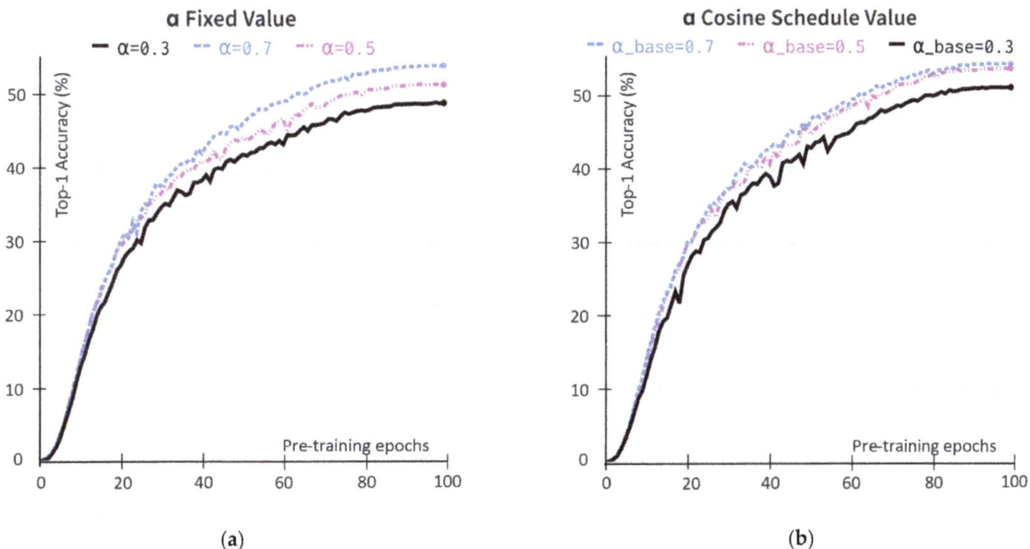

Figure 4. The impact of weighting coefficient α value to the obtained representation during the self-supervised pretraining stage with the ResNet-50 backbone. The evaluation during pretraining uses the ImageNet linear evaluation protocol in Top-1 accuracy (in%). (**a**) The α value is the fixed value; (**b**) The α value follows the cosine function scheduler.

5.2.2. Hybrid Loss

The objective combines whole image representation embedding v_1 and v_2 together with object-level representation embedding mask loss described in Equation (1). v_1 and v_2

are extracted from the two augmented views x and x′ and are denoted as $v_1 \triangleq_\theta \ {}^\circ g_\theta {}^\circ q_\theta(x) \ \mathbb{R}^d$ and $v_2 \triangleq_\xi \ {}^\circ g_\xi(x') \ \mathbb{R}^d$. The hybrid loss minizines the negative cosine similarity with weighting coefficient λ:

$$\mathcal{L}_\theta^{hybrid} = -\left[\lambda \cdot \frac{v_1}{\|v_1\|_2} \cdot \frac{v_2}{\|v_2\|_2} + (1-\lambda) \cdot \mathcal{L}_\theta^{Maskloss}\right], \tag{4}$$

where v_1 and v_2 are the whole image latent representation; $\mathcal{L}_\theta^{Maskloss}$ is the distance loss computed from the foreground and background latent representation described in Equation (1); $\|.\|_2$ is ℓ_2-norm; and λ is the weighting coefficient in the range [0–1].

To study the impact of weighting coefficient λ, we use a cosine scheduling value similar to α in the mask loss section. In our experiment, the weighting coefficient λ cosine scheduling sweeping over four λ_{base} values {0.3, 0.6, 0.7, 0.9}. We report the Top-1 accuracy of the ImageNet linear evaluation protocol on the validation set during the self-supervised pretraining stage, shown in Figure 5. We found using the weighting coefficient λ_{base} value of 0.7 obtains the consistent learned representation when transferring to downstream tasks.

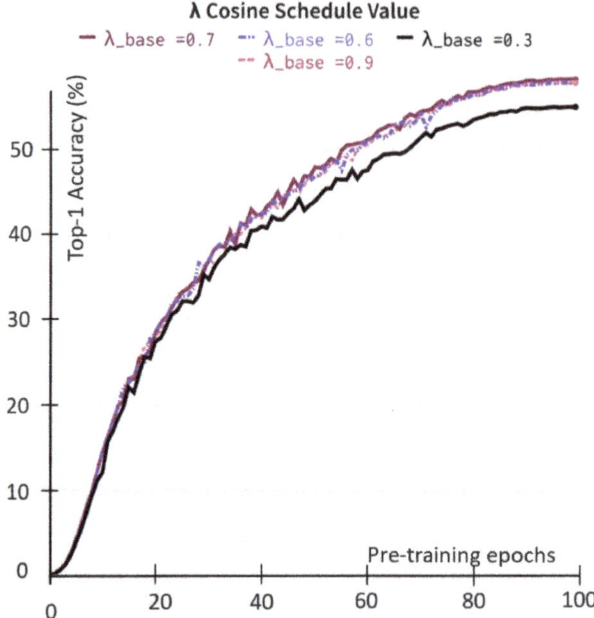

Figure 5. The impact of the weighting coefficient λ value to the obtained representation of the pre-trained encoder (ResNet-50) during the self-supervised pretraining stage on the ImageNet linear evaluation protocol in Top-1 accuracy (in%).

5.2.3. Mask Loss versus Hybrid Loss

We compare the obtained representation using mask loss and hybrid loss on self-supervised pretraining. To do so, we implement the HARL framework with both loss objectives on self-supervised pretraining. We use the cosine schedule function to control the weighting coefficient α and λ sweeping on three different initial values {0.3, 0.5, 0.7} for both coefficients. We evaluate the obtained representation of the pre-trained encoder using ResNet-50 backbone in Top-1 and Top-5 accuracy (in%) on ImageNet linear evaluation protocol, as shown in Table 4. According to the experimental result, using the hybrid loss incorporated between global and object-level latent representation yields better representation learning during self-supervised pretraining.

Table 4. The comparison obtained representation of HARL framework using mask loss and hybrid loss objective. We report Top-1 and Top-5 (in %) accuracy on ImageNet linear evaluation from 100 epochs pre-trained ResNet-50 backbone on ImageNet 1000 classes.

Method	Top-1 Accuracy	Top-5 Accuracy
Mask Loss		
α_base = 0.3	51.3	77.4
α_base = 0.5	53.9	79.4
α_base = 0.7	54.6	79.8
Hybrid Loss		
λ_base = 0.3	55.0	79.4
λ_base = 0.5	57.8	81.7
λ_base = 0.7	58.2	81.8

5.3. The Impact of Heuristic Mask Quality

In our work, the HARL objective uses two different image segmentation techniques. Which ones lead to the best representation? We first consider the heuristics mask retrieving from the computer vision DRFI [44] approach by varying the two hyperparameters (the Gaussian filter variance σ and the minimum cluster size s) as described in detail in Appendix C.1. In our implementation, we generate a diverse set of binary masks by different combinations of $\sigma \in \{0.2, 0.4, 0.8\}$ and c $\in \{1000, 1500\}$. The sets of the generated masks are shown in Figure 6. We found that the setting of $\sigma = 0.8$ and s = 1000 generate more stable mask quality than other combinations. Following the deep learning technique, we use the pre-trained deep convolution neural network as the feature extractor and design a saliency head prediction on top of the feature extractor output's representation in the following three steps described in Appendix C.2. The generated masks are dependent on the pixel saliency threshold, which determines the foregroundness and backgrounness of the pixel. In our implementation, we tested the saliency threshold value ranging in $\{0.4, 0.6, 0.7\}$ as shown in Figure 7. We choose the threshold value equal to 0.5 for generating masks in the ImageNet dataset. After choosing the best configure of the two techniques, we generate the mask for the whole training set of the ImageNet [24] dataset. We evaluate the mask quality generated by computing the mean Intersection-Over-Union (mIoU) between masks generated with the ImageNet ground-truth mask annotated by humans from Pixel-ImageNet [60]. The mIoU of the deep learning masks achieves 0.485 over 0.398 of DRFI masks on the subset of 0.485 million images (946/1000 classes of ImageNet). We found that in a complex scene, where multiple objects exist in a single image, the mask generated from the DRFI technique is noisier and less accurate than the deep learning masks, as illustrated in Figure 8.

To fully evaluate the impact of representation learning on downstream performance, we inspect the obtained representational quality with the transfer learning performance on the object detection and segmentation shown in Table 5. The result indicates that for most object detection and segmentation tasks, HARL learning based on masks with deep learning outperforms the one with DRFI masks, although the difference is very small. It shows that the quality of the mask used for HARL does have a small impact on the performance of the downstream task.

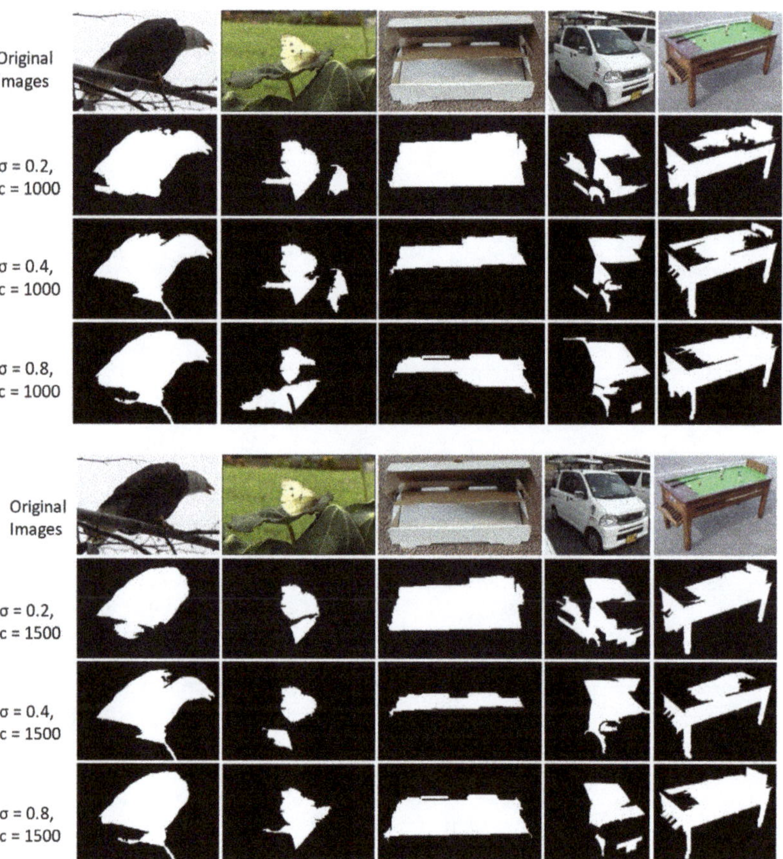

Figure 6. The heuristic binary masks are generated using DRFI with σ = {0.2, 0.4, 0.8} with c = {1000, 1500}.

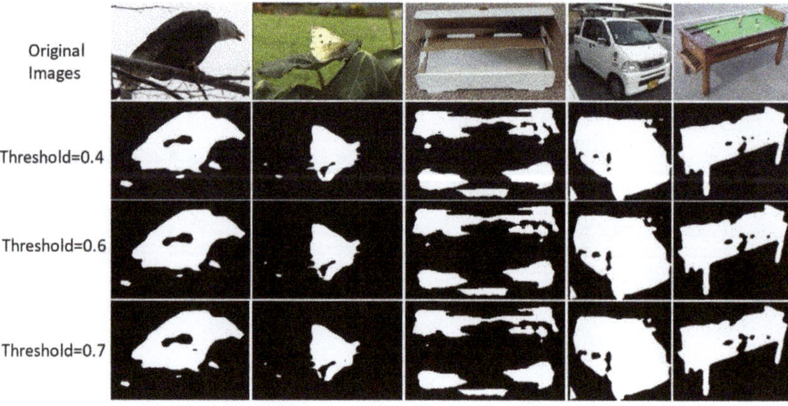

Figure 7. The heuristic binary masks are generated using an unsupervised deep learning encoder with saliency threshold values.

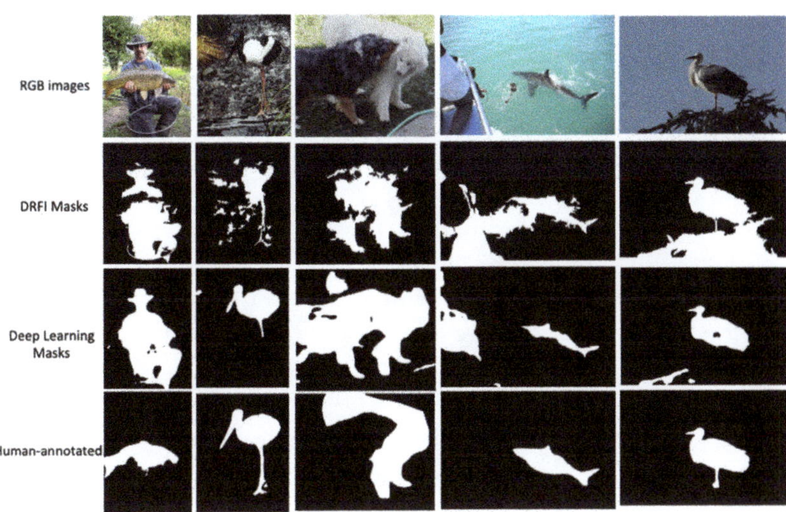

Figure 8. The inspection examples of the generated heuristic binary masks between DRFI and deep learning.

Table 5. The impact of mask quality on HARL framework performance on the downstream object detection and instance segmentation task. We use Faster R-CNN with C4 backbone for object detection and Mask-RCNN with FPN backbone for instance segmentation. Object detection and instance segmentation backbones are initialized with the 100-epoch pre-trained ResNet-50 backbone on ImageNet dataset. The best result is bolded.

Method	Object Detection						Instance Segmentation		
	VOC07 + 12 Detection			COCO Detection			COCO Segmentation		
	AP_{50}	AP	AP_{75}	AP_{50}	AP	AP_{75}	AP_{50}^{mask}	AP^{mask}	AP_{75}^{mask}
HARL (DRFI Masks)	**82.3**	55.4	61.2	44.2	24.6	24.8	41.8	24.3	25.1
HARL (Deep Learning Masks)	82.1	**55.5**	**61.7**	**44.7**	**24.7**	**25.3**	**42.3**	**24.6**	**25.2**

6. Conclusions and Future Work

We introduce the HARL framework, a new self-supervised visual representation learning framework, by leveraging visual attention with the heuristic binary mask. As a result, HARL manages higher-quality semantical information that considerably improves representation learning of self-supervised pretraining compared to previous state-of-the-art methods [9,13,17,18] on semi-supervised and transfers learning on various benchmarks. The two main advantages of the proposed method include: (i) the early attention mechanism that can be applied across different natural image datasets because we use unsupervised techniques to generate the heuristic mask and do not rely on external supervision; (ii) the entire framework can transfer and adapt quickly either to self-supervised contrastive or non-contrastive learning framework. Furthermore, our method will apply and accelerate the currently self-supervised learning direction on pixel-level objectives. Our object-level abstract will make this objective more efficient than the existing work based on computing pixel distance [61].

In our HARL framework, the heuristic binary mask is critical. However, the remaining challenge of estimating accurate masks is suitable for datasets with one primary object, such as the ImageNet dataset. The alternative is mining the object proposal of the image in the complex dataset which contains multiple things by producing heuristic semantic segmentation masks. Designing the new self-supervised framework to solve the remaining challenge of datasets which contain multiple objects is an essential next step and exciting research direction for our future work.

Supplementary Materials: The following documents support our experimental results reported in this study. Our code implementation on PyTorch implementation (https://github.com/TranNhiem/Heuristic_Attention_Represenation_Learning_SSL_Pytorch accessed on 19 September 2021) and TensorFlow (https://github.com/TranNhiem/Heuristic_Attention_Representation_Learning_SSL_Tensorflow accessed on 26 November 2021). Our experimental results and report included in different sections can be downloaded at: https://www.hh-ri.com/2022/05/30/heuristic-attention-representation-learning-for-self-supervised-pretraining/ (accessed on 30 May 2021).

Author Contributions: Conceptualization, Y.-H.L.; methodology, Y.-H.L. and V.N.T.; software, S.-H.L. and V.N.T.; validation, S.-H.L. and V.N.T.; writing—original draft preparation, Y.-H.L. and V.N.T.; writing—review and editing, Y.-H.L. and V.N.T.; supervision, Y.-H.L. and J.-C.W.; project administration, Y.-H.L. and J.-C.W.; funding acquisition, Y.-H.L. All authors have read and agreed to the published version of the manuscript.

Funding: This research received no external funding.

Data Availability Statement: In this study, we construct two novel sets of heuristic binary mask datasets for the ImageNet ILSVRC training set, which can be found here: https://www.hh-ri.com/2022/05/30/heuristic-attention-representation-learning-for-self-supervised-pretraining/ (accessed on 30 May 2022).

Acknowledgments: The authors would like to thank the following people for their help throughout the process of building this study: Timothy Ko for helping create the heuristic binary mask on the DRFI technique for the ImageNet ILSVRC-2012 dataset and Kai-Lin Yang who provided the setting virtual machines for experiments and gave suggestions during the development of the project.

Conflicts of Interest: The authors declare no conflict of interest.

Appendix A. Implementation Detail

Appendix A.1. Implementation Data Augmentation

HARL data augmentation pipeline starts with the standard inception-style random cropping [62]. These cropping views continue to transform using the same set of image augmentation as in SimCLR [9], consisting of the arbitrary sequence composition transformation (color distortion, grayscale conversion, gaussian blur, solarization).

Each RGB image and the heuristic binary mask corresponding to each image are transformed through the augmentation pipeline composed of the following operations described below. First, we utilize the image with the random crop with resizing and random flipping. For the binary mask, these masks apply, only cropping and flipping the underlying RGB image which corresponds. Then, these crop images used give the probability of color distortion (color jittering, color dropping), random Gaussian blur and solarization.

1. Random cropping with resizes: a random patch of the image is selected. In our pipeline, we use the inception-style random cropping [62], whose area crop is uniformly sampled in [0.08 to 1.0] of the original image, and the random aspect ratio is logarithmically sampled in [3/4, 4/3]. The patch is then resized to 224×224 pixels using bicubic interpolation;
2. Optional horizontal flipping (left and right);
3. Color jittering: the brightness, contrast, saturation and hue are shifted by a uniformly distributed offset;
4. Optional color dropping: the RGB image is replaced by its greyscale values;
5. Gaussian blurring with a 224×224 square kernel and a standard deviation uniformly sampled from [0.1, 2.0];
6. Optional solarization: a point-wise color transformation $x \mapsto x \cdot \mathbb{1}_{x<0.5} + (1 - x) \cdot \mathbb{1}_{x<0.5}$ for pixel values in the range [0–1].

The two views augmented image's x, x′ and mask pair m, m′ results from augmentations sample from distributions T, T′, M and M′, respectively. These distributions apply the primitives described above with different probabilities and magnitudes shown in Table A1. The following table specifies these parameters' inherence from the BYOL framework [13] without modification.

Table A1. Parameters used to generate image augmentations.

Parameter	T	T′	M	M′
Inception-style random crop probability	1.0	1.0	1.0	1.0
Flip probability	0.5	0.5	0.5	0.5
Color jittering probability	0.8	0.8	-	-
Brightness adjustment max intensity	0.4	0.4	-	-
Contrast adjustment max intensity	0.4	0.4	-	-
Saturation adjustment max intensity	0.2	0.2	-	-
Hue adjustment max intensity	0.1	0.1	-	-
Color dropping probability	0.2	0.2	-	-
Gaussian blurring probability	1.0	0.1	-	-
Solarization probability	0.0	0.2	-	-

Appendix A.2. Implementation Masking Feature

The masking feature step of the HARL framework is essential to leverage the objective-level information from the heuristic binary mask. The masking features method is composed of three steps. The first step is taking the spatial feature map output $7 \times 7 \times 2048$, which is the final layer before the global average output pooling of the ResNet architecture. Second, in our training loop design, the mask image is directly resized to $7 \times 7 \times 3$ to match the size of the output spatial feature maps without passing through the encoder. Then, the resized mask indexes the feature, one encodes for the foreground feature and zero encodes for the background feature. In the end, we multiply the indexing mask with the spatial features maps to separate the foreground and background features (the corresponding output is two spatial features maps of $7 \times 7 \times 2048$ for foreground and background features). Then, these two spatial feature map outputs apply global average pooling and further reduce dimension with non-linear multi-layer perceptron (MLP) architecture.

Appendix B. Evaluation on the ImageNet and Transfer Learning

Appendix B.1. Implementation Masking Feature Linear Evaluation Semi-Supervised Protocol on ImageNet

Our data preprocessing procedure is described as follows: At training time, the images apply the simple augmentations strategies, including random flip and crops with resizing to 224×224 pixels. At testing time, all images applied are resized to 256 pixels along the shorter side using bicubic resampling, which took a 224×224 center crop. Images are normalized by color channel in training time and testing and divided by standard deviation computed on ImageNet ([9,13] provide a similar pipeline for data processing).

Linear evaluation: We train a linear classifier on top of the frozen pre-trained encoder representation in the linear evaluation without updating the network parameters and the batch statistics. In design and configuration protocol, we follow the standard on ImageNet as in [9,51,54,55]. To train and optimize the linear classifier, we use the SGD optimizer to optimize the cross-entropy loss with the Nesterov momentum over 100 epochs using a batch size of 1024 and a momentum of 0.9. without regularization methods such as weight decay, gradient clipping [63], etc. We report the test set's accuracy (the public validation set of the original ILSVRC2012 ImageNet [24] dataset).

Semi-supervised evaluation: We fine-tuned the network parameters of the pre-trained encoder representation following the semi-supervised learning protocol and procedure as in [9]. Data preprocessing and augmentation strategies at training and testing time for 1% and 10% follow a similar procedure of linear evaluation (described in Appendix C.1) except that with a larger batch size of 2048 and trained over 60 epochs for 1% labeled data and 30 epochs for 10% labeled data. In Table 1 of Section 5.1, we report that the result fine-tuned the representation over the 1% and 10% ImageNet splits from [9] with ResNet-50 ($1\times$) architectures.

Datasets: We followed previous works [9,13] to transfer the representation on the linear classification and fine-tuned it on six different natural image datasets. These datasets are namely Food-101 [64], CIFAR-10 [65] and CIFAR-100 [65], the SUN397 scene dataset [66], Stanford Cars [67] and the Describable Textures Dataset (DTD) [68]. The detail of each dataset is described in Table A2. We use the training set and validation set, which are specified by the dataset creators, to select hyperparameters. On datasets without a test set or validation set, we use the validation examples as a test set or hold out a subset of the training examples we use as the validation set, as described in Table A2.

Table A2. The different image datasets used in transfer learning. When an official test split with labels is not publicly available, we use the official validation split as a test set and create a held-out validation set from the training examples.

Dataset	Classes	Original Training Examples	Training Examples	Validation Examples	Test Examples	Accuracy Measure	Test Provided
Food101	101	75,750	68,175	7575	25,250	Top-1 accuracy	-
CIFAR-10	10	50,000	45,000	5000	10,000	Top-1 accuracy	-
CIFAR-100	100	50,000	44,933	5067	10,000	Top-1 accuracy	-
Sun397 (split 1)	397	19,850	15,880	3970	19,850	Top-1 accuracy	-
Cars	196	8144	6494	1650	8041	Top-1 accuracy	-
DTD (split 1)	47	1880	1880	1880	1880	Top-1 accuracy	Yes

Standard evaluation metrics: To evaluate HARL transfer learning on different datasets and other vision tasks, we use the standard evaluation metrics of each dataset to assess and benchmark our results on these datasets as described in Top-1, AP, AP_{50} and AP_{75}.

- Top-1: We compute the proportion of correctly classified examples.
- AP, AP_{50} and AP_{75}: We compute the average precision as defined in [56].

Appendix B.2. Transfer via Linear Classification and Fine-Tuning

Transfer linear classification: We initialize the network parameters and freeze the pre-trained encoder without updating the network parameters and batch statistics. The standard linear evaluation protocol follows [9,51,55]. In training and testing, the images are resized to 224×224 along the shorter side using bicubic resampling and then normalized with ImageNet statistics without data augmentation. Both phase images normalized the color channels by subtracting the average color and dividing by the standard deviation. We train a regularized multinomial logistic regression classifier on top of the frozen representation. We optimize cross-entropy loss ℓ_2—regularization with the parameters from a range of 45 logarithmically spaced values between 10^{-6} and 10^5 (similar to the optimization procedure of [13]). The model is retrained on the training and validation set combined. The model accuracy performance is reported for the test set.

Transfer fine-tuning: We follow fine-tuning protocol as in [9,51,69] to initialize the network with the parameters of the pre-trained representation. At both phase training and testing time, we follow the image preprocessing and data augmentation strategies to the linear evaluation procedure in Appendix B.1. To fine-tune the network, we optimized the cross-entropy loss using SGD optimizer with a Nesterov momentum value of 0.9 and trained over 20,000 steps with a batch size of 256. We set a hyperparameter including the momentum parameter for batch statistics, learning rate and weight decay selection method, same as in [9,13]. After selecting the optimal hyperparameters configured for the validation set, the model is retrained on the combined training and validation set together, using the specified parameters. The absolute accuracy is reported for the test set.

Appendix B.3. Transfer Learning to Other Vision Tasks

Object detection and instance segmentation: We followed previous works [13,17] for the standard setup transferring procedure on Pascal object detection. We use a Faster R-CNN [57] with the R50-C4 backbone. We fine-tune with the training and validation set (16K images) and report the results for the test set of the PASCAL VOC07 + 12 [56] dataset. The backbone is initialized with our pre-trained ResNet50. We use the SGD optimizer to optimize network parameters for 24K iterations with a batch size of 16. We use the initial learning rate of 0.08, then it is reduced to 10^{-2} at 18K and 10^{-3} at 22K with a linear warmup of the slope 0.3333 for 1000 iterations and the region proposal loss weight of 0.2. Then, we report the final results of AP, AP_{50} and AP_{75} metrics for the test set. For instance, regarding the segmentation task on the COCO [58] dataset, we use Mask R-CNN with FPN backbone to iterate over 90K iterations with a batch size of 16. We initialize the learning rate at 0.03 and reduce it by 10 at the 60K and 80K iterations with warmup iterations of 50.

Appendix C. Heuristic Mask Proposal Methods

In our HARL framework, to generate the heuristic binary mask we investigated various supervised and unsupervised techniques from conventional machine learning to deep-learning-based approaches. The benchmark qualitative and quantitatively state-of-the-art approaches use computer vision methods [70]. The comprehensive literature survey and benchmark [71] offer multiple supervised deep-learning-based methods for salient object detection on multi-level supervision, network architectures and learning paradigms. Several works of the unsupervised deep learning method [72,73] used predictions obtained with the hand-crafted prior as the pseudo label to train the deep neural network.

Appendix C.1. Heuristic Binary Mask Generates Using DRFI

Our first approach uses the conventional machine learning method to generate binary masks by adopting the DRFI [44] technique. This method detects a salient object inside an image by carrying out three main steps: multi-level segmentation that segments an image into regions and regional saliency computation that maps the features extracted from each area to a saliency score, which is predicted by a random forest based on three elements: regional contrast, regional property and regional backgrounds. Additionally, at last, multi-level saliency fusion combines the saliency maps over all the layers of segmentation to obtain the final saliency map. To obtain a binary mask, we generate the saliency map of an image. Then, we define a threshold of 40% (top 40% saliency score) to determine what regions are considered salient objects. Any area that is not a salient object will be regarded as background. We generate a diverse set of binary masks by varying the two hyperparameters σ and the minimum cluster size c. Using $\sigma \in \{0.2, 0.4, 0.8\}$ and $c \in \{1000, 1500\}$ in our implementation, we defined $\sigma = 0.8$ and $c = 1000$ for generating masks in the ImageNet dataset. Additionally, the different configuration hyperparameters experimented with sweeping sigma values $\sigma = \{0.2, 0.4, 0.8\}$ and component sizes of $c = \{1000, 1500\}$ are shown in Figure 6.

Appendix C.2. Heuristic Binary Mask Generates Using Unsupervised Deep Learning

The second approach in our mask-generated techniques is based on a self-supervised pre-trained feature extractor from previous works [9,17,39,42]. We design a new saliency head prediction with pre-trained encoder representation to generate the binary masks. The design is to obtain a binary mask by carrying out three main steps. First, we take the output feature maps from a pre-trained ResNet-50 encoder [9,42]. Second, we pass the output feature map into a 1×1 convolutional classification layer for saliency prediction. The classification layer predicts the saliency or "foregroundness" of a pixel. Finally, we take the classification layer's output values and set a threshold to decide which pixels belong to the foreground. The pixel saliency value more significant than the threshold is determined as a foreground object. In our implementation, we defined a threshold value equal to 0.5 for generating masks in the ImageNet dataset. We further experiment with several threshold values in {0.4, 0.6, 0.7}; all these configure mask-generated examples in Figure 7.

References

1. Shu, Y.; Kou, Z.; Cao, Z.; Wang, J.; Long, M. Zoo-tuning: Adaptive transfer from a zoo of models. *arXiv* **2021**, arXiv:2106.15434.
2. Yang, Q.; Zhang, Y.; Dai, W.; Pan, S.J. *Transfer Learning*; Cambridge University Press: Cambridge, UK, 2020.
3. You, K.; Kou, Z.; Long, M.; Wang, J. Co-Tuning for Transfer Learning. *Adv. Neural Inf. Process. Syst.* **2020**, *33*, 17236–17246.
4. Misra, I.; Shrivastava, A.; Gupta, A.; Hebert, M. Cross-Stitch Networks for Multi-task Learning. In Proceedings of the 2016 IEEE Conference on Computer Vision and Pattern Recognition (CVPR), Las Vegas, NV, USA, 27–30 June 2016; pp. 3994–4003.
5. Li, X.; Xiong, H.; Xu, C.; Dou, D. SMILE: Self-distilled mixup for efficient transfer learning. *arXiv* **2021**, arXiv:2103.13941.
6. Tishby, N.; Zaslavsky, N. Deep learning and the information bottleneck principle. In Proceedings of the 2015 IEEE Information Theory Workshop (ITW), Jeju Island, Korea, 11–15 October 2015; pp. 1–5.
7. Shwartz-Ziv, R.; Tishby, N. Opening the black box of deep neural networks via information. *arXiv* **2017**, arXiv:1703.00810.
8. Amjad, R.A.; Geiger, B.C. Learning representations for neural network-based classification using the information bottleneck principle. *IEEE Trans. Pattern Anal. Mach. Intell.* **2020**, *42*, 2225–2239. [CrossRef]
9. Chen, T.; Kornblith, S.; Norouzi, M.; Hinton, G.E. A simple framework for contrastive learning of visual representations. *arXiv* **2020**, arXiv:2002.05709.
10. Goyal, P.; Caron, M.; Lefaudeux, B.; Xu, M.; Wang, P.; Pai, V.; Singh, M.; Liptchinsky, V.; Misra, I.; Joulin, A.; et al. Self-supervised Pretraining of Visual Features in the Wild. *arXiv* **2021**, arXiv:2103.01988.
11. Misra, I.; Maaten, L.v.d. Self-supervised learning of pretext-invariant representations. In Proceedings of the 2020 IEEE/CVF Conference on Computer Vision and Pattern Recognition (CVPR), Seattle, WA, USA, 13–19 June 2020; pp. 6706–6716.
12. Ermolov, A.; Siarohin, A.; Sangineto, E.; Sebe, N. Whitening for self-supervised representation learning. In Proceedings of the International Conference on Machine Learning ICML, Virtual, 18–24 July 2021.
13. Grill, J.-B.; Strub, F.; Altch'e, F.; Tallec, C.; Richemond, P.H.; Buchatskaya, E.; Doersch, C.; Pires, B.v.; Guo, Z.D.; Azar, M.G.; et al. Bootstrap your own latent: A new approach to self-supervised learning. *arXiv* **2020**, arXiv:2006.07733.
14. Caron, M.; Touvron, H.; Misra, I.; J'egou, H.e.; Mairal, J.; Bojanowski, P.; Joulin, A. Emerging properties in self-supervised vision transformers. *arXiv* **2021**, arXiv:2104.14294.
15. Caron, M.; Misra, I.; Mairal, J.; Goyal, P.; Bojanowski, P.; Joulin, A. Unsupervised learning of visual features by contrasting cluster assignments. *arXiv* **2020**, arXiv:2006.09882.
16. Bromley, J.; Bentz, J.W.; Bottou, L.; Guyon, I.; LeCun, Y.; Moore, C.; Säckinger, E.; Shah, R. Signature verification using a "Siamese" time delay neural network. In Proceedings of the 13th International Joint Conference on Artificial Intelligence, Chambéry, France, 28 August–3 September 1993.
17. He, K.; Fan, H.; Wu, Y.; Xie, S.; Girshick, R.B. Momentum contrast for unsupervised visual representation learning. In Proceedings of the 2020 IEEE/CVF Conference on Computer Vision and Pattern Recognition (CVPR), Seattle, WA, USA, 14–19 June 2020; pp. 9726–9735.
18. Chen, X.; He, K. Exploring simple siamese representation learning. In Proceedings of the IEEE/CVF Conference on Computer Vision and Pattern Recognition CVPR, Nashville, TN, USA, 20–25 June 2021.
19. Hayhoe, M.M.; Ballard, D.H. Eye movements in natural behavior. *Trends Cogn. Sci.* **2005**, *9*, 188–194. [CrossRef] [PubMed]
20. Borji, A.; SihiteDicky, N.; Itti, L. Quantitative analysis of human-model agreement in visual saliency modeling. *IEEE Trans. Image Process.* **2013**, *22*, 55–69. [CrossRef] [PubMed]
21. Benois-Pineau, J.; Callet, P.L. Visual content indexing and retrieval with psycho-visual models. In *Multimedia Systems and Applications*; Springer: Cham, Switzerland, 2017.
22. Awh, E.; Armstrong, K.M.; Moore, T. Visual and oculomotor selection: Links, causes and implications for spatial attention. *Trends Cogn. Sci.* **2006**, *10*, 124–130. [CrossRef] [PubMed]
23. Tian, Y.; Chen, X.; Ganguli, S. Understanding self-supervised learning dynamics without contrastive Pairs. *arXiv* **2021**, arXiv:2102.06810.
24. Russakovsky, O.; Deng, J.; Su, H.; Krause, J.; Satheesh, S.; Ma, S.; Huang, Z.; Karpathy, A.; Khosla, A.; Bernstein, M.S.; et al. ImageNet large scale visual recognition challenge. *Int. J. Comput. Vis.* **2015**, *115*, 211–252. [CrossRef]
25. Vincent, P.; Larochelle, H.; Bengio, Y.; Manzagol, P.-A. Extracting and composing robust features with denoising autoencoders. In Proceedings of the 25th International Conference on Machine Learning ICML, Helsinki, Finland, 5–9 July 2008.
26. Bojanowski, P.; Joulin, A. Unsupervised learning by predicting noise. *arXiv* **2017**, arXiv:1704.05310.
27. Larsson, G.; Maire, M.; Shakhnarovich, G. Colorization as a proxy task for visual understanding. In Proceedings of the 2017 IEEE Conference on Computer Vision and Pattern Recognition (CVPR), Honolulu, HI, USA, 21–26 July 2017; pp. 840–849.
28. Iizuka, S.; Simo-Serra, E. Let there be color!: Joint end-to-end learning of global and local image priors for automatic image colorization with simultaneous classification. *ACM Trans. Graph. (ToG)* **2016**, *35*, 1–11. [CrossRef]
29. Pathak, D.; Krähenbühl, P.; Donahue, J.; Darrell, T.; Efros, A.A. Context encoders: Feature learning by inpainting. In Proceedings of the IEEE Conference on Computer Vision and Pattern Recognition, Las Vegas, NV, USA, 26 June–1 July 2016; pp. 2536–2544.
30. Gidaris, S.; Singh, P.; Komodakis, N. Unsupervised representation learning by predicting image rotations. *arXiv* **2018**, arXiv:1803.07728.

31. Noroozi, M.; Favaro, P. Unsupervised learning of visual representations by solving jigsaw puzzles. In Proceedings of the European Conference on Computer Vision ECCV, Amsterdam, The Netherlands, 8–16 October 2016.
32. Zhang, R.; Isola, P.; Efros, A.A. Split-brain autoencoders: Unsupervised learning by cross-channel prediction. In Proceedings of the 2017 IEEE Conference on Computer Vision and Pattern Recognition (CVPR), Honolulu, HI, USA, 21–26 July 2017; pp. 645–654.
33. Mundhenk, T.N.; Ho, D.; Chen, B.Y. Improvements to context based self-supervised learning. In Proceedings of the 2018 IEEE/CVF Conference on Computer Vision and Pattern Recognition, Salt Lake City, UT, USA, 18–23 June 2018; pp. 9339–9348.
34. Donahue, J.; Krähenbühl, P.; Darrell, T. Adversarial feature learning. *arXiv* **2017**, arXiv:1605.09782.
35. Goodfellow, I.J.; Pouget-Abadie, J.; Mirza, M.; Xu, B.; Warde-Farley, D.; Ozair, S.; Courville, A.C.; Bengio, Y. Generative adversarial nets. In Proceedings of the Neural Information Processing Systems NIPS, Montreal, QC, Canada, 8–13 December 2014.
36. Donahue, J.; Simonyan, K. Large scale adversarial representation learning. In Proceedings of the Neural Information Processing Systems NeurIPS, Vancouver, BC, Canada, 8–14 December 2019.
37. Bansal, V.; Buckchash, H.; Raman, B. Discriminative auto-encoding for classification and representation learning problems. *IEEE Signal Process. Lett.* **2021**, *28*, 987–991. [CrossRef]
38. Kingma, D.P.; Welling, M. Auto-encoding variational bayes. *arXiv* **2013**, arXiv:1312.6114.
39. Chen, T.; Kornblith, S.; Swersky, K.; Norouzi, M.; Hinton, G.E. Big Self-supervised models are strong semi-supervised learners. *arXiv* **2020**, arXiv:2006.10029.
40. Jaiswal, A.; Babu, A.R.; Zadeh, M.Z.; Banerjee, D.; Makedon, F. A Survey on contrastive self-supervised learning. *Technologies* **2020**, *9*, 2. [CrossRef]
41. Hinton, G.E.; Vinyals, O.; Dean, J. Distilling the knowledge in a neural network. *arXiv* **2015**, arXiv:1503.02531.
42. Van Gansbeke, W.; Vandenhende, S.; Georgoulis, S.; Gool, L.V. unsupervised semantic segmentation by contrasting object mask proposals. *arXiv* **2021**, arXiv:2102.06191.
43. Zhang, X.; Maire, M. Self-Supervised visual representation learning from hierarchical grouping. *arXiv* **2020**, arXiv:2012.03044.
44. Jiang, H.; Yuan, Z.; Cheng, M.-M.; Gong, Y.; Zheng, N.; Wang, J. Salient object detection: A discriminative regional feature integration approach. *Int. J. Comput. Vis.* **2013**, *123*, 251–268.
45. He, K.; Zhang, X.; Ren, S.; Sun, J. Deep residual learning for image recognition. In Proceedings of the 2016 IEEE Conference on Computer Vision and Pattern Recognition (CVPR), Las Vegas, NV, USA, 27–30 June 2016; pp. 770–778.
46. Ioffe, S.; Szegedy, C. Batch normalization: Accelerating deep network training by reducing internal covariate shift. *arXiv* **2015**, arXiv:1502.03167.
47. Nair, V.; Hinton, G.E. Rectified linear units improve restricted boltzmann machines. In Proceedings of the 27th International Conference on Machine Learning ICML, Haifa, Israel, 21–24 June 2010.
48. You, Y.; Gitman, I.; Ginsburg, B. Scaling SGD batch size to 32K for imageNet training. *arXiv* **2017**, arXiv:1708.03888.
49. Loshchilov, I.; Hutter, F. SGDR: Stochastic gradient descent with warm restarts. *arXiv* **2017**, arXiv:1608.03983.
50. Goyal, P.; Dollár, P.; Girshick, R.B.; Noordhuis, P.; Wesolowski, L.; Kyrola, A.; Tulloch, A.; Jia, Y.; He, K. Accurate, large Minibatch SGD: Training ImageNet in 1 hour. *arXiv* **2017**, arXiv:1706.02677.
51. Kolesnikov, A.; Zhai, X.; Beyer, L. Revisiting self-supervised visual representation learning. In Proceedings of the 2019 IEEE/CVF Conference on Computer Vision and Pattern Recognition (CVPR), Long Beach, CA, USA, 15–20 June 2019; pp. 1920–1929.
52. Ye, M.; Zhang, X.; Yuen, P.; Chang, S.-F. Unsupervised embedding learning via invariant and spreading instance feature. In Proceedings of the 2019 IEEE/CVF Conference on Computer Vision and Pattern Recognition (CVPR), Long Beach, CA, USA, 15–20 June 2019; pp. 6203–6212.
53. Hjelm, R.D.; Fedorov, A.; Lavoie-Marchildon, S.; Grewal, K.; Trischler, A.; Bengio, Y. Learning deep representations by mutual information estimation and maximization. *arXiv* **2019**, arXiv:1808.06670.
54. Chen, X.; Fan, H.; Girshick, R.B.; He, K. Improved baselines with momentum contrastive learning. *arXiv* **2020**, arXiv:2003.04297.
55. Kornblith, S.; Shlens, J.; Le, Q.V. Do Better ImageNet models transfer better? In Proceedings of the 2019 IEEE/CVF Conference on Computer Vision and Pattern Recognition (CVPR), Long Beach, CA, USA, 15–20 June 2019; pp. 2656–2666.
56. Everingham, M.; Gool, L.V.; Williams, C.K.I.; Winn, J.M.; Zisserman, A. The pascal visual object classes (VOC) challenge. *Int. J. Comput. Vis.* **2009**, *88*, 303–338. [CrossRef]
57. Ren, S.; He, K.; Girshick, R.B.; Sun, J. Faster R-CNN: Towards real-time object detection with region proposal networks. *IEEE Trans. Pattern Anal. Mach. Intell.* **2015**, *39*, 1137–1149. [CrossRef]
58. Lin, T.-Y.; Maire, M.; Belongie, S.J.; Hays, J.; Perona, P.; Ramanan, D.; Dollár, P.; Zitnick, C.L. Microsoft COCO: Common objects in context. In Proceedings of the European Conference on Computer Vision ECCV, Zurich, Switzerland, 6–12 September 2014.
59. He, K.; Gkioxari, G.; Dollár, P.; Girshick, R.B. Mask R-CNN. *IEEE Trans. Pattern Anal. Mach. Intell.* **2020**, *42*, 386–397. [CrossRef]
60. Zhang, S.; Liew, J.H.; Wei, Y.; Wei, S.; Zhao, Y. Interactive object segmentation with inside-outside guidance. In Proceedings of the 2020 IEEE/CVF Conference on Computer Vision and Pattern Recognition (CVPR), Seattle, WA, USA, 13–19 June 2020; pp. 12231–12241.
61. Xie, Z.; Lin, Y.; Zhang, Z.; Cao, Y.; Lin, S.; Hu, H. Propagate yourself: Exploring pixel-level consistency for unsupervised visual representation learning. In Proceedings of the 2021 IEEE/CVF Conference on Computer Vision and Pattern Recognition (CVPR), Nashville, TN, USA, 20–25 June 2021; pp. 16679–16688.

62. Szegedy, C.; Liu, W.; Jia, Y.; Sermanet, P.; Reed, S.E.; Anguelov, D.; Erhan, D.; Vanhoucke, V.; Rabinovich, A. Going deeper with convolutions. In Proceedings of the 2015 IEEE Conference on Computer Vision and Pattern Recognition (CVPR), Boston, MA, USA, 7–12 June 2015; pp. 1–9.
63. Pascanu, R.; Mikolov, T.; Bengio, Y. On the difficulty of training recurrent neural networks. In Proceedings of the International Conference on Machine Learning ICML, Atlanta, GA, USA, 16–21 June 2013.
64. Bossard, L.; Guillaumin, M.; Gool, L.V. Food-101-mining discriminative components with random forests. In Proceedings of the European Conference on Computer Vision ECCV, Zurich, Switzerland, 6–12 September 2014.
65. Krizhevsky, A. Learning Multiple Layers of Features from Tiny Images. Master's Thesis, University of Toronto, Toronto, ON, Canada, 2009. Available online: https://www.cs.toronto.edu/~{}kriz/learning-features-2009-TR.pdf (accessed on 8 April 2009).
66. Xiao, J.; Hays, J.; Ehinger, K.A.; Oliva, A.; Torralba, A. SUN database: Large-scale scene recognition from abbey to zoo. In Proceedings of the 2010 IEEE Computer Society Conference on Computer Vision and Pattern Recognition, San Francisco, CA, USA, 13–18 June 2010; pp. 3485–3492.
67. Krause, J.; Stark, M.; Deng, J.; Fei-Fei, L. 3D Object Representations for fine-grained categorization. In Proceedings of the 2013 IEEE International Conference on Computer Vision Workshops, Sydney, Australia, 2–8 December 2013; pp. 554–561.
68. Cimpoi, M.; Maji, S.; Kokkinos, I.; Mohamed, S.; Vedaldi, A. Describing textures in the wild. In Proceedings of the 2014 IEEE Conference on Computer Vision and Pattern Recognition, Columbus, OH, USA, 23–28 June 2014; pp. 3606–3613.
69. HÈnaff, O.J.; Srinivas, A.; Fauw, J.D.; Razavi, A.; Doersch, C.; Eslami, S.M.A.; Oord, A.R.V.D. Data-efficient image recognition with contrastive predictive coding. *arXiv* **2020**, arXiv:1905.09272.
70. Borji, A.; Cheng, M.-M.; Jiang, H.; Li, J. Salient object detection: A benchmark. *IEEE Trans. Image Process.* **2015**, *24*, 5706–5722. [CrossRef]
71. Wang, W.; Lai, Q.; Fu, H.; Shen, J.; Ling, H. Salient object detection in the deep learning era: An in-depth survey. *IEEE Trans. Pattern Anal. Mach. Intell.* **2021**, *44*, 3239–3259. [CrossRef]
72. Zou, W.; Komodakis, N. HARF: Hierarchy-associated rich features for salient object detection. In Proceedings of the 2015 IEEE International Conference on Computer Vision (ICCV), Santiago, Chile, 7–13 December 2015; pp. 406–414.
73. Zhang, J.; Zhang, T.; Dai, Y.; Harandi, M.; Hartley, R.I. Deep unsupervised saliency detection: A multiple noisy labeling perspective. In Proceedings of the 2018 IEEE/CVF Conference on Computer Vision and Pattern Recognition, Salt Lake City, UT, USA, 18–23 June 2018; pp. 9029–9038.

sensors

Article

Hyper-Parameter Optimization of Stacked Asymmetric Auto-Encoders for Automatic Personality Traits Perception

Effat Jalaeian Zaferani [1], Mohammad Teshnehlab [1], Amirreza Khodadadian [2,*], Clemens Heitzinger [3,4], Mansour Vali [1], Nima Noii [5] and Thomas Wick [2]

1   Electrical & Computer Engineering Faculty, K. N. Toosi University of Technology, Tehran 19967-15433, Iran
2   Institute of Applied Mathematics, Leibniz University of Hannover, 30167 Hannover, Germany
3   Institute of Analysis and Scientific Computing, TU Wien, 1040 Vienna, Austria
4   Center for Artificial Intelligence and Machine Learning (CAIML), TU Wien, 1040 Vienna, Austria
5   Institute of Continuum Mechanics, Leibniz University of Hannover, 30823 Garbsen, Germany
*   Correspondence: khodadadian@ifam.uni-hannover.de

Citation: Jalaeian Zaferani, E.; Teshnehlab, M.; Khodadadian, A.; Heitzinger, C.; Vali, M.; Noii, N.; Wick, T. Hyper-Parameter Optimization of Stacked Asymmetric Auto-Encoders for Automatic Personality Traits Perception. *Sensors* 2022, 22, 6206. https://doi.org/10.3390/s22166206

Academic Editor: Jing Tian

Received: 17 July 2022
Accepted: 16 August 2022
Published: 18 August 2022

Abstract: In this work, a method for automatic hyper-parameter tuning of the stacked asymmetric auto-encoder is proposed. In previous work, the deep learning ability to extract personality perception from speech was shown, but hyper-parameter tuning was attained by trial-and-error, which is time-consuming and requires machine learning knowledge. Therefore, obtaining hyper-parameter values is challenging and places limits on deep learning usage. To address this challenge, researchers have applied optimization methods. Although there were successes, the search space is very large due to the large number of deep learning hyper-parameters, which increases the probability of getting stuck in local optima. Researchers have also focused on improving global optimization methods. In this regard, we suggest a novel global optimization method based on the cultural algorithm, multi-island and the concept of parallelism to search this large space smartly. At first, we evaluated our method on three well-known optimization benchmarks and compared the results with recently published papers. Results indicate that the convergence of the proposed method speeds up due to the ability to escape from local optima, and the precision of the results improves dramatically. Afterward, we applied our method to optimize five hyper-parameters of an asymmetric auto-encoder for automatic personality perception. Since inappropriate hyper-parameters lead the network to over-fitting and under-fitting, we used a novel cost function to prevent over-fitting and under-fitting. As observed, the unweighted average recall (accuracy) was improved by 6.52% (9.54%) compared to our previous work and had remarkable outcomes compared to other published personality perception works.

Keywords: big five personality traits; cultural algorithm; deep learning; hyper-parameter optimization; personality perception

1. Introduction

Whether deep or shallow, the operation of artificial neural networks (ANNs) depends on their hyper-parameters and parameters [1–3]. Certain variables of ANNs are called hyper-parameters, such as the number of layers [2], or control the training process, such as the learning rate [4]. In contrast, the trainable variables pertaining to layer connections and tuned during the training process, which are weights and biases, are called parameters [5–7]. Although parameter tuning may yield good results, it does not yield notable results without hyper-parameter tuning (HPT).

The importance of HPT became more manifest than before with the development of deep learning algorithms. Deep learning is a type of machine learning (ML) technique with diverse hyper-parameters that severely affect its performance [8–10]. Since HPT is an arduous task and requires data and network knowledge [11,12], it is often acquired by empirical methods (trial-and-error), which is time-consuming and does not guarantee significant results in terms of efficient algorithms and overall cost complexity. Therefore, studies based on applying optimization methods to ANNs have gained attention.

Accordingly, the usage of optimization algorithms is divided into three groups, as follows:

1. HPT with the classical method and parameter optimization [13–16]: The fine-tuning of weights and biases (parameters) can provide useful information about the problem, but their size and initial value rely on HPT. Moreover, the number of parameters in deep neural networks (DNNs) and high dimensional datasets is enormous, and calculating the optimum value of these parameters is complicated, not easily implemented, and requires computational systems with remarkable capabilities.

2. Hyper-parameter and parameter optimization [17–19]: Adaptive hyper-parameters are obtained by parameter training. The critical disadvantage is that with each possible vector of hyper-parameters, the parameters must be optimized, which causes runtime errors in the computational system and requires expensive training and large storage capacity to save the best parameters value over epochs. Additionally, all possible combinations of hyper-parameters are computationally infeasible. Hence, this method is not applicable in a large model such as deep learning [20,21].

3. Hyper-parameter optimization (HPO) and parameter tuning with back-propagation [4,11,22]: The main drawback is that although optimization methods are efficient in finding global optima, the gradient may vanish when back-propagating. As a result, not all network parameters are tuned well, which impacts results [23]. To tackle the poor-tuning process of deep neural network parameters, an asymmetric auto-encoder ($\mathrm{Asy_{AE}}$) was presented in our previous work for automatic personality perception (APP) from speech [24]. We showed that $\mathrm{Asy_{AE}}$ could improve the model outcome results compared with conventional auto-encoders by semi-supervised training of parameters, and it can be effectively employed in deep learning. However, the stacked asymmetric auto-encoder ($\mathrm{SA_{AE}}$) hyper-parameters were chosen by trial-and-error, which was time-consuming, and two personality traits achieved lower accuracy than other prior research [24].

Thus, the aims of the present work were to (1) propose a novel optimization method based on cultural evolution and parallel computing, (2) obtain the near-optimal values of hyper-parameters of $\mathrm{SA_{AE}}$, and (3) classify five personality traits.

The rest of the article is organized as follows. In Section 2, some related works of HPO in deep learning and APP are explained. In Section 3, the dataset is introduced, and the summary of the feature extraction method is presented in Section 4. The new optimization method is proposed in Section 5. The simulation results of the new method, which is applied to three benchmark functions of finding global optima, are presented in Section 6. In addition, this section discusses the outcomes of applying the proposed method to $\mathrm{SA_{AE}}$ for automatic personality perception classification.

2. Related Works

Given that this article examines HPO methods in order to find a proper one to optimize the hyper-parameters of $\mathrm{SA_{AE}}$ for automatic personality trait perception, the related works section is divided into two parts. The focus of the first part is on recently published methods of neural network hyper-parameter tuning, regardless of the application in which it is used. Thus, the works related to the investigation of HPO in ML are summarized in the first part. Since the aim of our research was HPT of $\mathrm{SA_{AE}}$ to classify five personality traits from speech, the second part is related to studying HPO in machine learning methods applied in the field of personality trait perception.

2.1. Hyper-Parameter Tuning in ML

Deep learning hyper-parameter types are vast and can be divided into three groups: integer, real, and categorical. The integer group consists of variables such as the number of layers (whether hidden or convolutional) [25], the number of neurons [8], the size of the kernel [26], the number of kernels [27], batch size, pooling size, and number of maximum epochs [9]. The real group includes the learning rate [25], dropout rate [25], regularization factor [25], network weight initialization [5], and momentum [4]. The categorical group comprises activation function type [8] and optimization method [8].

Considering that a change in the value of each hyper-parameter changes the values of the neural network parameters that affect the output of the network, and also that examination of any possible combination of hyper-parameters is time-consuming, expensive and practically impossible, studies have investigated the effect of adjusting and optimizing some of the most important hyper-parameters.

In this regard, the article in [4] employed the HPO method for bearing fault diagnosis in mechanical equipment. Parallel computing was used to find hyper-parameters of the deep belief network (DBN). The learning rate and momentum were optimized, while other hyper-parameters were predefined and kept constant. Additionally, Wu Deng et al. used quantum-inspired differential evolution (DE) to optimize DBN parameters. Results showed an improvement in global search and avoiding premature convergence for fault classification [28].

The numbers of hidden neurons as a hyper-parameter and of the weights/biases as parameters were optimized in a feed-forward ANN by Gray wolf optimizer in [18]. Feed-forward ANNs (not back-propagation) were used because adjusted parameters were achieved by the optimization method.

Y. Peng et al. proposed an HPO method based on a fuzzy system in [8]. They optimized the number of hidden layers and the number of neurons in each layer of a DNN. The activation function type and optimization method, including Genetic Algorithms (GA), Bayesian search, grid search, random search, and quasi-random search, were selected automatically during HPO. For preventing over-fitting, the dropout technique was used. The proposed method was tested in three rainfall prediction datasets.

The authors of [29] suggested a distributed particle swarm optimization (PSO) for the HPO of a convolution neural network (CNN). They were concerned about the time-consuming population search based on distributed PSO, and parallel computing was employed to speed up the algorithm. They optimized the number and size of the kernels, the type of pooling (max or average) for two convolutional layers, the activation function type in convolutional layers, the number of neurons, learning rate, and the dropout rate of the fully connected layers.

Time-series prediction of congestion in highway systems based on long short-term memory (LSTM) was investigated in [9]. To obtain the proper model and structure, the authors recommended an HPO method by applying the Bayesian optimization (BO) method. Five hyper-parameters were automatically obtained, including learning rate, the number of hidden layers, the number of neurons in each layer, batch size, and dropout rate.

The intention of [25] was to examine the robustness of one HPO method over six benchmarks, contrary to other works that designed an algorithm that fit one problem. In other work, the authors used BO as an old HPO method in CNN [1] and applied four strategies to alleviate the drawbacks of BO. They tuned the hyper-parameters of two convolutional layers and two fully connected layers in this way.

In [26], an intuitive architecture design using GA was proposed for CNN. The obtained model was evaluated on a CNN with a single convolutional layer and a fully connected layer. Additionally, some hyper-parameters, including maximum epochs, batch size, initial learning rate, regularization, and momentum were optimized by PSO to prepare a CNN for expression recognition in [30].

Since the success of neural networks depends on their structure, the article in [31] proposed a micro-canonical optimization algorithm for overcoming large parameter spaces and optimizing hyper-parameters of a CNN. Hyper-parameters were the number of convolution layers, activation function type, batch size, pooling type, and dropout rate. The method was evaluated by six image recognition datasets and exhibited accuracy improvement.

State-of-health estimation and remaining usable life prediction in battery prognosis were examined in [32] by a deep convolution neural network. The authors addressed hyper-parameter tuning that affected DNN performance. They improved the algorithm by using the BO method.

Anjir A. Chowdhury et al. concentrated on the role of hyper-parameter optimization in the performance and reliability of deep learning outcomes [33]. They compared several HPO algorithms to obtain better validation accuracy in DNNs and concluded that most of them are computationally expensive. Finally, a greedy approach-based HPO algorithm was proposed for enabling faster computing on edge devices for on-the-fly learning applications. The VGG and ResNet architectures were used, and their hyper-parameters such as epochs, number of hidden layers, number of units per layer, activation function, dropout rate, batch size, and learning rate were optimized.

The Gray wolf optimization was employed to optimize the parameters of the kernel extreme learning machine to realize a hyperspectral image classification method in [34].

2.2. Automatic Personality Perception

In psychology, the big five inventory (BFI) is a well-known theory of personality with five traits, including openness to experience (Ope.), conscientiousness (Con.), extraversion (Ext.), agreeableness (Agr.), and neuroticism (Neu.). These traits are in an individual simultaneously by different scores and can be measured by a BFI questionnaire in general [35,36].

Due to the importance of personality in daily life, computer science researchers have investigated personality trait identification by multimodal media (audio, text, video, image) recently. Here, we focus on studies structured by deep learning methods.

A multimodal approach for perceiving personality traits was proposed by employing well-known deep structures (ResNet-v2-101 and VGGish) [37]. The LSTM network for using temporal information was added at the end. The authors optimized only the learning rate, while other hyper-parameters were configured manually. It is clear that the structure of the mentioned deep methods is fixed, and the weights and biases are pre-trained. Therefore, HPO or HPT does not tune according to each dataset in these networks.

Given the fact that personality traits can influence appearance, MobileNetv2 and ResNeSt50 networks were employed in [38] to extract facial features and classification. Results specified that one pre-trained network such as MobileNetv2 is inappropriate for classifying all five personality traits. It indicated that each trait must classify by a specific model, which means different hyper-parameters are necessary. However, the authors did not mention it directly and applied a combination of two pre-trained deep networks to build a complex deep model.

Onno Kampman et al. examined feature extraction and the classification of five personality traits by applying a one-dimensional CNN to a raw audio dataset. The HPT of the deep network containing regularization factors and kernel size was performed manually [39].

One of the personality detection applications is discovering interpersonal communication skills. Article [40] investigated this aspect from a video interview using a semisupervised CNN in which HPT was performed by trial-and-error. The authors concentrated on video processing, and a fixed hyper-parameter set to utilize for all traits.

The study in [41] analyzed the acoustic and lexical features of a speech signal that were affected by BFI traits. Additionally, it designed six models based on recurrent neural networks for classifying those traits. Hyper-parameters such as hidden size, learning rate, batch size, and dropout percentage were defined, but tuning them was not discussed.

3. Dataset

The SSPNet speaker personality corpus (SPC) is a well-known automatic personality perception dataset introduced in 2010. This dataset originally contained 640 recorded speech signals of 322 native French speakers. There is one speaker in each clip recorded for 10 s. Due to the studies on the effect of mental factors on speech signals [42], the collected clips were emotionally neutral, and to confirm that lexical content did not affect the personality scores, evaluators who were foreign to the French language were selected. Therefore, eleven assessors who did not understand French evaluated each clip based on the BFI questionnaire. The average score of these assessors was considered as the final score for each clip. Hence, five scores were obtained for each clip [43].

Although the SPC dataset has been applied in several works and is a proper dataset for comparison with the new methods, the number of samples is uses is low to train the enormous number of parameters of a DNN. This important challenge was addressed in our previous work [24], and we proved that the sample size of speech signals could be enhanced with data augmentation methods based on a spectrogram so that the prosodic content of speech could be preserved. Data augmentation is a popular technique to expand the size of the dataset artificially and is widely used in image processing. However, using this technique in speech is not as easy as using an image. In other words, we needed to choose transformations that maintain the speaker's personality, and we had to be confident that such manipulations in the spectrogram do not interfere with the extracted features related to personality traits. In this regard, frequency masking and time warping were selected as data augmentation methods, and the number of clips increased up to 640,000. For more details, please see [24].

4. Feature Extraction

Despite DNN's ability to perform automatic feature extraction from raw speech signals, deep learning methods have been generally applied to manually extracted hand-crafted audio features. This is mainly because of the large volume of data required for deep learning methods to outperform. Nevertheless, building a dataset with large available labeled samples is costly, time-consuming, and laborious work in the automatic personality perception field, which restricts various methods. Therefore, previous studies have used handcrafted features for the DNN input [44].

These handcrafted features contain 6373 statistical features extracted from 130 low-level descriptions (LLD) [45]. Table 1 contains 65 LLD features and 65 first derivatives of LLD (ΔLLD), for a total of 130 LLD features.

For the LLD feature extraction process, each clip was divided into 60 ms frames with a 20 ms overlap in the time domain and 20 ms frames with a 10 ms overlap in the frequency domain by the Opensmile2.3 toolkit.

Table 1. The 130 LLD features, including 65 LLD and 65 ΔLLD features [46].

4 Energy Related LLD	Group
Sum of Auditory Spectrum (Loudness)	Prosodic
Sum of RASTA-Style Filtered Auditory Spectrum	Prosodic
RMS Energy, Zero-Crossing Rate	Prosodic

55 Spectral LLD	Group
RASTA-Style Auditory Spectrum, Bands 1–26 (0–8 kHz)	Spectral
MFCC 1-14	Cepstral
Spectral Energy 250–650 Hz, 1 k–4 kHz	Spectral
Spectral Roll Off Point 0.25, 0.50, 0.75, 0.90	Spectral
Spectral Flux, Centroid, Entropy, Slope, Harmonicity	Spectral
Spectral Psychoacoustic Sharpness	Spectral
Spectral Variance, Skewness, Kurtosis	Spectral

6 Voicing Related LLD	Group
F0 (SHS & Viterbi Smoothing)	Prosodic
Probability of Voicing	Sound Quality
Log. HNR, Jitter (Local, Delta), Shimmer (Local)	Sound Quality

Table 1. *Cont.*

4 Energy Related LLD	Group
Mean Values Arithmetic Mean A^{Δ}, B, Arithmetic Mean of Positive Values $^{A\Delta,\,B}$, Root-Quadratic Mean, Flatness	
Moments: Standard Deviation, Skewness, Kurtosis	
Temporal Centroid A^{Δ}, B	
Percentiles Quartiles 1–3, Inter-Quartile Ranges 1–2, 2–3, 1–3, 1%—tile, 99%—tile, Range 1–99%	
Extrema Relative Position of Maximum and Minimum, Full Range (Maximum–Minimum)	
Peaks and Valleys ^{A} Mean of Peak Amplitudes, Difference of Mean of Peak Amplitudes to Arithmetic Mean, Peak to Peak Distances: Mean and Standard Deviation, Peak Range Relative to Arithmetic Mean, Range of Peak Amplitude Values, Range of Valley Amplitude Values Relative to Arithmetic Mean, Valley-Peak (Rising) Slopes: Mean and Standard Deviation, Peak-Valley (Falling) Slopes: Mean and Standard Deviation	
Up-Level Times: 25%, 50%, 75%, 90%	
Rise and Curvature Time Relative Time in which Signal is Rising, Relative Time in which Signal has Left Curvative	
Segment Lengths ^{A} Mean, Standard Deviation, Minimum, Maximum	
Regression A^{Δ}, B Linear Regression: Slope, Offset, Quadratic Error, Quadratic Regression: Coefficients a and b, Offset c, Quadratic Error	
Linear Prediction LP Analysis Gain (Amplitude Error), LP Coefficients 1–5	

^{A} Functionals applied only to energy related and spectral LLDs (group A)
^{B} Functionals applied only to voicing related LLDs (group B)
$^{\Delta}$ Functionals applied only to ΔLLDs
$^{\Delta}$ Functionals not applied only to ΔLLDs

5. Proposed Method

This section is divided into two parts. In the first part, we thoroughly describe the new optimization method mathematically. In order to apply our optimization method to the SA_{AE}, we had to address several problems. The second part deals with this issue and its solution.

5.1. The Proposed Optimization Method

HPO of deep learning is a time-consuming task in practice that depends on the network depth, the size of parameters, processor system, and optimization algorithm speed [5]. Applying HPO to deep learning is challenging. It can be (1) the unsupervised learning of most deep learning methods that causes trouble for optimization and imperfect tuning of parameters [47], (2) a large model with enormous trainable parameters that lead the processing system to runtime errors [5,8], and (3) an intricate search space created by different types of hyper-parameter domains (categorical, continuous, and integer value), causing inherent computational complexity [5]. A larger search space gives rise to a longer search time.

Parallel evaluation can partly reduce optimization time [48], and culture speeds up the population's evolution more than chromosomes (each chromosome represents a solution in the population space) [49]. Accumulated experience that is potentially accessible to all individuals is called culture, which is used in problem-solving activities [50]. The knowledge extracted by identifying patterns in the population's problem-solving experiences influences the generation of new solutions [51]. Therefore, the combination of CA and parallel computing can facilitate the discovery of the search space [52]. In this regard, researchers are interested in combining CA with other optimization algorithms. Sun et al. combined a cultural algorithm and two PSO populations and shared their belief space. It indicated that sharing knowledge of belief space can improve performance by avoiding local optima [53]. A single population and multi-population based on CA was proposed in [54]. A PSO population-based method with interactive belief space was introduced by [49]. A hybrid evolutionary optimization method coupling CA with GAs was defined in [55]. Fuzzy operations were employed to exchange individuals between belief space and population space in [56].

From this perspective, we proposed a four-island approach based on the parallel evaluation and CA.

Although CA and parallel computing can perform better than the basic optimization algorithms [57], they do not provide enough convergence speed alone for deep learning. Thus, three driving force factors were applied to population space for creating interactive space between four island population spaces. Creating interactive population space causes interactive belief space, which can determine the direction and step size faster than traditional optimization methods. In this regard, our proposed method is called the multi-island interactive cultural (MIC) algorithm.

The MIC method is illustrated in Figure 1. In this method, control parameters are configured firstly. The initial population X[m, D] is generated randomly in the feasible space. The variable m indicates the population size (the number of chromosomes or individuals), and D is chromosome dimension (the number of genes).

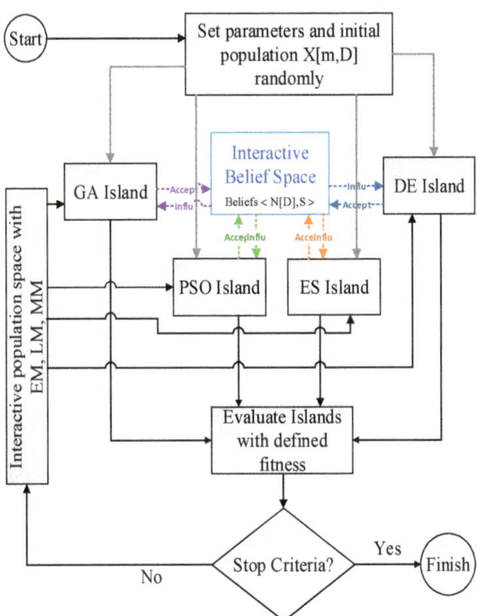

Figure 1. Flowchart of the MIC algorithm.

After preparing the random initial population, it transfers into the four islands in parallel (gray lines): GA, PSO, DE, and evaluation strategy (ES). The GA and PSO are the optimization algorithms widely applied to HPO studies in deep learning [1,8]. GA is far more successful in complex networks such as CNNs, but eliminates previous information by changing the population every iteration [50]. PSO shares information between the particles and is popular on the smaller networks [29]. The DE algorithm is utilized in optimization problems due to the high convergence speed and low control parameters when searching global optima. It is suitable for nonlinear search spaces [28]. The ES is less popular among the global optimization algorithms because it is a simple mutation-selection method, but it is helpful in making small changes [48]. It should be noticed that in the first iteration, the population of the four islands is the same.

The four islands were evaluated individually and in parallel. Then, some individuals of each island were randomly selected to transfer into an interactive belief space (InBS) through an acceptance function (colored arrows). Here, the acceptance function was 25% of the best individuals of each island. So, the belief space size was y[m, D].

The InBS consisted of normative (N[D]) and situational knowledge (S) of all islands. Knowledge of different islands in the belief space causes the chromosomes to move away from unwanted regions and get closer to the optimal points by using different experiences faster than previously published works. InBS can be used effectively to prune the population space.

Normative knowledge represents the range of the best solutions by determining the upper and lower bands of each gene of a chromosome and is used to influence the direction of the search efforts within the promising ranges. In other words, it computes the range of each gene that leads the individual to a good solution.

The offspring affected by normative knowledge are generated by Equation (1) as

$$
y_{p+i,j}^{t+1} = \begin{cases} y_{i,j}^{t} + \left| (u_j^t - l_j^t) * N(0,1) \right| & \text{if } y_{i,j}^{t} < l_j^t, \\ y_{i,j}^{t} - \left| (u_j^t - l_j^t) * N(0,1) \right| & \text{if } y_{i,j}^{t} > u_j^t, \\ y_{i,j}^{t} + \beta \left| (u_j^t - l_j^t) \right| * N(-1,1) & \text{otherwise,} \end{cases} \tag{1}
$$

where u_j is the upper and l_j is the lower band of InBS for jth gene, respectively, β is a constant value, t is the current iteration, and $N(0,1)$ is the normal distribution.

For each gene, the structure contains the upper band (u_j^t), the lower bound (l_j^t), the upper band value (U_j^t), and the lower bound value (L_j^t), which are obtained by Equations (2)–(5), respectively.

$$
L_j^{t+1} = \begin{cases} f(y_i) & \text{if } y_{i,j} \leq l_j^t \ \text{Or} \ f(y_i) < L_j^t, \\ L_j^t & \text{otherwise,} \end{cases} \tag{2}
$$

$$
l_j^{t+1} = \begin{cases} y_{i,j}^t & \text{if } y_{i,j}^t \leq l_j^t \ \text{Or} \ f(y_i^t) < L_j^t, \\ l_j^t & \text{otherwise,} \end{cases} \tag{3}
$$

$$
U_j^{t+1} = \begin{cases} f(y_i) & \text{if } y_{i,j} \geq u_j^t \ \text{Or} \ f(y_i) < U_j^t, \\ U_j^t & \text{Otherwise,} \end{cases} \tag{4}
$$

$$
u_j^{t+1} = \begin{cases} y_{i,j}^t & \text{if } y_{i,j}^t \geq u_j^t \ \text{Or} \ f(y_i^t) < U_j^t, \\ u_j^t & \text{otherwise,} \end{cases} \tag{5}
$$

where $y_{i,j}$ is the jth gene in the ith individual of InBS, and the $f(y_i)$ is the value of the individual y_i calculated by the fitness function. A fitness function (loss function) evaluated individuals of each island separately. The problem description determines the fitness function.

The situational knowledge, as seen in Equation (6), adjusts the mutation step size relative to the distance between the current best individual and the other individuals. The greater the distance between ith individual, y_i, and the current best individual, the greater the step size and vice versa.

Updating the situational knowledge adds the InBS's best individual to the situational knowledge if it outperforms the current best individual, as described in Equation (6).

Here, y^t_{best} is the best individual in the InBS at iteration t. The influence rule can be represented with Equation (7) (for $i = 1, \ldots, m$ and $j = 1, \ldots, D$).

$$< E^{t+1}_1, E^{t+1}_2, \ldots, E^{t+1}_e > = \begin{cases} < y^t_{best}, E^t_2, \ldots, E^t_e > & \text{if } f(y^t_{best}) > f(E^t_1), \\ < y^t_{best} > & \text{if change detected,} \\ < E^t_1, E^t_2 \ldots, E^t_e > & \text{otherwise,} \end{cases} \quad (6)$$

$$y^{t+1}_{p+i,j} = \begin{cases} y^t_{i,j} + \left| (y^t_{i,j} - E^t_{i,j}) \cdot N_{i,j}(0,1) \right| & \text{if } y^t_{i,j} < E^t_j, \\ y^t_{i,j} - \left| (y^t_{i,j} - E^t_{i,j}) \cdot N_{i,j}(0,1) \right| & \text{if } y^t_{i,j} > E^t_j, \\ y^t_{i,j} + \beta \left| (y^t_{i,j} - E^t_{i,j}) \right| \cdot N_{i,j}(0,1) & \text{otherwise,} \end{cases} \quad (7)$$

where E_j is the jth gene in the best individual, β is a constant factor, $N(0, 1)$ is the normal distribution, and $y_{p+i,j}$ is the offspring of the individual $y_{i,j}$.

After updating InBS with new generations, some individuals are transferred into each island population space by influence function. There is no doubt that the individuals of InBS contain the knowledge of all of the islands. This is the ability of the proposed method. Various studies have shown that the efficiency of optimization methods is altered for different problems. In other words, choosing an optimization method for a problem is a challenge that some researchers consider as a kind of hyper-parameter that needs to be tuned. Hence, 25% of the best individuals of InBS were replaced with 25% of the worst population on each island. Offspring generation processing is started in each island separately and evaluated through fitness function.

If the algorithm reaches the stopping criterion, the process will be stopped. Otherwise, interactive population space is created by three driving forces in order to promote cooperation among the islands and increase diversity.

The three driving-force methods are named the elitism method (EM), merge method (MM), and lambda method (LM).

In interactive population space, all individuals of each island are considered. In EM, the best individuals with size *m* are preserved and replaced with the old population on each island. As we use this method, the populations of the next generation for each island are the same. This driving force method forces the four basic algorithms to create interactive space only by the best individuals of four islands.

In MM, after considering all individuals of each island, a random number $a, a \in (0, 1)$, is produced. The $a \times m$ ($a * sizeofpopulation$) of the best individuals are merged with $(a - 1) \times m$ of the old population on each island. It is clear that each island has a unique new population in this interactive space.

In LM, two of the islands are selected randomly, according to two random numbers $\mu, \mu \in (0, 1)$, and $\lambda, \lambda \in (0, 1)$, representing emigration and immigration, respectively. The random numbers of individuals based on μ and λ of each island indicate which individuals can immigrate to and emigrate from another random island. This method forces islands to cooperate with the best individual and the worst one to create interactive space.

Due to the interaction and sharing of individuals among the four islands, if one algorithm traps in local optima, others can lead MIC into global optima because the result is not dependent on a single algorithm. This feature allows the MIC to be used for various global optimization problems to escape local optima efficiently.

The MIC strategy is presented step by step below (Algorithm 1).

Algorithm 1: Implementation of MIC

Step 1: Set the MIC parameters randomly.
Step 2: Generate the initial population randomly.
Step 3: Transfer 25% of the best individuals of each island into InBS (Accept).
Step 4: Update Belief space whith Equations (1)–(7).
Step 5: Transfer 25% of offspring into each island (Influ).
Step 6: **If** stop criterion $< \zeta$
      Stop algoriyhm.
    Else
      Go to Step 7.
Step 7: Create Interactive population space by using the following three methods:
      EM: m of the best individuals of four islands are selected and replaced with an old population.
      MM: The $a \times m$ ($a * size of population$) of the best individuals are selected and merged with $(a - 1) \times m$, which is obtained from the old population in islands.
      LM: According to two random numbers, μ and λ, some individuals of a random island can immigrate to and emigrate from another random island.
Step 8: Go to Step 3.

5.2. Stacked Asymmetric Auto-Encoder HPO Using MIC

Since our work aimed to obtain the SA_{AE} near-optimal structure, a brief overview of this method is presented below.

(1)   **Stacked asymmetric auto-encoder**

The Asy_{AE} is a semi-supervised DNN that poses the curse of dimensionality. The schematic of the Asy_{AE} is illustrated in Figure 2.

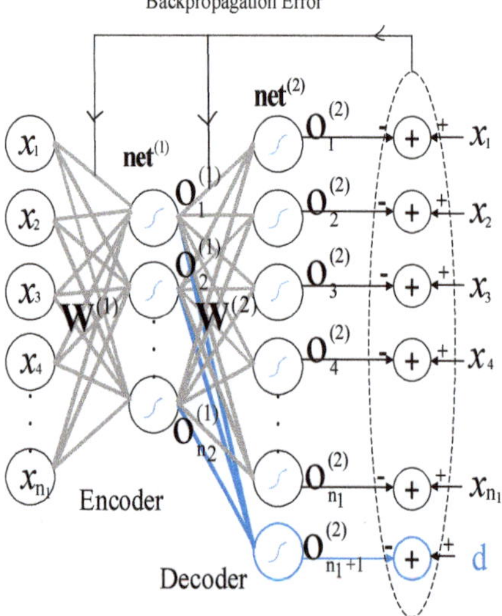

Figure 2. Schematic of the asymmetric auto-encoder [24].

In this type, one neuron is added in the decoder part of the conventional auto-encoder with the desired value of the problem, which is the studied personality score in our field. The symmetry of the encoder and decoder parts is disrupted by this single neuron and made asymmetric.

The feed-forward equations of the Asy_{AE} are similar to the conventional one as follows. For representing encoder and decoder layers, superscripts of 1 and 2 were used, respectively.

$$\mathbf{net}^{(1)} = \mathbf{W}^{(1)}\mathbf{X}, \tag{8}$$

$$\mathbf{O}^{(1)} = f\left(\mathbf{net}^{(1)}\right), \tag{9}$$

where $\mathbf{W}^{(1)}$ indicates the encoder weight matrix, $\mathbf{X}$ displays the input matrix, $\mathbf{O}^{(1)}$ is the encoder output matrix, and f is the activation function.

$$\mathbf{net}^{(2)} = \mathbf{W}^{(2)}\mathbf{O}^{(1)}, \tag{10}$$

$$\mathbf{O}^{(2)} = f\left(\mathbf{net}^{(2)}\right), \tag{11}$$

where $\mathbf{W}^{(2)}$ and $\mathbf{O}^{(2)}$ are the weight and output matrixes of the decoder layer, respectively.

The error back-propagation related to the encoder and decoder weights matrixes is calculated by Equation (12).

$$E := \frac{1}{k}\sum_{i=1}^{k} \log(\cosh(e_t)), \tag{12}$$

where e_t is the error vector of Asy_{AE} at time t, which is described by Equation (13), and k is the neuron size of decoder layer output.

$$\mathbf{e}_t := \mathbf{d}_t - \mathbf{o}_t^{(2)}. \tag{13}$$

The desired output vector at time t is presented by $\mathbf{d}_t$, which belongs to the matrix $\mathbf{D}$. It is the desired output matrix of Asy_{AE}, which is produced by the combination of desired labels and Asy_{AE} input.

$$\mathbf{D} = \begin{bmatrix} x_{11} & x_{12} & \cdots & x_{1n_0} & \mathcal{L} \\ x_{21} & x_{22} & \cdots & x_{2n_0} & \mathcal{L} \\ \vdots & \vdots & \vdots & \vdots & \vdots \\ x_{m1} & x_{m2} & \cdots & x_{mn_0} & \mathcal{L} \end{bmatrix}.$$

Here, x_{ij} is the Asy_{AE} input matrix element, and $\mathcal{L}$ is the desired label of the problem. A stacked asymmetric auto-encoder is a result of putting several Asy_{AES} together.

(2) **Optimizing some hyper-parameters of a stacked asymmetric auto-encoder**

Given the fact that the number of DNN hyper-parameters is significantly large, the simultaneous optimization of all of them complicates the computation and requires high-performance computing systems. Hence, we compromised between MIC and expertise for calculating the six critical DNN hyper-parameters as follows:

1. number of neurons in each hidden layer
2. learning rate value
3. initial parameters
4. number of hidden layers
5. maximum epoch of network training
6. preventing over-fitting and under-fitting

For HPO of SA_{AE}, the following principles come after. Figure 3 illustrates the flowchart of the proposed method in detail.

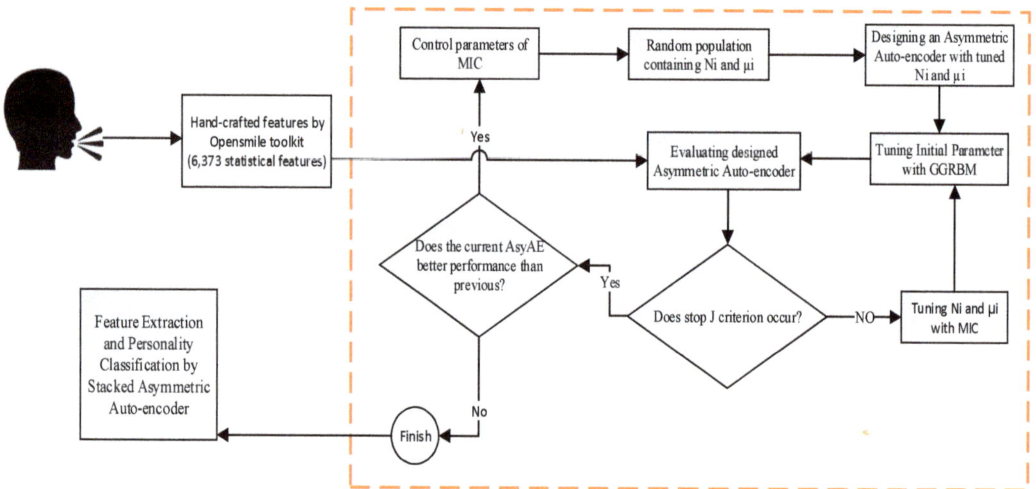

Figure 3. Flowchart of SA$_{AE}$ hyper-parameter optimization.

Determining the number of neurons in each hidden layer: In our work, N$_i$ indicates the number of neurons in the ith hidden layer that will be optimized by the MIC method. So, the first variable of MIC is N$_i$, which is an integer value, N$_i \in [1, m]$ where m value is equal to the input size of Asy$_{AE}$. It forces the Asy$_{AE}$ to be an incomplete network. It means the encoder layer has fewer neurons than the input layer.

Determining the learning rate in each hidden layer: µ$_i$ specifies the learning rate in the ith hidden layer, which will be optimized by the MIC method. Therefore, the second variable of the MIC population is a real value between zero and one, µ$_i \in (0, 1)$. It should be mentioned that we set the decimal digit of µ$_i$ equal to 5 to examine its effect on SA$_{AE}$ performance.

Initial value of trainable parameters: Although deep learning methods have good performance in various problems, they are complicated tasks. Because there are huge factors that strongly influence them, one of the critical factors is initialization.

The DNN parameters need a starting point in the feasible area to be trained. The proper initial parameters can accelerate the convergence. Contrarily, random initialization can trap the network in the local optima.

Optimization algorithms such as GA and PSO can be used in this field. However, the number of DNN parameters (weights and biases) is vast, e.g., 10^{15}, and producing the chromosomes with these dimensions causes a memory error in the processor system and is not efficient in practice. Another method, suggested by Hinton et al., applies the restricted Boltzman machine (RBM) network to tune the auto-encoder's initial parameters [58,59].

According to the ANN-base of an Asy$_{AE}$ and RBM, the Asy$_{AE}$ can be interpreted as two consecutive RBMs illustrated in Figure 4. The input layer is the visible unit, and the encoder layer is the hidden unit for the RBM$_1$. In the RBM$_2$, the encoder layer is the visible unit, and the decoder layer is the hidden unit.

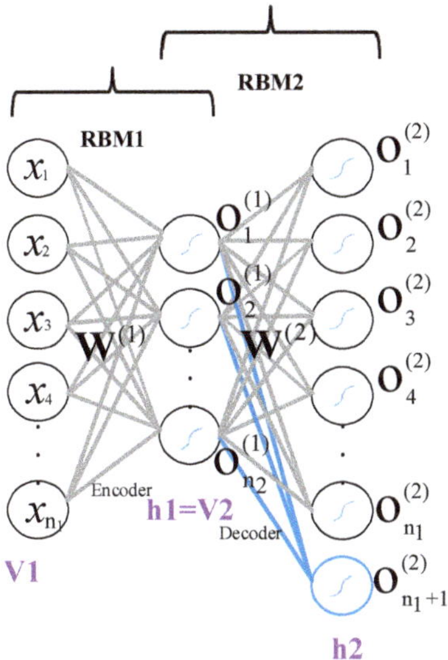

Figure 4. Converting auto-encoder to two RBMs for tuning the initial weights of the encoder and decoder layers.

The conventional RBM is based on binary visible and hidden units, called Bernoulli-Bernoulli RBM (BBRBM). If both visible and hidden units have a Gaussian distribution, the Gaussian-Gaussian RBM (GGRBM) is employed [60]. Since the Asy_{AE} input and parameters are real values, we used the GGRBM equations.

The energy function of the GGRBM is defined as Equation (14), where v presents visible units and h shows hidden units. It should be noted that the Asy_{AE} input and the encoder output are the visible units of RBM_1 and RBM_2, respectively.

$$E(v,h) = -\sum_{i=1}^{g_v}\sum_{j=1}^{g_h} W_{i,j}\frac{v_i h_j}{\sigma_i \sigma_j} - \sum_{i=1}^{g_v}\frac{(v_i - a_i)^2}{2\sigma_i^2} - \sum_{j=1}^{g_h}\frac{(h_j - b_j)^2}{2\sigma_j^2}, \tag{14}$$

where a_i and b_j are visible and hidden units biases, respectively, σ_i and σ_j are their standard deviations. $W_{i,j}$ is the weight between the visible and hidden units. A probability value is assigned to each possible visible and hidden unit by Equation (15),

$$P(v,h) = \frac{1}{Z}\exp(-E(v,h)). \tag{15}$$

Here, Z is the normalization constant calculated by Equation (16).

$$Z = \sum_v \sum_h \exp(-E(v,h)). \tag{16}$$

Equation (17) shows the loss function, which must be maximized,

$$\text{maximize}_{\{W_{i,j},a_i,b_j\}} \frac{1}{c}\sum_{L=1}^{c} \log\left(P\left(v^L, h^L\right)\right), \tag{17}$$

The updating functions are

$$\Delta W_{i,j} = \zeta\left(< v_i h_j >_{data} - < v_i h_j >_{model}\right), \tag{18}$$

$$\Delta a_i = \zeta\left(< v_i >_{data} - < v_i >_{model}\right), \tag{19}$$

$$\Delta b_j = \zeta(<h_j>_{\text{data}} - <h_j>_{\text{model}}),\tag{20}$$

where $<\bullet>_{\text{data}}$ and $<\bullet>_{\text{model}}$ are expanded values of sample data and model probabilistic distribution, and ζ is the learning rate.

We described GGRBM briefly, and this is the time to use it. For a traditional auto-encoder, first, the initial parameters of the encoder layer are randomly selected and then trained by the GGRBM method. The trained parameters are considered the encoder layer's initial parameters, and its transposition is employed for the decoder layer. However, in the Asy_{AE}, the encoder and decoder parameters are not symmetric and have to be obtained individually. So, the above principle is applied to the decoder layer to obtain the initial parameters.

The number of hidden layers: The value of this hyper-parameter is dependent on the performance of Asy_{AE}s. The classification performance of each Asy_{AE} is examined in MIC for each pair of (N_i, μ_i). For the next Asy_{AE}, the performance has to be better than that of the previous one. If the performance of $\text{Asy}_{\text{AE}(i+1)}$ is better than that of $\text{Asy}_{\text{AE}(i)}$, the MIC algorithm is continued.

The performance criterion is different from one problem to another. The Unweighted Average (UA) recall criterion frequently used in personality perception studies is calculated by Equation (21),

$$\text{UA recall} = \frac{1}{2}\left(\text{recall}_{\text{Low}} + \text{recall}_{\text{High}}\right),\tag{21}$$

The $\text{recall}_{\text{Low}}$ means the recall of detecting the low degree of studied personality, and the $\text{recall}_{\text{High}}$ indicates the recall of detecting its high degree.

The maximum epoch of network training: Generally, the DNN training process proceeds to reach maximum epoch (updating time) [40]. As discussed in [24], proper data separation does not occur in the maximum epoch. Thus, a J variation is employed as a stopping criterion to finish the training process in the epoch in which the maximum separation is achieved.

J is calculated as follows,

$$J = \frac{\det(\mathbf{S_B})}{\det(\mathbf{S_W})},\tag{22}$$

where S_W is a within-class scattering matrix, and S_B is a between-class scattering matrix [61]. **det** represents the determinant of a matrix.

$$\mathbf{S_w} = \sum_{i=1}^{c}\sum_{x\in c_i}(\mathbf{X}-\mathbf{\mu_i})(\mathbf{X}-\mathbf{\mu_i})^{\text{T}},\tag{23}$$

$$\mathbf{S_B} = \sum_{i=1}^{c} n_i(\mathbf{\mu_i}-\mathbf{\mu})(\mathbf{\mu_i}-\mathbf{\mu})^{\text{T}}.\tag{24}$$

Here, n_i is the instance number of ith class, $\mathbf{X}$ is the encoder output matrix, and c is the number of classes, μ is the matrix for average all instances, and μ_i is the class average matrix of ith class.

Preventing over-fitting and under-fitting problems: The over-fitting problem happens when a model trains properly on the training dataset but performs poorly on the testing dataset. The under-fitting problem occurs when a model performs poorly on both the training and testing samples.

The number of layers and the neurons in each layer can excessively lead a model into over-fitting or under-fitting. This can be easily changed by changing the structure. More neurons and layers complicate the model, but fewer cannot pursue the data pattern. Therefore, this is one of the problems that has to be dealt with in designing an optimum structure. So, a new loss function is defined to guide the model toward good fitting.

$$\text{Loss} = \frac{\text{UA}_{\text{train}}}{a} * \frac{\text{UA}_{\text{val}}}{b},\tag{25}$$

where a is the training threshold, and b is the validation threshold. We already discussed the UA recall criterion used chiefly in personality perception. We applied the loss function defined in Equation (25) instead of Equation (21). The aim is the maximization of Equation (25). We set a = 0.8 and b = 0.6 because a UA_{train} of more than 80% and UA_{val} of more than 60% are acceptable. The loss value can be in the range of [2.08, 0]. So, the set of (N_i, μ_i) is acceptable to be maximized in Equation (25).

Final algorithm: The pseudo-code of optimizing SA_{AE} hyper-parameters is described in Algorithm 2.

Algorithm 2: Optimizing SAAE hyperparameters

Set the initial parameters Old_max = 0, G_max = 2.08 (upper band of Loss), OldEv_Asy = 0 (the first Asy_{AE} performance) and the other randomly.
Set the input matrix of Asy_{AE}.
Set i = 1 (i indicates the number of hidden layer)
Set NewEv_Asy = 1 (the (i + 1)th Asy_{AE} performance)
While NewEv_Asy> OldEv_Asy
    OldEv_Asy = NewEv_Asy
    While (G_max-Old_max) > 0.1
        Optimize (N_i, μ_i) with MIC.
        Initialize the Asy_{AE} parameter randomly.
        Tune Asy_{AE} initial parameters with GGRBM.
        Train Asy_{AE} while J increases.
        Evaluate Equation (25).
        If the value of Equation (25) $\geq$ Old_max
          Old_max = the current value of Equation (25),
          NewEv_Asy = the value of Equation (25).
        End if
        Set the encoder layer output of ith Asy_{AE} as the input of (i + 1)th Asy_{AE}.
        i = i + 1.
    End while
End while

---

6. Simulations and Results

In this section, firstly, the results of the MIC method on three benchmarks and comparison with other published methods will be discussed. Then, the MIC will be used to design the structure of five individual DNNs for classifying five personality traits. A final comparison can be found at the end of this section.

6.1. The Results of the MIC on Three Optimization Benchmarks

Three well-known, multimodal, continuous, and non-separable benchmark functions that have a global minimum value of zero, called Rastrigin [52], Ackley [62], and Griewang [62], are used to validate the MIC method.

The multimodal property means having many local optima or peaks in the function, which can test the ability of an algorithm to avoid being stuck in a local minimum. Non-separable refers to the independence of obtained solution variables. If all variables are independent, they can be optimized independently, and the function will be optimized [62]. Therefore, these three functions are complex problems in evaluating the performance of any new optimization algorithm.

The formula, feasible range of variables, and the global optima points of three functions are summarized in Table 2.

Table 2. Description of Three Benchmark Functions.

Name	Formula	Range	Optimal $f(x)$
Rastrigin	$f(x) = 10n + \sum_{i=1}^{n} x_i^2 - 10\cos(2\pi x_i)$	$-5.12 < x_i < 5.12$	0
Ackley	$f(x) = -20\exp\left(-0.02\sqrt{\frac{1}{n}\sum_{i=1}^{n} x_i^2}\right) - \exp\left(\frac{1}{n}\sum_{i=1}^{n}\cos(2\pi x_i)\right) + 20 + \exp(1)$	$-32 \le x_i \le 32$	0
Griewang	$f(x) = \frac{1}{4000}\sum_{i=1}^{n} x_i^2 - \prod_{i=1}^{n}\cos\left(\frac{x_i}{\sqrt{i}}\right) + 1$	$-600 < x_i < 600$	0

Here, n indicates the dimension of the function, which is $n \ge 2$ for all mentioned functions.

Figure 5 shows the shape of the functions described in Table 2. As can be seen, all three functions have many local optima and are suitable to show the ability of optimization methods to escape from being stuck in local optima.

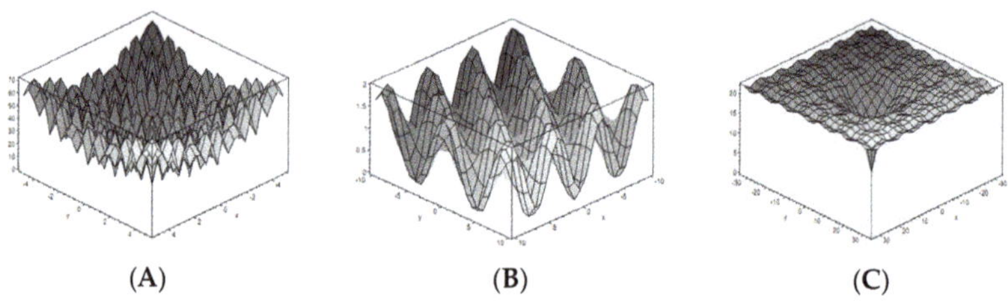

(A)   (B)   (C)

Figure 5. Benchmark functions (**A**) Rastrigin, (**B**) Ackley, and (**C**) Griewang.

In order to show the performance of MIC against the conventional optimization methods, the comparison results of the mentioned four islands and MIC are reported in Table 3.

Given the fact that the problem complexity increases with increasing dimensionality, increasing the number of the variables (dimension) grows the search space, which makes exploring the best solution difficult [62]. To investigate the effect of dimension on searching quality in MIC, we compared our results with 30D and 10D in Table 3.

For a fair comparison, all parameters and initial populations for the basic algorithms and MIC were set to the same values.

The following six criteria were utilized for a more reliable analysis. It should be mentioned that these criteria are common in optimization problems.

- The average of iterations where the stop criterion is reached for examining convergence speed (AvI).
- The average of obtained best optima point (AvP).
- The smallest iteration at which the stop criterion occurs (SI).
- The best-obtained optima point (BOP).
- Calculating the standard deviation (SD) for proving the efficiency and robustness of the algorithm.
- The number of successful runs divided by the total number of runs called success rate (SR).

It was concluded by AvI numerical results that MIC can reach more accurate solutions with a faster convergence speed than traditional algorithms in n = 10. Although for the n = 30, the MM performance diminishes, LM and EM preserve their performance with increasing complexity. It is demonstrated that LM and EM improve solutions steadily for a long time without getting stuck in local minima. It is clear that MIC is more powerful than the four basic algorithms alone when it comes to solving global optimization problems.

Table 3. The Results of MIC Compared with Traditional GA, DE, PSO, and ES in Three Benchmark Functions (10D and 30D).

Benchmark Functions	Optimization Algorithm	AvI	AvP	SI	BOP	SD	SR (%)	AvI	AvP	SI	BOP	SD	SR (%)
					n=10						n=30		
Rastrigin	MIC by LM	83.4	6.7×10^{-5}	61	7.2×10^{-5}	4.7×10^{-5}	100	307.1	7.8×10^{-4}	254	4.8×10^{-4}	2.4×10^{-4}	100
	MIC by EM	120.5	7.1×10^{-5}	101	8.8×10^{-6}	5.8×10^{-5}	100	321.5	7.4×10^{-4}	187	1.8×10^{-4}	5.6×10^{-4}	100
	MIC by MM	572.5	9.9×10^{-3}	131	4.9×10^{-3}	4.3×10^{-3}	100	2000	0.46	2000	6.1×10^{-4}	1.32	60
	GA	617.1	1.2×10^{-5}	324	6.8×10^{-4}	4.6×10^{-3}	40	1178.4	1.34	926	6.8×10^{-4}	3.20	50
	DE	1000	0.22	1000	0.14	0.47	0	2000	4.72	2000	0.10	2.55	0
	ES	1000	3.48	1000	1.94	2.39	0	2000	37.4	2000	24.7	14.4	0
	PSO	985.7	0.82	857	6.8×10^{-4}	0.71	20	2000	10.9	2000	3.13	5.60	0
Ackley	MIC by LM	467.2	4.4×10^{-15}	355	4.4×10^{-15}	0	100	1039.6	4.4×10^{-15}	956	4.4×10^{-15}	0	100
	MIC by EM	788.4	4.4×10^{-15}	462	4.4×10^{-15}	0	100	1154.8	3.1×10^{-14}	937	4.4×10^{-15}	1.9×10^{-15}	100
	MIC by MM	725.2	1.4×10^{-9}	324	3.5×10^{-10}	9.2×10^{-10}	100	2000	2.48	2000	2.24	0.20	0
	GA	957.5	7.3×10^{-2}	565	7.2×10^{-3}	1.58	70	1895.1	2.86	951	0.01	1.53	10
	DE	1000	1.69	1000	1.24	4.47	0	2000	5.35	2000	4.34	0.67	0
	ES	1000	4.96	1000	3.20	3.09	0	2000	5.43	2000	5.23	1.9×10^{-1}	0
	PSO	557.5	8.6×10^{-4}	344	6.2×10^{-4}	4.1×10^{-4}	100	839.5	4.5×10^{-3}	162	8.8×10^{-4}	7.9×10^{-3}	100
Griewang	MIC by LM	154.7	6.2×10^{-14}	38	1.2×10^{-14}	2.8×10^{-14}	100	106	8.4×10^{-14}	92	8.7×10^{-14}	4.4×10^{-14}	100
	MIC by EM	171.2	8.4×10^{-14}	43	6.4×10^{-14}	1.5×10^{-14}	100	489	1.1×10^{-13}	94	9.1×10^{-14}	2.5×10^{-14}	100
	MIC by MM	775.4	3.3×10^{-13}	146	9.6×10^{-14}	3.2×10^{-13}	100	2000	0.27	2000	9.1×10^{-13}	0.16	20
	GA	909.8	0.09	84	0.8×10^{-3}	0.18	10	2000	0.21	2000	0.09	1.1×10^{-1}	0
	DE	337.4	9.1×10^{-3}	44	7.3×10^{-3}	1.1×10^{-3}	100	993	0.01	588	7.9×10^{-3}	8.8×10^{-1}	70
	ES	1000	0.36	1000	1.2×10^{-1}	0.20	0	2000	0.81	2000	0.76	0.19	0
	PSO	555.2	0.02	258	6.6×10^{-2}	2.8×10^{-2}	30	2000	0.77	2000	0.37	3.1×10^{-2}	0

Table 3 shows the simulation outcomes of MIC and four basic optimization algorithms.

According to the AvP values in n = 10 and n = 30, traditional algorithms are often unsuccessful in finding favorable solutions in comparison to MIC, especially EM. Additionally, it can be concluded from AvP that the MIC speeds up the convergence to the global optima. The AvP values in n = 30 in comparison to n = 10 decreased about 0.1 in Rastrigin and remained constant for the other two functions in LM and EM. The change in the AvP values in MM is meaningful, which indicates getting stuck in the local optimum with the increase in the complexity of the problem, like the traditional methods.

Our SI outcomes show that the MIC method, especially EM and LM, reaches the stop criterion in a few iterations. It means the MIC method speeds up convergence. Moreover, the SI criterion shows that although the MM method performs better than the basic optimization methods in simpler functions (n = 10), its performance drops in complex functions (n = 30). LM and EM not only show their effectiveness in simple functions, but also perform well in complex problems compared to other methods.

The evaluation results of criterion BOP show that the LM and EM methods achieve the global optimal value more accurately than the basic methods in n = 10 and n = 30. However, MM implementation results decrease with increasing complexity.

It can be seen that the SD values of MIC, except for MM, are very small in comparison to those of the four basic algorithms in n = 10 and n = 30, which means the repeatability and robustness of the new algorithm are due to pruning search space.

The SR results prove that the MIC is very promising in bringing higher reliability than traditional algorithms because the number of times that LM, EM, and MM reached the desired value of the function was 100% in n = 10. As can be seen, as the complexity of the function increases (n = 30), the LM and EM methods are still successful in reaching the desired value.

From Table 3, it is concluded that despite increasing dimensions, the implementation outcomes of all algorithms decrease, except EM and LM.

Our study indicates that the quality of the solutions found using our proposed method for widespread global optima functions is higher than that of the solutions provided by traditional algorithms. This is due to a more appropriate tradeoff between exploring new individuals and exploiting highly fit individuals found at the parallelism level. By means of three widespread test functions, it is demonstrated that the new method has great potential for substantial improvement in search performance.

Due to the wide usage of these benchmarks, a comparison with other published works is presented in Table 4. It can be observed that LM and EM achieved the best solution in Ackley and Griewang functions (30D).

Table 4. Comparison with Other Published Methods in 30D. N/A means not available.

Methods	Benchmarks								
	Rastrigin			Ackley			Griewang		
	AvP	SD	SR%	AvP	SD	SR%	AvP	SD	SR%
Xin Zhao et al., 2022 [55]	2.1×10^{-13}	4.1×10^{-14}	100	8.2×10^{-15}	1.3×10^{-15}	100	3.78×10^{-13}	1.7×10^{-13}	100
Chentoufi et al., 2021 [49]	0.99	1.31	100	1.0×10^{-15}	6.4×10^{-16}	43	8.3×10^{-4}	5.4×10^{-4}	67
MIC_LM	7.8×10^{-4}	2.4×10^{-4}	100	4.4×10^{-15}	0	100	8.4×10^{-14}	4.4×10^{-14}	100
MIC_EM	7.4×10^{-4}	5.6×10^{-4}	100	3.1×10^{-14}	1.9×10^{-15}	100	1.1×10^{-13}	2.5×10^{-14}	100
MIC_MM	0.46	1.32	60	2.48	0.20	0	0.27	0.16	20

6.2. The Results of Personality Perception with The MIC Method

After the successful outcomes with the MIC method to find the global optima of three complex benchmark functions, we applied our novel method to find the near-optimal values of hyper-parameters for classifying five personality traits. We used "near-optimal" instead of "optimal" structure because tuning of MIC hyper-parameters such as mutation and crossover rating is chosen randomly.

Taking into account that different personality traits have different effects on speech characteristics [24,42], using the same DNN structure for all traits to extract features is not recommended. Assuming the five personality traits were independent, five separate neural networks were designed and trained to classify the five personality traits.

Hence, the network's depth was determined by classifying the output of each Asy_{AE} encoder layer by the SVM with radial basis function kernel. The Asy_{AE} with higher classification results is considered as the output layer of the S_{AE}.

Table 5 shows the comparison results of our proposed method with other works in the SPC dataset in terms of UA recall and accuracy. In our previous work, the structure of SA_{AE} was chosen by trial-and-error, which was time-consuming, and two traits (extraversion and openness) achieved lower accuracy than reported by other research [24]. N/A means not available.

Table 5. Comparison Results of Our Proposed Method with Other Works in the SPC Dataset in Terms of UA Recall % (Accuracy %).

Methods	Traits				
	Neu.	**Ext.**	**Ope.**	**Agr.**	**Con.**
Mohammadi et al., 2010 [43]	N/A (63)	N/A (76.3)	N/A (57.9)	N/A (63)	N/A (72)
Mohammadi et al., 2012 [63]	N/A (65.9)	N/A (73.5)	N/A (60.1)	N/A (63.1)	N/A (71.3)
Chastagnol et al., 2012 [64]	58 (N/A)	75.5 (N/A)	73.4 (N/A)	65 (N/A)	62.2 (N/A)
Mohammadi et al., 2015 [65]	N/A (66.1)	N/A (71.4)	N/A (58.6)	N/A (58.8)	N/A (72.5)
Solera-Urena et al., 2017 [66]	65.1 (64.7)	75 (75.1)	59.1 (58.2)	60.3 (60.2)	75.7 (75.6)
Carbonneau et al., 2017 [67]	70.8 (N/A)	75.2 (N/A)	56.3 (N/A)	64.9 (N/A)	63.8 (N/A)
Zhen-Tao Liu et al., 2020 [68]	N/A (69.2)	N/A (76.3)	N/A (74.7)	N/A (65.3)	N/A (73.3)
Our privuse work 2021 [24]	77.1 (76.9)	76.6 (72.9)	81.2 (70.4)	80.7 (68.7)	78.5 (69.5)
Proposed method	89.8 (80.5)	82.2 (83.4)	87.1 (84.7)	85.8 (76.2)	81.8 (72.6)

In the present study, not only were the accuracy of extraversion and openness improved, but UA recalls were also increased more than before. This evidences that the performance and robustness of trained models are highly dependent on their hyper-parameter settings.

7. Conclusions

Since HPT is the most challenging aspect of ANN studies, it is mostly obtained by trial-and-error, affecting its performance. This article proposed a new approach based on cultural evolution and parallel computing to achieve a near-optimal structure of SA_{AE} in a reasonable time for automatic personality perception. We used the concept of parallelism and information on different regions of the search space to improve the search spaces in MIC and exchanged them between islands to provide greater population diversity. The proposed approach was implemented on three complex benchmarks, and six criteria evaluated our method's performance in comparison with four basic optimization methods. The results showed that our approach outperforms other traditional optimization and newly published algorithms in four aspects: (1) convergence speed, (2) precision, (3) escaping from entrapment in local optima, and (4) repeatability. As an indication of our method's performance, we increased the problem complexity by increasing the number of variables up to 30. The outcomes demonstrated the reliability of the MIC method, especially LM and EM. Subsequently, five hyper-parameters of SA_{AE} were optimized. Since the tuning of hyper-parameters affects over-fitting and under-fitting, we introduced a new cost function to control them during the optimization process.

In comparison with the results of our previous published work, the outcomes of applying MIC to SA_{AE} indicated 3.3% (3.1%) for consciousness, 5.1% (7.5%) for agreeableness, 5.9% (14.3%) for openness, 5.6% (10.1%) for extraversion, and 12.7% (3.6%) for neuroticism.

Author Contributions: Writing—review & editing, E.J.Z., M.T., A.K., C.H., M.V., N.N. and T.W. All authors have read and agreed to the published version of the manuscript.

Funding: This research received no external funding.

Institutional Review Board Statement: Not applicable.

Informed Consent Statement: Not applicable.

Data Availability Statement: Not applicable.

Conflicts of Interest: The authors declare no conflict of interest.

References

1. Bischl, B.; Binder, M.; Lang, M.; Pielok, T.; Richter, J.; Coors, S.; Thomas, J.; Ullmann, T.; Becker, M.; Boulesteix, A.-L. Hyperparameter optimization: Foundations, algorithms, best practices and open challenges. *arXiv* **2021**, arXiv:107.05847.
2. Szepannek, G.; Lübke, K. Explaining Artificial Intelligence with Care. In *KI-Künstliche Intell.*; 2022; Volume 16, pp. 1–10. [CrossRef]
3. Khodadadian, A.; Parvizi, M.; Teshnehlab, M.; Heitzinger, C. Rational Design of Field-Effect Sensors Using Partial Differential Equations, Bayesian Inversion, and Artificial Neural Networks. *Sensors* **2022**, *22*, 4785. [CrossRef]
4. Guo, C.; Li, L.; Hu, Y.; Yan, J. A Deep Learning Based Fault Diagnosis Method With hyperparameter Optimization by Using Parallel Computing. *IEEE Access* **2020**, *8*, 131248–131256. [CrossRef]
5. Feurer, M.; Hutter, F. Hyperparameter optimization. In *Automated Machine Learning*; Springer: Cham, Switzerland, 2019; pp. 3–33.
6. Yang, L.; Shami, A. On hyperparameter optimization of machine learning algorithms: Theory and practice. *Neurocomputing* **2020**, *415*, 295–316. [CrossRef]
7. Wu, D.; Wu, C. Research on the Time-Dependent Split Delivery Green Vehicle Routing Problem for Fresh Agricultural Products with Multiple Time Windows. *Agriculture* **2022**, *12*, 793. [CrossRef]
8. Peng, Y.; Gong, D.; Deng, C.; Li, H.; Cai, H.; Zhang, H. An automatic hyperparameter optimization DNN model for precipitation prediction. *Appl. Intell.* **2021**, *52*, 2703–2719. [CrossRef]
9. Yi, H.; Bui, K.-H.N. An automated hyperparameter search-based deep learning model for highway traffic prediction. *IEEE Trans. Intell. Transp. Syst.* **2020**, *22*, 5486–5495. [CrossRef]
10. Kinnewig, S.; Kolditz, L.; Roth, J.; Wick, T. *Numerical Methods for Algorithmic Systems and Neural Networks*; Institut für Angewandte Mathematik, Leibniz Universität Hannover: Hannover, Germany, 2022. [CrossRef]
11. Han, J.-H.; Choi, D.-J.; Park, S.-U.; Hong, S.-K. Hyperparameter optimization using a genetic algorithm considering verification time in a convolutional neural network. *J. Electr. Eng. Technol.* **2020**, *15*, 721–726. [CrossRef]
12. Yao, R.; Guo, C.; Deng, W.; Zhao, H. A novel mathematical morphology spectrum entropy based on scale-adaptive techniques. *ISA Trans.* **2022**, *126*, 691–702. [CrossRef] [PubMed]
13. Raji, I.D.; Bello-Salau, H.; Umoh, I.J.; Onumanyi, A.J.; Adegboye, M.A.; Salawudeen, A.T. Simple deterministic selection-based genetic algorithm for hyperparameter tuning of machine learning models. *Appl. Sci.* **2022**, *12*, 1186. [CrossRef]
14. Harichandana, B.; Kumar, S. LEAPMood: Light and Efficient Architecture to Predict Mood with Genetic Algorithm driven Hyperparameter Tuning. In Proceedings of the 2022 IEEE 16th International Conference on Semantic Computing (ICSC), Laguna Hills, CA, USA, 26–28 January 2022; pp. 1–8.
15. Guido, R.; Groccia, M.C.; Conforti, D. Hyper-Parameter Optimization in Support Vector Machine on Unbalanced Datasets Using Genetic Algorithms. In *Optimization in Artificial Intelligence and Data Sciences*; Springer: Berlin/Heidelberg, Germany, 2022; pp. 37–47.
16. Thavasimani, K.; Srinath, N.K. Hyperparameter optimization using custom genetic algorithm for classification of benign and malicious traffic on internet of things-23 dataset. *Int. J. Electr. Comput. Eng.* **2022**, *12*, 4031–4041. [CrossRef]
17. Awad, M. Optimizing the Topology and Learning Parameters of Hierarchical RBF Neural Networks Using Genetic Algorithms. *Int. J. Appl. Eng. Res.* **2018**, *13*, 8278–8285.
18. Faris, H.; Mirjalili, S.; Aljarah, I. Automatic selection of hidden neurons and weights in neural networks using grey wolf optimizer based on a hybrid encoding scheme. *Int. J. Mach. Learn. Cybern.* **2019**, *10*, 2901–2920. [CrossRef]
19. Li, A.; Spyra, O.; Perel, S.; Dalibard, V.; Jaderberg, M.; Gu, C.; Budden, D.; Harley, T.; Gupta, P. A generalized framework for population based training. In Proceedings of the 25th ACM SIGKDD International Conference on Knowledge Discovery & Data Mining, Anchorage, AK, USA, 4–8 August 2019; pp. 1791–1799.
20. Luo, G. A review of automatic selection methods for machine learning algorithms and hyper-parameter values. *Netw. Model. Anal. Health Inform. Bioinform.* **2016**, *5*, 18. [CrossRef]
21. An, Z.; Wang, X.; Li, B.; Xiang, Z.; Zhang, B. Robust visual tracking for UAVs with dynamic feature weight selection. *Appl. Intell.* **2022**. [CrossRef]
22. Bhandare, A.; Kaur, D. Designing convolutional neural network architecture using genetic algorithms. *Int. J. Adv. Netw. Monit. Control* **2021**, *6*, 26–35. [CrossRef]
23. Tan, H.H.; Lim, K.H. Vanishing gradient mitigation with deep learning neural network optimization. In Proceedings of the 2019 7th International Conference on Smart Computing & Communications (ICSCC), Sarawak, Malaysia, 28–30 June 2019; pp. 1–4.
24. Zaferani, E.J.; Teshnehlab, M.; Vali, M. Automatic Personality Traits Perception Using Asymmetric Auto-Encoder. *IEEE Access* **2021**, *9*, 68595–68608. [CrossRef]
25. Cho, H.; Kim, Y.; Lee, E.; Choi, D.; Lee, Y.; Rhee, W. Basic enhancement strategies when using bayesian optimization for hyperparameter tuning of deep neural networks. *IEEE Access* **2020**, *8*, 52588–52608. [CrossRef]

26. Sun, Y.; Xue, B.; Zhang, M.; Yen, G.G.; Lv, J. Automatically designing CNN architectures using the genetic algorithm for image classification. *IEEE Trans. Cybern.* **2020**, *50*, 3840–3854. [CrossRef]
27. Cabada, R.Z.; Rangel, H.R.; Estrada, M.L.B.; Lopez, H.M.C. Hyperparameter optimization in CNN for learning-centered emotion recognition for intelligent tutoring systems. *Soft Comput.* **2020**, *24*, 7593–7602. [CrossRef]
28. Deng, W.; Liu, H.; Xu, J.; Zhao, H.; Song, Y. An improved quantum-inspired differential evolution algorithm for deep belief network. *IEEE Trans. Instrum. Meas.* **2020**, *69*, 7319–7327. [CrossRef]
29. Guo, Y.; Li, J.-Y.; Zhan, Z.-H. Efficient hyperparameter optimization for convolution neural networks in deep learning: A distributed particle swarm optimization approach. *Cybern. Syst.* **2020**, *52*, 36–57. [CrossRef]
30. Ozcan, T.; Basturk, A. Static facial expression recognition using convolutional neural networks based on transfer learning and hyperparameter optimization. *Multimed. Tools Appl.* **2020**, *79*, 26587–26604. [CrossRef]
31. Gülcü, A.; Kuş, Z. Hyper-parameter selection in convolutional neural networks using microcanonical optimization algorithm. *IEEE Access* **2020**, *8*, 52528–52540. [CrossRef]
32. Kong, D.; Wang, S.; Ping, P. State-of-health estimation and remaining useful life for lithium-ion battery based on deep learning with Bayesian hyperparameter optimization. *Int. J. Energy Res.* **2022**, *46*, 6081–6098. [CrossRef]
33. Chowdhury, A.A.; Hossen, M.A.; Azam, M.A.; Rahman, M.H. Deepqgho: Quantized greedy hyperparameter optimization in deep neural networks for on-the-fly learning. *IEEE Access* **2022**, *10*, 6407–6416. [CrossRef]
34. Chen, H.; Miao, F.; Chen, Y.; Xiong, Y.; Chen, T. A hyperspectral image classification method using multifeature vectors and optimized KELM. *IEEE J. Sel. Top. Appl. Earth Obs. Remote Sens.* **2021**, *14*, 2781–2795. [CrossRef]
35. Phan, L.V.; Rauthmann, J.F. Personality computing: New frontiers in personality assessment. *Soc. Personal. Psychol. Compass* **2021**, *15*, e12624. [CrossRef]
36. Koutsombogera, M.; Sarthy, P.; Vogel, C. Acoustic Features in Dialogue Dominate Accurate Personality Trait Classification. In Proceedings of the 2020 IEEE International Conference on Human-Machine Systems (ICHMS), Rome, Italy, 7–9 September 2020; pp. 1–3.
37. Aslan, S.; Güdükbay, U.; Dibeklioğlu, H. Multimodal assessment of apparent personality using feature attention and error consistency constraint. *Image Vis. Comput.* **2021**, *110*, 104163. [CrossRef]
38. Xu, J.; Tian, W.; Lv, G.; Liu, S.; Fan, Y. Prediction of the Big Five Personality Traits Using Static Facial Images of College Students With Different Academic Backgrounds. *IEEE Access* **2021**, *9*, 76822–76832. [CrossRef]
39. Kampman, O.; Siddique, F.B.; Yang, Y.; Fung, P. Adapting a virtual agent to user personality. In *Advanced Social Interaction with Agents*; Springer: Berlin/Heidelberg, Germany, 2019; pp. 111–118.
40. Suen, H.-Y.; Hung, K.-E.; Lin, C.-L. Intelligent video interview agent used to predict communication skill and perceived personality traits. *Hum.-Cent. Comput. Inf. Sci.* **2020**, *10*, 1–12. [CrossRef]
41. Liam Kinney, A.W.; Zhao, J. Detecting Personality Traits in Conversational Speech. Stanford University: Stanford, CA, USA, 2017; Available online: https://web.stanford.edu/class/cs224s/project/reports_2017/Liam_Kinney.pdf (accessed on 15 June 2022).
42. Jalaeian Zaferani, E.; Teshnehlab, M.; Vali, M. Automatic personality recognition and perception using deep learning and supervised evaluation method. *J. Appl. Res. Ind. Eng.* **2022**, *9*, 197–211.
43. Mohammadi, G.; Vinciarelli, A.; Mortillaro, M. The voice of personality: Mapping nonverbal vocal behavior into trait attributions. In Proceedings of the 2nd international workshop on Social signal processing, Firenze, Italy, 29 October 2010; pp. 17–20.
44. Rosenberg, A. Speech, Prosody, and Machines: Nine Challenges for Prosody Research. In Proceedings of the 9th International Conference on Speech Prosody 2018, Poznań, Poland, 13–16 June 2018; pp. 784–793.
45. Junior, J.C.S.J.; Güçlütürk, Y.; Pérez, M.; Güçlü, U.; Andujar, C.; Baró, X.; Escalante, H.; Guyon, I.; van Gerven, M.; van Lier, R. First impressions: A survey on computer vision-based apparent personality trait analysis. *arXiv* **2019**, arXiv:1804.08046v1.
46. Schuller, B.; Weninger, F.; Zhang, Y.; Ringeval, F.; Batliner, A.; Steidl, S.; Eyben, F.; Marchi, E.; Vinciarelli, A.; Scherer, K. Affective and behavioural computing: Lessons learnt from the first computational paralinguistics challenge. *Comput. Speech Lang.* **2019**, *53*, 156–180. [CrossRef]
47. Hutter, F.; Kotthoff, L.; Vanschoren, J. *Automated Machine Learning: Methods, Systems, Challenges*; Springer: Berlin/Heidelberg, Germany, 2019.
48. Harada, T.; Alba, E. Parallel genetic algorithms: A useful survey. *ACM Comput. Surv.* **2020**, *53*, 1–39. [CrossRef]
49. Chentoufi, M.A.; Ellaia, R. A novel multi-population passing vehicle search algorithm based co-evolutionary cultural algorithm. *Comput. Sci.* **2021**, *16*, 357–377.
50. Liu, Z.-H.; Tian, S.-L.; Zeng, Q.-L.; Gao, K.-D.; Cui, X.-L.; Wang, C.-L. Optimization design of curved outrigger structure based on buckling analysis and multi-island genetic algorithm. *Sci. Prog.* **2021**, *104*, 368504211023277. [CrossRef]
51. Shah, P.; Kobti, Z. Multimodal fake news detection using a Cultural Algorithm with situational and normative knowledge. In Proceedings of the 2020 IEEE Congress on Evolutionary Computation (CEC), Glasgow, UK, 19–24 July 2020; pp. 1–7.
52. Al-Betar, M.A.; Awadallah, M.A.; Doush, I.A.; Hammouri, A.I.; Mafarja, M.; Alyasseri, Z.A.A. Island flower pollination algorithm for global optimization. *J. Supercomput.* **2019**, *75*, 5280–5323. [CrossRef]
53. Sun, Y.; Zhang, L.; Gu, X. A hybrid co-evolutionary cultural algorithm based on particle swarm optimization for solving global optimization problems. *Neurocomputing* **2012**, *98*, 76–89. [CrossRef]
54. da Silva, D.J.A.; Teixeira, O.N.; de Oliveira, R.C.L. *Performance Study of Cultural Algorithms Based on Genetic Algorithm with Single and Multi Population for the MKP. Bio-Inspired Computational Algorithms and Their Applitions*; IntechOpen: London, UK, 2012; pp. 385–404.

55. Zhao, X.; Tang, Z.; Cao, F.; Zhu, C.; Periaux, J. An Efficient Hybrid Evolutionary Optimization Method Coupling Cultural Algorithm with Genetic Algorithms and Its Application to Aerodynamic Shape Design. *Appl. Sci.* **2022**, *12*, 3482. [CrossRef]
56. Muhamediyeva, D. Fuzzy cultural algorithm for solving optimization problems. *J. Phys. Conf. Ser.* **2020**, *1441*, 012152. [CrossRef]
57. Xu, W.; Wang, R.; Zhang, L.; Gu, X. A multi-population cultural algorithm with adaptive diversity preservation and its application in ammonia synthesis process. *Neural Comput. Appl.* **2012**, *21*, 1129–1140. [CrossRef]
58. Hinton, G.E.; Salakhutdinov, R.R. Reducing the dimensionality of data with neural networks. *Science* **2006**, *313*, 504–507. [CrossRef] [PubMed]
59. Cho, K.H.; Raiko, T.; Ilin, A. Gaussian-bernoulli deep boltzmann machine. In Proceedings of the 2013 International Joint Conference on Neural Networks (IJCNN), Dallas, TX, USA, 4–9 August 2013; pp. 1–7.
60. Ogawa, S.; Mori, H. A gaussian-gaussian-restricted-boltzmann-machine-based deep neural network technique for photovoltaic system generation forecasting. *IFAC-Pap.* **2019**, *52*, 87–92. [CrossRef]
61. Tharwat, A.; Gaber, T.; Ibrahim, A.; Hassanien, A.E. Linear discriminant analysis: A detailed tutorial. *AI Commun.* **2017**, *30*, 169–190. [CrossRef]
62. Jamil, M.; Yang, X.-S. A literature survey of benchmark functions for global optimisation problems. *Int. J. Math. Model. Numer. Optim.* **2013**, *4*, 150–194. [CrossRef]
63. Mohammadi, G.; Vinciarelli, A. Automatic personality perception: Prediction of trait attribution Based Prosodic Features. *IEEE Trans. Affect. Comput.* **2012**, *3*, 273–284. [CrossRef]
64. Chastagnol, C.; Devillers, L. Personality traits detection using a parallelized modified SFFS algorithm. *Computing* **2012**, *15*, 16.
65. Mohammadi, G.; Vinciarelli, A. Automatic personality perception: Prediction of trait attribution based on prosodic features extended abstract. In Proceedings of the 2015 International Conference on Affective Computing and Intelligent Interaction (ACII), Xi'an, China, 21–24 September 2015; pp. 484–490.
66. Solera-Ureña, R.; Moniz, H.; Batista, F.; Cabarrão, R.; Pompili, A.; Astudillo, R.; Campos, J.; Paiva, A.; Trancoso, I. A semi-supervised learning approach for acoustic-prosodic personality perception in under-resourced domains. In Proceedings of the 18th Annual Conference of the International Speech Communication Association, INTERSPEECH 2017, Stockholm, Sweden, 20–24 August 2017; pp. 929–933.
67. Carbonneau, M.-A.; Granger, E.; Attabi, Y.; Gagnon, G. Feature learning from spectrograms for assessment of personality traits. *IEEE Trans. Affect. Comput.* **2017**, *11*, 25–31. [CrossRef]
68. Liu, Z.-T.; Rehman, A.; Wu, M.; Cao, W.; Hao, M. Speech personality recognition based on annotation classification using log-likelihood distance and extraction of essential audio features. *IEEE Trans. Multimed.* **2020**, *23*, 3414–3426. [CrossRef]

Article

Multiple Attention Mechanism Graph Convolution HAR Model Based on Coordination Theory

Kai Hu [1,2,*], Yiwu Ding [1], Junlan Jin [1], Min Xia [1,2] and Huaming Huang [1]

[1] School of Automation, Nanjing University of Information Science and Technology, Nanjing 210044, China; 20191223014@nuist.edu.cn (Y.D.); 20201249090@nuist.edu.cn (J.J.); xiamin@nuist.edu.cn (M.X.); 850028@nuist.edu.cn (H.H.)
[2] Jiangsu Collaborative Innovation Center of Atmospheric Environment and Equipment Technology (CICAEET), Nanjing University of Information Science and Technology, Nanjing 210044, China
* Correspondence: 001600@nuist.edu.cn

Abstract: Human action recognition (HAR) is the foundation of human behavior comprehension. It is of great significance and can be used in many real-world applications. From the point of view of human kinematics, the coordination of limbs is an important intrinsic factor of motion and contains a great deal of information. In addition, for different movements, the HAR algorithm provides important, multifaceted attention to each joint. Based on the above analysis, this paper proposes a HAR algorithm, which adopts two attention modules that work together to extract the coordination characteristics in the process of motion, and strengthens the attention of the model to the more important joints in the process of moving. Experimental data shows these two modules can improve the recognition accuracy of the model on the public HAR dataset (NTU-RGB + D, Kinetics-Skeleton).

Keywords: human action recognition; graph neural network; attention module

Citation: Hu, K.; Ding, Y.; Jin, J.; Xia, M.; Huang, H. Multiple Attention Mechanism Graph Convolution HAR Model Based on Coordination Theory. *Sensors* **2022**, *22*, 5259. https://doi.org/10.3390/s22145259

Academic Editor: Jing Tian

Received: 15 June 2022
Accepted: 11 July 2022
Published: 14 July 2022

Publisher's Note: MDPI stays neutral with regard to jurisdictional claims in published maps and institutional affiliations.

1. Introduction

With the rapid development of artificial intelligence algorithms, motion-recognition technology, which is an important part of artificial intelligence, is being studied for its application in many fields, such as human–computer interaction, video surveillance, film and television production, and other areas [1–3]. Many researchers [4–6] have invested a great deal of energy in this field and designed many excellent algorithms. Among them, most of the traditional algorithms use manual feature extraction, and these algorithms have made a breakthrough [7]. With the rapid development of machine learning and deep learning, many end-to-end motion recognition algorithms have appeared. These methods do not need to consume a lot of manpower and can achieve high recognition accuracy [8,9].

On the one hand, with deep learning and the rapid development of computer hardware, especially GPU, the performance of action-recognition algorithms is getting better and better. These algorithms can recognize more and more complex actions. Action-recognition algorithms based on deep learning can be roughly divided into the following two categories.

(1) The first category is the motion-recognition algorithm based on traditional CNN, RNN, and LSTM networks, for example, two-stream [10], C3D [11], and LSTM [12], and so on. These algorithms use end-to-end methods to train the model, which can effectively reduce the number of parameters and improve the accuracy of model recognition. Karens et al. [13] designed a two-stream model, which can extract the features of space and time latitude at the same time. They creatively fused the models of the two branches, effectively improving the recognition accuracy of the model. Du et al. [11] applied 3D convolution to action-recognition tasks. The model proposed by them can effectively extract the features of spatial and temporal latitude, and proved that $3 \times 3 \times 3$ convolution is more suitable for action-recognition tasks through experiments. Jeff et al. [12] applied the

LSTM model to the action-recognition task and proved through experiments that LSTM is more prominent in the features with time series.

(2) Secondly, with the rise of the graph convolution model, a large number of bone-based motion-recognition models have emerged. These models use human bones as the data of the training model. This type of data is not affected by environmental occlusion, complex background, and optical flow interference, which makes the model more robust. Yan et al. [14] used the graph convolution network in the task of action recognition for the first time. They used the graph convolution model to extract the features in the human skeleton map, and combined with the time convolution to extract the features in the time latitude. Kalpit et al. [15] proposed a bone-partition strategy. They use a partition strategy, which effectively fits the task of local graph convolution. Shi et al. [16] creatively proposed an adaptive graph convolution method based on the spatio–temporal graph convolution, which can adaptively learn bone features and further extract the hidden length, direction, and other features in bones.

On the other hand, coordination is not only the key to improving athletes' technical ability, but also is an essential part of everyday human physical activities. Coordination refers to the ability of each part of an organism to cooperate with other parts in time and space, and to complete actions in an effective manner. Coordination ability can make movements more accurate and subtle, especially periodic movements. Therefore, athletes attach great importance to the training of coordination ability and regard it as an indispensable and important physical quality to develop in order to more effectively compete and improve. Body coordination also includes three categories: force coordination, movement coordination, and space coordination. First, force coordination refers to the coordination ability of each muscle during tension and contraction. The coordination among the active, antagonistic, and supportive muscles is an important factor in muscle tension and contraction. Therefore, strength coordination training is mainly performed to improve the ability of the nervous system, to get more athletes to participate, to improve the degree of muscle fiber synchronization, to improve the coordination of muscles, and to make athletes exert their maximum potential when exerting strength. Secondly, movement coordination refers to the coordination ability that all humans shows when completing a certain action. Strengthening coordination training can improve human sports performance. Therefore, athletes with good movement coordination ability demonstrate the timeliness and economy of sports technology when they complete technical movements. Finally, spatial coordination refers to the body's coordination and adaptability with regard to its ability to maintain balance when changing its position. The training of spatial coordination ability is mainly performed to improve people's adaptability to their three-dimensional sense of space (up and down, left and right, front and back), so as to enhance their spatial awareness or position perception [17]. In terms of coordination in motion theory, we associate coordination features with motion-recognition algorithms. Therefore, this paper proposes a coordinated attention module based on coordination theory.

Through the research and learning of the existing algorithms, the author found the following two problems:

(1). According to the theory of human body-motion balance, the body will produce a coordination feature to maintain balance in the process of moving. Learning about this coordination feature was very helpful for understanding action, but the existing models did not make full use of this feature.

(2). Although the graph convolution neural network was successful in the field of action recognition, the limitation of its adjacency matrix led to the model that can only extract features at the neighbor nodes, and cannot extract features from the global perspective.

To solve the above problems, we improve the Two-Stream Adaptive Graph Convolutional Network (2S-AGCN) algorithm and propose a novel multiple attention mechanism graph convolution action-recognition model based on coordination theory (MA-CT). In this paper, a coordinated attention module (CAM) and an important attention module (IAM) are proposed. The important takeaways from these developments are as follows.

(1). The CAM effectively extracts coordination features generated during motion, and simulate the coordination of human movement through the covariance matrix. This module could effectively improve the accuracy of the basic model.

(2). In addition, the IAM directly started from the feature level, captured the changes of features on nodes, and gave more weight to the more important joints. The module could realize plug and play and effectively improve the accuracy of the basic model.

The structure of this paper is as follows. In the first section, this paper briefly introduces the development of action recognition and the previous methods. Section 2 briefly introduces the graph convolution neural network and the related knowledge of attention mechanism. Section 3 introduces the graph convolution action recognition model based on multiple attention modules, and introduces the details of the two attention modules in detail. In Section 4, experiments are carried out on two large public datasets to verify the effectiveness of the module proposed in this paper, and the model in this paper is compared with the existing model. Section 5 is the summary and prospect of this paper.

2. Related Works

2.1. Graph Convolution Neural Network

The graph convolution neural network (GCN) [18–21] summarized the convolution operation from grid image data to graph data with a topological structure. Its main idea was to aggregate the characteristics of its nodes and the characteristics of neighbor nodes, coupled with the natural constraints of the topological graph so that new node characteristics could be generated. The motivation of GCN comes from the combination of convolutional neural networks (CNN) [22–24] and topological graphs. With the further development of GCN, graph convolution neural networks could be divided into graph convolution neural networks based on spectral method and graph convolution neural networks based on the spatial method. Kipf et al. proposed a convolution formula combined with a graph Laplacian under the background of spectrum graph theory; however, the spatial-based method was intended to directly convolute the structure of the graph and its neighborhood, and then extract and normalize it according to the manually designed rules. After that, more and more scholars devoted themselves to the task of studying graph convolution neural networks. The fundamental reason is human bone data is topology type data, whereas CNN can only deal with two-dimensional grid data-like images, which is not competent for most tasks in human life. Therefore, in the field of action recognition, more and more people are engaged in the research of graph neural networks because the skeleton data is represented as a topological graph structure rather than a sequence or 2D grid structure.

2.2. Study on Action Coordination

Sports cannot be played without the intensively cultivated body coordination of athletes. To improve sports performance, athletes also need to carry out coordination training. Existing algorithms in the field of action recognition do not make full use of the coordination features of the body. Therefore, after consulting many books and papers on basic theories and training methods related to coordination, we chose the skeleton-based action-recognition dataset to deeply study the specific expression of body coordination. Among our findings, we learned that the coordination of the human body requires the sense of space when moving. This sense of space refers to the orientation of each part of the body when moving. Take running as an example. As shown in Figure 1, when a human is running, his hands and legs always swing alternately one after the other, and the arms and legs on the same side must be one after the other. According to the characteristics of the sense of motion space in the coordination of body movement, we studied how to extract the coordination features in the process of movement. To this end, we roughly divide the human body into five areas, including the left arm, the right arm, the left leg, the right leg, and the trunk, which includes the head. The position feature is expressed through appropriate expression, and the position relationship between two pairs is calculated. The coordination feature generated by human motion is calculated through this relationship. In this paper, the local center of gravity theory in physics and the covariance matrix in mathematics are used to express the coordination characteristics of the body.

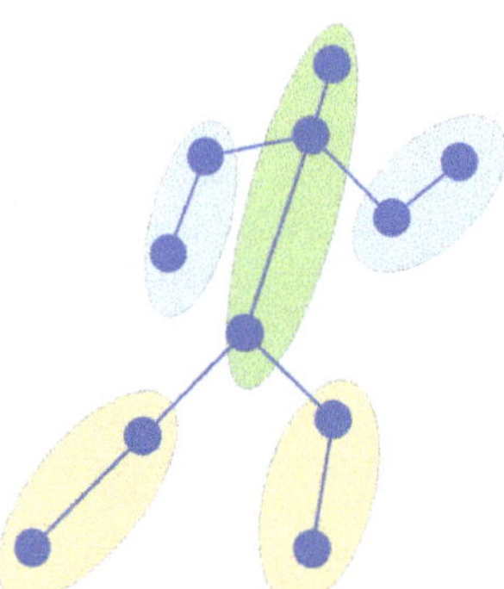

Figure 1. Running: position diagram of arms, legs and body.

3. Proposed Methods

In recent years, GCN has been used successfully in the field of motion recognition. On the one hand, because human bone data is not affected by interference information such as optical flow and occlusion, the data is purer. On the other hand, the topology of human bone data is a beat set with a graph neural network. The first section investigates the advantages and disadvantages of the existing algorithms in detail. When dealing with the task of human motion recognition based on bone data, these models ignore the coordination features of human action and cannot pay good attention to the more important joints in the process of motion due to the limitation of GCN. On the one hand, the theory of human movement balance [17] describes how the body acts in order to prevent the act of falling and the body's need to constantly adjust its posture to keep the position of the center of gravity unchanged. In particular, athletes can maintain their balance by swinging their arms and stretching their legs. For ordinary people, everyday actions are also needed to maintain balance, and the cooperation of limbs and trunk is needed to ensure that people will not fall to the ground. Therefore, in the process of completing a certain action, people's limbs have roughly fixed movement tracks. For example, in the action of running, when the left foot moves forward, the right arm must swing back to keep the position of the body's center of gravity unchanged; otherwise, there will be a risk of falling. On the other hand, the importance of different joints in different human actions is different, and these more important joints often number more than one. The existing models fail to pay good attention to the extraction of this part of the features. In addition, due to the fixity of the physical connection of the human body, the GCN is often fixed when extracting features and fails to pay better attention to the mutual features of several more important joints from a global perspective. These joints are often not connected in most actions. For example, in the action of clapping hands, from the perspective of the human skeleton map, the nodes of both hands are not directly connected and are far apart. However, both hands are an important part of the action of clapping hands, and the changes of various characteristics also focus on both hands. To solve the above problems, we propose two attention modules, namely the coordination attention module and the importance attention module, to solve the above two problems.

3.1. Multiple Attention Mechanism Graph Convolution Action-Recognition Model Based on Action Coordination Theory

Based on the 2S-AGCN algorithm, we propose a multiple attention mechanism graph convolution action-recognition model based on action coordination theory (MA-CT). The model solves some problems and helps the model to better identify the categories of human actions. On the one hand, the coordinated attention module (CAM) is mainly used to extract the coordination features generated in the process of human movement, and use this coordination feature to further strengthen the input of the model. On the other hand, the importance of attention module (IAM) aims to solve the problem that the model is limited by the graph convolution neural network, which makes the model unable to observe the more important joints in the movement process through the global field of vision. This section mainly introduces the original adaptive graph convolution model structure, the multiple attention mechanism graph convolution action-recognition model structure based on action coordination theory, and the structure of two attention modules.

3.1.1. Adaptive Graph Convolution Module

We take 2S-AGCN as the basic model that was introduced in detail in our other paper [5]. This article will briefly introduce the prominent contents. As shown in Figure 2, an adaptive graph convolution network is used to stack the above adaptive graph convolution modules. There are nine modules in total. The numbers of output channels of each module are 64, 64, 64, 128, 256, 256, and 256. Before the beginning of the network, add a BN layer to standardize the input data, add global average pooling after the ninth module, and finally input the results into the softmax layer to obtain the predicted result. The calculation formula of adaptive graph convolution is shown in Equation (1),

$$f_{out} = \sum_k^{K_v} W_k f_{in}(A_k + B_k + C_k),\tag{1}$$

where K_v is the kernel size of the spatial dimension and set to 3, W_k is the weight matrix. A_k, B_k, and C_k is three kinds of the adjacency matrix.

Here we will focus on the calculation process of C_k. C_k can learn a unique graph for each sample. To determine whether there is a connection between two adjacent nodes and how strong the connection is, we use the normalized Gaussian embedding function to calculate the similarity of the two nodes, as shown in Equation (2):

$$f(v_i, v_j) = \frac{e^{\theta}(v_i)^T \Phi(v_j)}{\sum\limits_{j=1}^{N} e^{\theta}(v_i)^T \Phi(v_j)}.\tag{2}$$

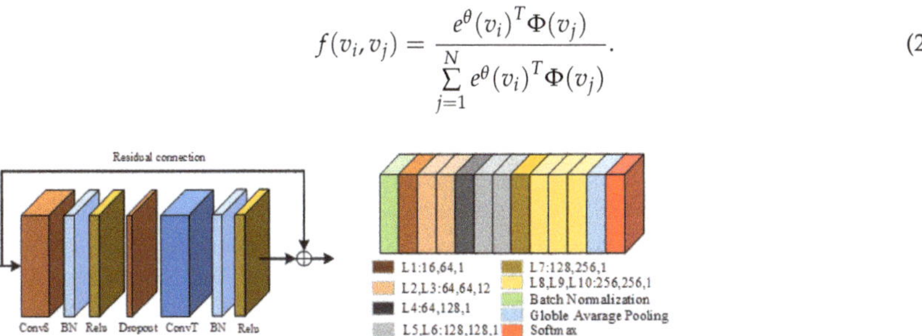

Figure 2. Original adaptive graph convolution module (**left**) and adaptive graph convolution model (**right**) [7].

3.1.2. Multiple Attention Mechanism Graph Convolution Action-Recognition Model Based on Action Coordination Theory

In this section, aiming at the existing models cannot effectively use the coordination characteristics of the body in the process of human movement, and due to the limitations of the graph convolution network, it is impossible to obtain the importance of joints from the global field of vision; therefore, a multiple attention mechanism graph convolution action-recognition model based on motion coordination theory is proposed. The overall framework of the model is shown in Figure 3. The light blue square in Figure 3 represents the CAM proposed in this paper, and the highlighted part in yellow represents the new adaptive graph convolution network after inserting IAM.

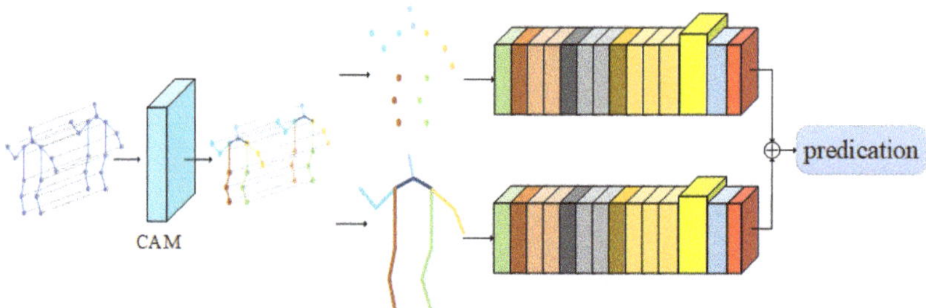

Figure 3. Multiple attention mechanism convolution action-recognition model based on action coordination theory (MA-CT).

The multi-attention mechanism graph convolution action-recognition model based on the action coordination theory proposed in this paper is an end-to-end training model. The overall framework can be roughly divided into three parts: coordination attention module, dual flow adaptive graph convolution model, and importance attention module. The model is based on the 2S-AGCN algorithm. After inputting the action sequence, the coordination attention module is used to preprocess the original data, mine the coordination characteristics of human action, and obtain a group of action sequences with coordination characteristics, which effectively integrates the concept of body coordination in human motion theory into the deep learning model. Then, according to the idea of the dual flow adaptive graph convolution model, the new action sequence is decomposed into two parts; one is a node feature, and the other is bone feature. Among them, the node characteristics include the coordinates on the node, confidence, and so on. Bone length, orientation, and other features are included. The two sets of data are used as the input of two identical adaptive graph convolutions for feature extraction. After the ninth layer of the adaptive graph convolution model, the features are input into the importance attention module, which can pay attention to the more important joints in the movement process, which effectively solves the deficiency that the existing models cannot obtain the important joints through the global field of view. Finally, through the softmax layer, two classification results are obtained, respectively. Finally, the two classification results are fused to obtain the final classification result of this model.

3.2. Coordination Attention Module

In the process of movement, people are always maintaining balance, which requires the cooperation of limbs and the trunk. Therefore, in the process of movement, the position and trajectory of each body part are roughly fixed. Inspired by this idea, the coordination of human motion is introduced into the action-recognition model. Therefore, this paper proposes a coordinated attention module, which is a computing unit, which is composed of the bone-partition strategy, matrix calculation, covariance matrix, and so on. The bone-partition strategy of the coordinated attention module proposed in this paper is shown in Figure 4. According to the structure of the human body, the human bone map is divided into five partitions, including the head, left arm, right arm, left leg, and right leg, and five subgraphs are obtained.

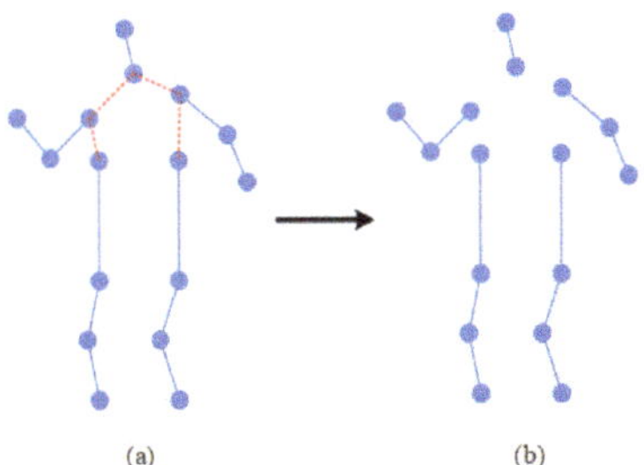

Figure 4. Partition strategy of human skeleton map. (**a**) shows the unprocessed human skeleton diagram, in which the red connecting part represents the divided connecting line, and (**b**) shows the human skeleton diagram after being divided into five partitions.

Then the model calculates the center-of-gravity point of each region. Mathematically and physically, it is stipulated that the center of gravity is closely related to the balance of the object, the motion of the object, and the internal force distribution of the constituent object. The author considers that to reduce the calculation amount of the model, the module uses the center of gravity of each region to calculate the coordination, which will be much less than the calculation amount of directly using joints, and can effectively avoid the problem of inconsistent nodes of each part. According to Equation (3), the center of gravity points on the five sub-graphs are calculated, respectively, and the center-of-gravity coordinates of each part are calculated to represent the general position of the area. The general motion trajectory of each area can be obtained by tracking the motion trajectory of the center of gravity. Let the center-of-gravity matrix be $(w_1, w_2, w_3, w_4, w_5)$. n in Equation (3) represents the number of nodes, and x_n represents the value of the abscissa of the nth node. Here, to simplify the expression, only the calculation formula of abscissa is shown, and the calculation of the other two coordinates is consistent with Equation (3):

$$w = \frac{(x_1 + x_2 + \ldots + x_n)}{n}, n = 1, 2, \ldots, n. \tag{3}$$

As shown in Figure 5, according to Equation (3), the center-of-gravity points in five zones can be obtained. Then calculate the body coordination matrix. Covariance is widely used in statistics and machine learning. Statistically, covariance is generally used to describe the similarity between two variables, and variance is a special case of covariance. The author believes that the covariance matrix can be used to calculate the similarity between various regions, and the similarity between two barycenters can be used to express the coordination of the body. The module introduces the covariance matrix into the action-recognition module to calculate the coordination relationship between two regions. The following will introduce the specific calculation methods of covariance and variance and rewrite the calculation of the covariance matrix according to the characteristics of the data used in this paper to make it more consistent with said data. The standard variance and covariance are calculated as shown in Equations (4) and (5).

$$s^2 = \frac{\sum\limits_{i-1}^{n} (X_i - \overline{X})^2}{n - 1}, i = 1, 2, \ldots, n \tag{4}$$

$$cov(X, Y) = \frac{\sum\limits_{i-1}^{n} (X_i - \overline{X})(Y_i - \overline{Y})}{n - 1}, i = 1, 2, \ldots, n \tag{5}$$

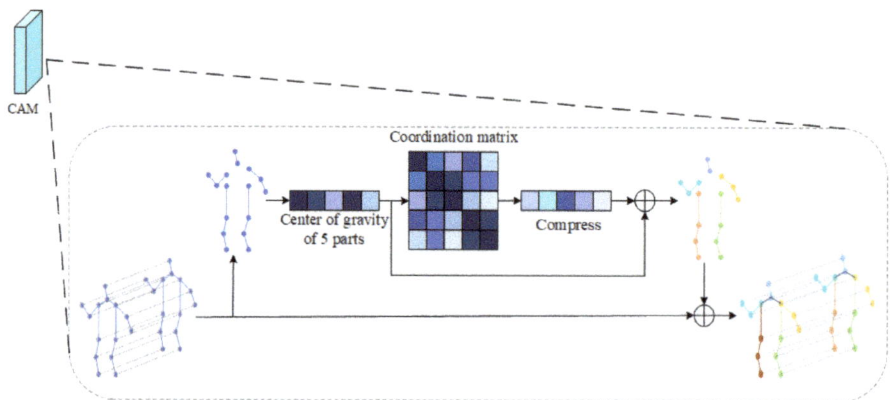

Figure 5. Coordination attention module.

Here, s represents variance, X and Y represent two groups of random variables, $cov(X, Y)$ represents the covariance of variables X and Y, i represents the ith variable in X or Y, and n represents the number of samples. According to the characteristics of the data in this paper, combined with Equations (4) and (5), we rewrite the covariance matrix into a form suitable for application in this paper. Here, n is set to 5, samples X and Y are set to the same sample, and the values are consistent, which is the center-of-gravity matrix. Let $X_i = Y_j = (w_1, w_2, w_3, w_4, w_5), i = j = 1, 2, 3, 4, 5$. Rewrite Equation (5) to obtain the calculation formula of coordination matrix used in this module, as shown in Equation (6):

$$cov(X, Y) = \frac{\sum_{i-1}^{5} (X_i - \overline{X})(Y_i - \overline{Y})}{4}, i = 1, 2, 3, 4, 5. \tag{6}$$

According to Equation (6) and the center-of-gravity matrix, the coordination matrix related to each other can be calculated. The matrix form is shown in Equation (7). Similarly, the coordination matrix of the remaining two coordinates can be calculated by using Equation (6).

$$cov(X, Y) =$$
$$\begin{bmatrix} cov(w_1, w_1) & cov(w_1, w_2) & cov(w_1, w_3) & cov(w_1, w_4) & cov(w_1, w_5) \\ cov(w_2, w_1) & cov(w_2, w_2) & cov(w_2, w_3) & cov(w_2, w_4) & cov(w_2, w_5) \\ cov(w_3, w_1) & cov(w_3, w_2) & cov(w_3, w_3) & cov(w_3, w_4) & cov(w_3, w_5) \\ cov(w_4, w_1) & cov(w_4, w_2) & cov(w_4, w_3) & cov(w_4, w_4) & cov(w_4, w_5) \\ cov(w_5, w_1) & cov(w_5, w_2) & cov(w_5, w_3) & cov(w_5, w_4) & cov(w_5, w_5) \end{bmatrix} \tag{7}$$

According to Equation (7), three groups of coordination matrices can be obtained. These three groups of coordination matrices are expressed as w_x, w_y, and w_z, respectively. These three groups of matrices can be used to represent the coordination characteristics of the body. Compress w_x, w_y, and w_z to the same size as the dimension of the center-of-gravity matrix. The compression method here is in the form of column-by-column addition, as shown in Equation (8). Take the first column as an example to illustrate the compression method.

$$X_i = cov(w_1, w_1) + cov(w_2, w_1) + cov(w_3, w_1) + cov(w_4, w_1) + cov(w_5, w_1) \tag{8}$$

Add the barycentric matrix and the compressed coordination matrix to obtain the barycentric matrix $(\tilde{w}_1, \tilde{w}_2, \tilde{w}_3, \tilde{w}_4, \tilde{w}_5)$ with coordination characteristics. Here, we consider the operation of matrix multiplication, but the coordinate values of most points are less than 1. If the matrix is multiplied, it will be smaller, and even lead to the loss of features.

Finally, the center-of-gravity matrix is added to each node according to the region, so that a set of bone data with coordination characteristics can be obtained.

3.3. Importance Attention Module

The graph volume model processes the data of the topology structure, which is in good agreement with the action-recognition task based on the human skeleton graph. At present, many models have achieved very good results. However, these models still have some shortcomings in the global field of view. Due to the limitation of human body topology, it is difficult for the graph volume model to learn the relationship between various end nodes, which is often an important part of the action. In addition, the deep graph convolution model easily leads to the phenomenon of excessive smoothing of features [25–27], so it is not suitable to use the deep model [28–30]. Inspired by the dual attention network (DA-net) [31,32], an attention module is proposed. DA-net can capture the global feature dependencies in both spatial and channel dimensions. The model uses the location attention module to learn the spatial interdependence of features and designs the channel attention module to simulate the interdependence between channels. Inspired by this idea, the location attention mechanism is embedded into the adaptive graph convolution model to obtain the important features of nodes in the feature graph and transfer them to the original feature graph. This paper proposes an important attention module. When extracting features, the module operates directly on the feature map, which can effectively overcome the limitations of the graph convolution neural network. The important attention module proposed in this paper is shown in Figure 6. The input of this module is the feature map obtained after spatial map convolution sampling and time convolution sampling, and the output is the feature map with attention characteristics.

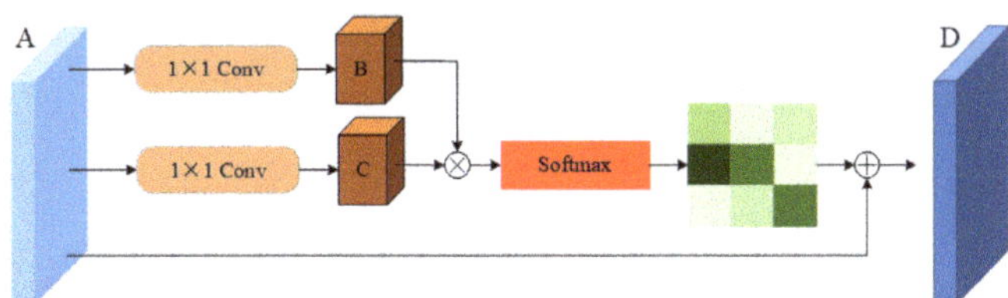

Figure 6. Importance attention module (IAM).

Because the number of channels in the ninth layer of the adaptive graph convolution model has reached 256, the value is too large, and the calculation size in the process of parameter transmission is large. To reduce the computational burden, use the convolution of 11 reduces the dimension of the feature channel, which effectively reduces the amount of calculation. First, the characteristic diagram in Figure·6 is divided into three branches, $A \in R^{(N \times M) \times C \times T \times V}$, where $(N \times M)$ represents the product of the batch size and the number of characters, C indicates the number of channels, T indicates the number of action frames, and V indicates the number of nodes. Then, A is sent into two convolution layers of 11 to obtain two new feature maps B and C, $\{B,C\} \in R^{(N \times M) \times C \times T \times V}$. Then the characteristic figure B and C are reconstituted into $R^{C \times D}$, $D = (NM)TV$, where D represents the number of feature points on each channel. Then the transposition of B and C is matrix-multiplied, and the position attention feature map S is calculated by the softmax layer, $S \in R^{D \times D}$. The calculation formula of the attention characteristic map is shown in Equation (9). where s_{ji} represents the influence of the ith position on the jth position:

$$S_{ji} = \frac{\exp(B_i \cdot C_j)}{\sum\limits_{i=1}^{N} \exp(B_i \cdot C_j)}.$$
(9)

At the same time, the feature map S is reorganized and multiplied by a scale coefficient α, which is added to the feature map a to obtain the final output D, $D \in R^{(N \times M) \times C \times T \times V}$.

The initial value of α is set to 0 and can gradually learn greater weight. The feature D of each position is the weighted sum of all position features and the original features. Therefore, it has a global vision and can selectively aggregate context information according to the spatial attention map:

$$E_j = \alpha \sum_{i=1}^{N} s_{ji} + A_j. \tag{10}$$

Here, the initial value of α is set to 0, and the corresponding weight can be gradually obtained through training.

The importance attention module proposed in this paper can realize plug and play. The author puts it after the space graph convolution and time convolution in the adaptive graph convolution model. As shown in Figure 7, the more important joints in the process of human motion are extracted from the space dimension and time dimension respectively.

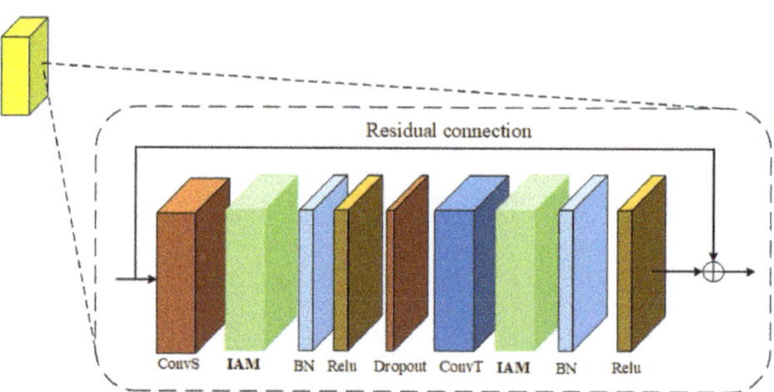

Figure 7. Adaptive graph convolution model with importance attention module.

In order to better explain the algorithm proposed in this paper, we simply provide an algorithm flow chart, as shown in Figure 8.

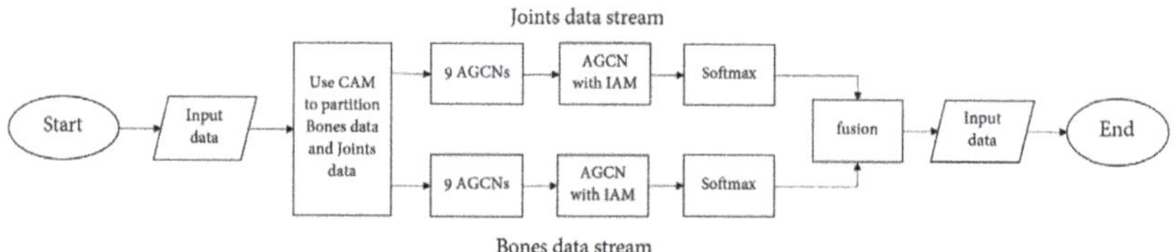

Figure 8. Flowchart of the methodology.

4. Experimental Results and Analysis

This section verifies the effectiveness of the coordination attention module and importance attention module proposed in this paper through experiments. To facilitate the comparison with the initial model 2S-AGCN, experimental verification is carried out on two large datasets: Kinetics-Skeleton and NTU-RGB + D. When verifying the coordination attention module, this section compares each branch of the two-stream network and then compares the results of the two-stream fusion. When verifying the importance attention module, because this paper inserts the importance attention module in two positions, to verify its effectiveness this section verifies the effectiveness of the importance attention module in space and time dimensions respectively. Then the two modules are fused to verify the effectiveness of the spatio–temporal importance attention module. Finally, the graph convolution motion recognition model based on multiple attention modules proposed in this paper is compared with the model on the same dataset to verify its effectiveness.

4.1. Datasets and Experimental Details

4.1.1. NTU-RGB + D

NTU-RGB + D [33] is one of the largest datasets in the human action-recognition task and contains 56,000 action clips in 60 action classes. Each action is taken with three cameras. The dataset gives the position information of nodes in each frame. There are 25 nodes in each frame. The author of this dataset two proposed benchmarks—cross-subject (X-Sub) and cross-view (X-View)—in his paper [33]. The former divides the training set and the test set according to the subject, and the latter divides the training set and the test set according to the camera number.

4.1.2. Kinetics-Skeleton

Kinetics [34] is another of the largest human action datasets, and contains 400 action categories. These video clips are taken from YouTube. We use the OpenPose toolbox to extract bone data from these videos, and extract bone data with 18 key points from the video sequence. In this paper, we use their released data (Kinetics-Skeleton) to evaluate the model in this paper. This dataset can be divided into a training set and a verification set. The training set has 240,000 segments, and the verification set includes 20,000 segments.

4.1.3. Training Details

All the experiments in this paper were completed under the same equipment. The hardware condition of the device was the ninth-generation Intel CPU, 64 g RAM and two 2080 Ti GPUs. The software condition was based on the Pytoch framework. The optimization algorithm was the stochastic gradient descent (SGD). Its momentum was set to 0.9. The cross-entropy loss function was used, and the initial learning rate was set to 0.1. For the NTU-RGBD and Kinetics-Skeleton datasets, due to the limitations of the experimental conditions in this paper, we set the batch size of the model to 16. The learning rate is set as 0.1 and is divided by 10 at the 30th epoch and the 40th epoch. The training process is ended at the 50th epoch [16]. For the Kinetics-Skeleton dataset, the size of the input tensor of Kinetics is set the same as [16], which contains 150 frames with two bodies in each frame. We perform the same data-augmentation methods as in [16]. In detail, we randomly choose 300 frames from the input skeleton sequence and slightly disturb the joint coordinates with randomly chosen rotations and translations. The learning rate is also set as 0.1 and is divided by 10 at the 45th epoch and 55th epoch. The training process is ended at the 65th epoch [16]. To enhance the accuracy of the experimental results, we did 10 experiments and took the average value as the final experimental results.

4.2. Ablation Experiment

4.2.1. Effectiveness Analysis of Coordination Attention Module

To verify the effectiveness of the coordination attention module (CAM) proposed in this paper, this section uses two large datasets—NTU-RGB + D and Kinetics-Skeleton—and compares the effectiveness of the coordination module by controlling variables. Under the same hardware conditions and the same parameter settings, the results shown in Table 1 are obtained, in which "J-Stream" and "B-Stream" respectively represent the joint stream and bone stream of the 2S-AGCN, and "CAM" represents the abbreviation of the coordinated attention module proposed in this paper. On the CV index of the NTU-RGB + D dataset, the accuracy of "J-Stream" of the initial 2S-AGCN is 93.1%, the accuracy of "B-Stream" is 93.3%, and the accuracy after two-stream fusion is 95.1%. In terms of CS index, the accuracy of "J-Stream" in the experimental environment is 86.3%, the accuracy of "B-Stream" is 86.7%, and the accuracy after two-stream fusion is 88.5%. In the Kinetics-Skeleton dataset, the accuracy of "J-Stream" in the experimental environment is 34.0%, that of "B-Stream" is 34.3%, and that of 2S-AGCN is 36.1%.

Table 1. Effectiveness analysis of coordination attention module on NTU-RGB + D and Kinetics-Skeleton datasets.

Methods	NTU-RGB + D		Kinetics-Skeleton (%)
	CV (%)	CS (%)	
J-Stream	93.1	86.3	34.0
B-Stream	93.3	86.7	34.3
2s-AGCN	95.1	88.5	36.1
CAM + J-Stream	94.0	86.9	35.4
CAM + B-Stream	93.5	87.5	34.5
CAM + 2s-AGCN	95.3	88.8	36.5

Under the same test conditions, the CAM proposed in this paper is inserted into the adaptive graph convolution model. In the CV index of the NTU-RGB + D dataset, the accuracy of "CAM + J-Stream" is 94%, which is 0.9% higher than the original accuracy. The accuracy of "CAM + B-Stream" is 93.5%, which is 0.2% higher than the original accuracy. The accuracy of two-stream fusion is 95.3%, which is 0.2% higher than the original accuracy. In terms of CS index, the accuracy of "CAM + J-Stream" is 86.9%, which is 0.6% higher than the original accuracy. The accuracy of "CAM + B-Stream" is 87.5%, which is 0.8% higher than the original accuracy. The accuracy of two-stream fusion is 88.8%, which is 0.3% higher than the original accuracy. In the Kinetics-Skeleton dataset, the accuracy of "CAM + J-Stream" is 35.4%, which is 1.4% higher than the original accuracy. The accuracy of "CAM + B-Stream" is 34.5%, which is 0.2% higher than the original accuracy. The accuracy of two-stream fusion is 36.5%, which is 0.4% higher than the original accuracy.

It can be seen from Table 1 that the performance of the adaptive graph convolution model combined with the coordinated attention module has improved in the two datasets. The module calculates the barycenter positions of the five partitions of the body, then calculates the relationship between the five locations by using the covariance matrix, and adds it to the features as a coordination matrix, which enriches the representation of the features. From the experimental results, this module can improve the accuracy of the model. After the two-stream fusion, the effect is better than the model without the coordination feature.

To further verify the effectiveness of the module, this section also records the changes in accuracy during model training and draws a curve to compare the changes in inaccuracy, as shown in Figures 9–11. It can be seen from the accuracy curve that the coordinated attention module proposed in this paper is better able to help the model understand the action semantics. In the two datasets, the initial accuracy of the model is higher than that of the original two-stream adaptive graph convolution model, and the oscillation amplitude of the accuracy is also small in the training process. When the final model tends to converge, the accuracy is also improved to a certain extent, which shows that the coordination attention module can effectively extract the coordination features of human bones, and provide help for the discrimination of action semantics.

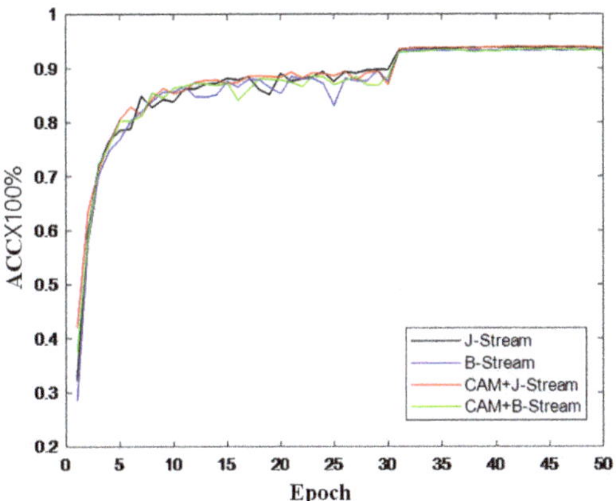

Figure 9. Accuracy change curve on CV index of NTU-RGB + D.

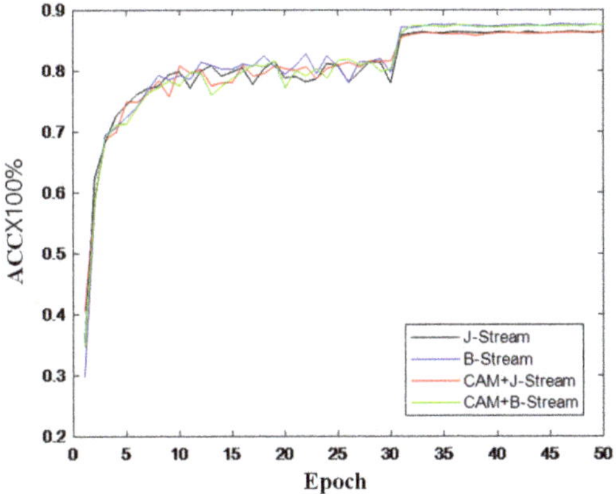

Figure 10. Accuracy change curve on CS index of NTU-RGB + D.

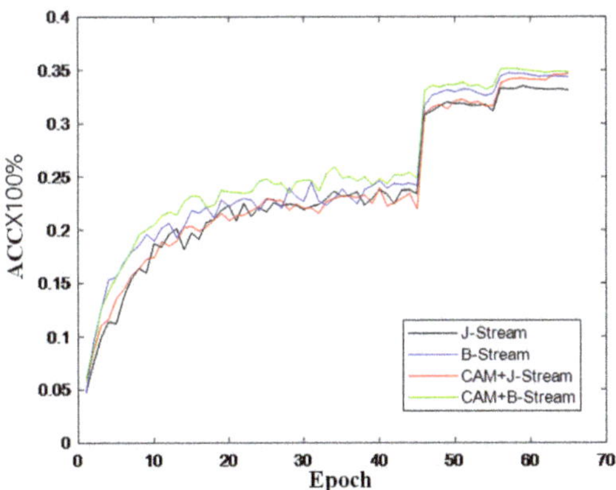

Figure 11. Accuracy change curve of Kinetics-Skeleton.

4.2.2. Effectiveness Analysis of Importance Attention Module

Aiming at the shortcomings of the existing models, this paper proposes an importance attention module (IAM). The module can observe the changes in joints from a global perspective and can calculate the dependencies between non-adjacent nodes from the topology. This module can realize plug and play. Because the adaptive graph convolution module extracts the features of the data in both space and time dimensions at the same time, this paper places the importance module after the spatial graph convolution layer and time convolution layer respectively. This section will verify and analyze the effectiveness of the module on two large public datasets. All data in Table 2 are completed under the same parameter settings and hardware conditions. "IAM-S" in the table indicates that the importance attention module is placed after spatial map convolution, "IAM-T" indicates that the importance attention module is placed after time convolution, and "IAM-ST" indicates that the importance attention module is placed at both locations. To facilitate the comparison with the 2S-AGCN, this section compares the accuracy of the importance attention module with the "J-Stream" and "B-Stream" of the initial model and verifies its effectiveness one by one. Then, the results of the two streams are fused to obtain the final classification result. From the experimental results in Table 2, it can be concluded that the IAM will improve the accuracy of the model after the spatial map convolution or time convolution. When the two positions are placed at the same time, the accuracy of the model will be further improved, which shows that the important attention module can effectively observe the joints that are more important for motion understanding from the perspective of global vision. In the CV index of the NTU-RGB + D dataset, after adding two groups of IAMs, the accuracy of the model is 95.7%, which is 0.6% higher than the initial 2S-AGCN. In the CS index, the accuracy of the model is improved by 0.4% compared with the initial model. In the Kinetics-Skeleton dataset, the accuracy of the model is improved by 0.9% compared with the initial model. The results in Table 2 illustrate the effectiveness of the importance attention module proposed in this paper.

Table 2. Effectiveness analysis of importance attention module in NTU-RGB + D and Kinetics-Skeleton.

Methods	NTU-RGB + D		Kinetics-Skeleton (%)
	CV (%)	CS (%)	
J-Stream	93.1	86.3	34.0
B-Stream	93.3	86.7	34.3
IAM-S + J-Stream	93.9	86.9	34.9
IAM-S + B-Stream	93.5	86.5	34.5
IAM-T + J-Stream	94.4	87.1	35.0
IAM-T + B-Stream	94.1	86.7	34.5
(IAM-ST) + J-Stream	94.6	86.9	34.8
(IAM-ST) + B-Stream	94.3	86.6	34.6
2s-AGCN	95.1	88.5	36.1
IAM-S + 2s-AGCN	95.2	88.6	36.3
IAM-T + 2s-AGCN	95.5	88.7	36.4
(IAM-ST) + 2s-AGCN	95.7	88.9	37.0

4.3. Comparison with Other Methods

This paper proposes a convolution action recognition model of multiple attention mechanism graphs based on action coordination theory. The experiments in Section 4.3 confirm the effectiveness of the two attention modules proposed in this paper. This section compares the MA-CT with some existing algorithms in the same datasets. The results of these two comparisons are shown in Tables 3 and 4, in these tables, bolded data is best. The methods used for comparison include the handcrafted feature-based method [35], RNN-based methods [36–38], CNN-based methods [39,40], and GCN-based methods [5,14,16,41–47]. The accuracy of MA-CT in CV index on NTU-RGB + D is 95.9%, and the accuracy in the CS index is 89.7%. Compared with the original 2S-AGCN, it is improved by 0.8% and 1.2%, respectively. In the Kinetics-Skeleton dataset, the accuracy of MA-CT reaches 37.3%, which is 1.2% higher than that of the original model. At the same time, compared with the model proposed in Section 3 it is improved by 0.2%. As can be seen from Table 3, in terms of the CV index, the model proposed in this paper is still inferior to the more advanced MV-IGNet. However, in terms of the CS index, the accuracy of the model proposed in this paper is 0.3% higher than that of MV-IGNet. It can be seen from Table 3 that the model proposed in this paper has improved upon the initial model, indicating that the coordination attention module and importance attention module can improve the accuracy of model recognition to a certain extent. In the Kinetics-Skeleton dataset, the accuracy of the proposed model in top-1 is 37.3%, which is 1.2% higher than the original 2S-AGCN. The accuracy of this model in top-1 is still not as good as 2S-AAGCN and 4S-AAGCN, but the accuracy of top-5 is 1% and 0.4% higher.

Table 3. Comparison of accuracy between ours model and other models on NTU-RGB + D dataset.

Methods	CV (%)	CS (%)
Deep LSTM [36]	67.3	60.7
Temporal ConvNet [39]	83.1	74.3
VA-LSTM [37]	87.6	79.4
Two-stream CNN [40]	89.3	83.2
GCA-LSTM [41]	82.8	74.4
ARRN-LATM [38]	89.6	81.8
MANs [42]	93.22	83.01
ST-GCN [14]	88.3	81.5
DPRL + GCNN [43]	89.8	83.5
2S-AGCN [16]	95.1	88.5
RA-GCN [44]	93.6	87.3
MV-IGNet [45]	96.3	89.2
MST-AGCN [5]	95.5	89.5
MA-CT (ours)	**95.9**	**89.7**

Table 4. Comparison of accuracy between ours model and other models on Kinetics-Skeleton dataset.

Methods	CV (%)	CS (%)
Feature Encoding [35]	14.9	25.8
Deep LSTM [36]	16.4	35.3
Temporal ConvNet [39]	20.3	40.0
ST-GCN [14]	30.7	52.8
2S-AGCN [16]	36.1	58.7
GCN-NAS [46]	37.1	60.0
1s-AAGCN [47]	36.0	58.4
2s-AAGCN [47]	37.4	60.4
4s-AAGCN [47]	37.8	61.0
MST-AGCN [5]	37.1	61.0
MA-CT (ours)	**37.3**	**61.4**

5. Discussion

With the rapid development of artificial intelligence and its application in various fields, HAR has become an important area of development through deep learning to identify human movement. There is still room for further improvement in the accuracy of current HAR algorithms before its best engineering applications can be achieved.

In the development of existing HAR algorithms, people are always accustomed to introducing newly developed deep learning algorithms into HAR algorithms, which has played a role in improving the accuracy. Compared with traditional machine learning, deep learning essentially uses complex networks for automatic learning data features. In order to achieve better learning of such features, the network of deep learning becomes more and more complex, which requires more expensive hardware, and the requirement is contradictory to engineering application. Therefore, if the network structure remains unchanged (the requirements for hardware also remain unchanged), artificial emphasis on some prior knowledge and enhancement of some features will enable the network to quickly grasp these important features, and improve the accuracy to become a better choice.

Based on the coordination theory in sports kinematics, and by combining the digital robot control theory and the attention mechanism, this study has some innovations in feature enhancement and model structure. For feature extraction, this study uses a two-channel scheme to extract joint and bones features, which are divided into two data streams for analysis. In the aspect of feature enhancement, the coordination attention module and the importance attention module are designed and used to focus on the correlation of upper and lower frames action coordination, and finally achieve the fusion output. This study improves the accuracy, which proves that the idea of HAR combined with the coordination theory is correct.

In addition, we also recognize that because the learning data and validation data of this algorithm come from generally accepted standard datasets, and most of these standard datasets are stable movements of healthy people, this is obviously a positive sample for whole data, and uncoordinated actions should also be the content of learning and analysis, which is one of the defects of this study. Of course, it is easy to imagine that if human movements were inconsistent and the center of gravity was unstable, the predictable results are falls, so this algorithm should be used to predict the action of human falls.

6. Conclusions

In this work, we propose a multiple attention mechanism graph convolution action recognition model based on coordination theory (MA-CT). It parameterizes the graph structure of the skeleton data and embeds it into the network to be jointly learned and updated with the model. This data-driven approach increases the flexibility of the graph convolutional network and is more suitable for the action recognition task. Furthermore, the existing methods do not make full use of the coordination features of human motion, and because of the existence of an adjacency matrix, the model cannot extract features from the global perspective. In this work, we propose a coordination attention module (CAM) and importance attention module (IAM). In this paper, experiments are carried out on two large public datasets. For the two indicators of NTU-RGB + D, the CAM improves the accuracy of the model by 0.2% and 0.3%, and the IAM improves the accuracy of the model by 0.6% and 0.4%. In the Kinetics dataset, the CAM improves the accuracy of the model by 0.4%, and the IAM improves the accuracy of the model by 0.9%. They are used to solve the problems of insufficient feature extraction and the capturing of key joints. The final model has achieved good results in NTU-RGB + D and Kinetics.

Author Contributions: Conceptualization, K.H. and Y.D.; methodology, K.H. and H.H.; software, Y.D.; validation, Y.D.; formal analysis, Y.D.; investigation, Y.D. and J.J.; resources, K.H. and Y.D.; data curation, K.H.; writing—original draft preparation, Y.D. and M.X.; writing—review and editing, Y.D.; visualization, Y.D. and J.J.; supervision, K.H.; project administration, K.H. All authors have read and agreed to the published version of the manuscript.

Funding: Research in this article is supported by the key special project of the National Key R&D Program (2018YFC1405703), and the financial support of Jiangsu Austin Optronics Technology Co., Ltd. is deeply appreciated.

Institutional Review Board Statement: Not applicable.

Informed Consent Statement: Not applicable.

Data Availability Statement: The data and code used to support the findings of this study are available from the corresponding author upon request (001600@nuist.edu.cn).

Acknowledgments: We would like to express my heartfelt thanks to the reviewers and editors who submitted valuable revisions to this article.

Conflicts of Interest: The authors declare no conflict of interest.

References

1. Hu, K.; Jin, J.; Zheng, F.; Weng, L.; Ding, Y. Overview of behavior recognition based on deep learning. *Artif. Intell. Rev.* **2022**, *preprint*. [CrossRef]
2. Dai, R.; Gao, Y.; Fang, Z.; Jiang, X.; Wang, A.; Zhang, J.; Zhong, C. Unsupervised learning of depth estimation based on attention model and global pose optimization. *Signal Process. Image Commun.* **2019**, *78*, 284–292. [CrossRef]
3. Pareek, P.; Thakkar, A. A survey on video-based human action recognition: Recent updates, datasets, challenges, and applications. *Artif. Intell. Rev.* **2021**, *54*, 2259–2322. [CrossRef]
4. Kipf, T.N.; Welling, M. Semi-supervised classification with graph convolutional networks. *arXiv* **2016**, arXiv:1609.02907.
5. Hu, K.; Ding, Y.; Jin, J.; Weng, L.; Xia, M. Skeleton Motion Recognition Based on Multi-Scale Deep Spatio-Temporal Features. *Appl. Sci.* **2022**, *12*, 1028. [CrossRef]
6. Hu, K.; Zheng, F.; Weng, L.; Ding, Y.; Jin, J. Action Recognition Algorithm of Spatio–Temporal Differential LSTM Based on Feature Enhancement. *Appl. Sci.* **2021**, *11*, 7876. [CrossRef]
7. Sun, Z.; Ke, Q.; Rahmani, H.; Bennamoun, M.; Wang, G.; Liu, J. Human action recognition from various data modalities: A review. *IEEE Trans. Pattern Anal. Mach. Intell.* **2022**, 1–20. [CrossRef]
8. Ma, L.; Wang, X.; Wang, X.; Wang, L.; Shi, Y.; Huang, M. TCDA: Truthful combinatorial double auctions for mobile edge computing in industrial Internet of Things. *IEEE Trans. Mob. Comput.* **2021**, 1. [CrossRef]
9. Ma, L.; Li, N.; Guo, Y.; Wang, X.; Yang, S.; Huang, M.; Zhang, H. Learning to optimize: Reference vector reinforcement learning adaption to constrained many-objective optimization of industrial copper burdening system. *IEEE Trans. Cybern.* **2021**, 1–14. [CrossRef]
10. Feichtenhofer, C.; Pinz, A.; Zisserman, A. Convolutional two-stream network fusion for video action recognition. In Proceedings of the IEEE Conference on Computer Vision and Pattern Recognition, Las Vegas, NV, USA, 27–30 June 2016; pp. 1933–1941.
11. Tran, D.; Bourdev, L.; Fergus, R.; Torresani, L.; Paluri, M. Learning spatiotemporal features with 3d convolutional networks. In Proceedings of the IEEE International Conference on Computer Vision, Santiago, Chile, 7–13 December 2015; pp. 4489–4497.
12. Donahue, J.; Anne Hendricks, L.; Guadarrama, S.; Rohrbach, M.; Venugopalan, S.; Saenko, K.; Darrell, T. Long-term Recurrent Convolutional Networks for Visual Recognition and Description. *Potential Anal.* **2019**, *49*, 225–245.
13. Karens, A.Z. Two-Stream Convolutional Networks for Action Recognition in Videos. In Proceedings of the 28th Annual Conference on Neural Information Processing Systems, Montreal, QC, Canada, 8–13 December 2014; pp. 568–576.
14. Yan, S.; Xiong, Y.; Lin, D. Spatial temporal graph convolutional networks for skeleton-based action recognition. In Proceedings of the Thirty-Second AAAI Conference on Artificial Intelligence, New Orleans, LA, USA, 2–7 February 2018.
15. Thakkar, K.; Narayanan, P.J. Part-based Graph Convolutional Network for Action Recognition. In Proceedings of the 29th British Machine Vision Conference, Cardiff, UK, 9–12 September 2019; BMVA Press: Newcastle, UK, 2019.
16. Shi, L.; Zhang, Y.; Cheng, J.; Lu, H. Two-stream adaptive graph convolutional networks for sk eleton-based action recognition. In Proceedings of the 32th IEEE Conference on Computer Vision and Pattern Recognition, Long Beach, CA, USA, 16–20 June 2019; IEEE Computer Society: Long Beach, CA, USA, 2019; pp. 12018–12027.
17. Winter, D.A. Human balance and posture control during standing and walking. *Gait Posture* **1995**, *3*, 193–214. [CrossRef]
18. Bruna, J.; Zaremba, W.; Szlam, A.; LeCun, Y. Spectral networks and locally connected networks on graphs. *arXiv* **2013**, arXiv:1312.6203.
19. Hu, K.; Wu, J.; Li, Y.; Lu, M.; Weng, L.; Xia, M. FedGCN: Federated Learning-Based Graph Convolutional Networks for Non-Euclidean Spatial Data. *Mathematics* **2022**, *10*, 1000. [CrossRef]
20. Xia, M.; Wang, Z.; Lu, M.; Pan, L. MFAGCN: A new framework for identifying power grid branch parameters. *Electr. Power Syst. Res.* **2022**, *207*, 107855. [CrossRef]
21. Wang, Z.; Xia, M.; Lu, M.; Pan, L.; Liu, J. Parameter Identification in Power Transmission Systems Based on Graph Convolution Network. *IEEE Trans. Power Deliv.* **2021**, 1. [CrossRef]
22. Lu, C.; Xia, M.; Lin, H. Multi-scale strip pooling feature aggregation network for cloud and cloud shadow segmentation. *Neural Comput. Appl.* **2022**, *34*, 6149–6162. [CrossRef]
23. Hu, K.; Zhang, D.; Xia, M.; Qian, M.; Chen, B. LCDNet: Light-weighted Cloud Detection Network for High-resolution Remote Sensing Images. *IEEE J. Sel. Top. Appl. Earth Obs. Remote Sens.* **2022**, *15*, 4809–4823. [CrossRef]
24. Qu, Y.; Xia, M.; Zhang, Y. Strip pooling channel spatial attention network for the segmentation of cloud and cloud shadow. *Comput. Geosci.* **2021**, *157*, 104940. [CrossRef]
25. Hu, K.; Zhang, D.; Xia, M. CDUNet: Cloud Detection UNet for Remote Sensing Imagery. *Remote Sens.* **2021**, *13*, 4533. [CrossRef]
26. Hu, K.; Li, M.; Xia, M.; Lin, H. Multi-Scale Feature Aggregation Network for Water Area Segmentation. *Remote Sens.* **2022**, *14*, 206. [CrossRef]

27. Lu, C.; Xia, M.; Qian, M.; Chen, B. Dual-branch Network for Cloud and Cloud Shadow Segmentation. *IEEE Trans. Geosci. Remote Sens.* **2022**, *60*, 5410012. [CrossRef]

28. Gao, J.; Weng, L.; Xia, M.; Lin, H. MLNet: Multichannel feature fusion lozenge network for land segmentation. *J. Appl. Remote Sens.* **2022**, *16*, 016513. [CrossRef]

29. Miao, S.; Xia, M.; Qian, M.; Zhang, Y.; Liu, J.; Lin, H. Cloud/shadow segmentation based on multi-level feature enhanced network for remote sensing imagery. *Int. J. Remote Sens.* **2022**, 1–21. [CrossRef]

30. Xia, M.; Qu, Y.; Lin, H. PADANet: Parallel asymmetric double attention network for clouds and its shadow detection. *J. Appl. Remote Sens.* **2021**, *15*, 046512. [CrossRef]

31. Fu, J.; Liu, J.; Tian, H.; Li, Y.; Bao, Y.; Fang, Z.; Lu, H. Dual attention network for scene segmentation. In Proceedings of the IEEE/CVF Conference on Computer Vision and Pattern Recognition, Long Beach, CA, USA, 15–20 June 2019; pp. 3146–3154.

32. Chen, B.; Xia, M.; Qian, M.; Huang, J. MANet: A multilevel aggregation network for semantic segmentation of high-resolution remote sensing images. *Int. J. Remote Sens.* **2022**, 1–21. [CrossRef]

33. Shahroudy, A.; Liu, J.; Ng, T.T.; Wang, G. NTU RGB + D: A Large Scale Dataset for 3D Human Activity Analysis. In Proceedings of the IEEE Conference on Computer Vision and Pattern Recognition, Las Vegas, NV, USA, 27–30 June 2016; IEEE Computer Society: Long Beach, CA, USA, 2016; pp. 1010–1019.

34. Kay, W.; Carreira, J.; Simonyan, K.; Zhang, B.; Hillier, C.; Vijayanarasimhan, S.; Viola, F.; Green, T.; Back, T.; Natsev, P.; et al. The Kinetics Human Action Video Dataset. *arXiv* **2017**, arXiv:1705.06950.

35. Du, Y.; Fu, Y.; Wang, L. Skeleton based action recognition with convolutional neural network. In Proceedings of the 2015 3rd IAPR Asian Conference on Pattern Recognition (ACPR), Kuala Lumpur, Malaysia, 3–6 November 2015; IEEE: Piscataway, NJ, USA, 2015; pp. 579–583.

36. Twin, A.P.; Alkan, E.O.; Gangi, A.; de Mathelin, M.; Padoy, N. Data-driven spatio-temporal RGBD feature encoding for action recognition in operating rooms. *Int. J. Comput. Assist. Radiol.* **2015**, *10*, 737–747.

37. Zhang, P.; Lan, C.; Xing, J.; Zeng, W.; Xue, J.; Zheng, N. View adaptive recurrent neural networks for high performance human action recognition from skeleton data. In Proceedings of the IEEE International Conference on Computer Vision, Venice, Italy, 22–29 October 2017; pp. 2117–2126.

38. Zheng, W.; Li, L.; Zhang, Z.; Huang, Y.; Wang, L. Relational network for skeleton-based action recognition. In Proceedings of the 2019 IEEE International Conference on Multimedia and Expo (ICME), Shanghai, China, 8–12 July 2019; IEEE: Piscataway, NJ, USA, 2019; pp. 826–831.

39. Gammulle, H.; Denman, S.; Sridharan, S.; Fookes, C. Two Stream LSTM: A Deep Fusion Framework for Human Action Recognition. In Proceedings of the 2017 IEEE Winter Conference on Applications of Computer Vision (WACV), Santa Rosa, CA, USA, 24–31 March 2017; IEEE: Piscataway, NJ, USA, 2017.

40. Zhao, R.; Ali, H.; Van der Smagt, P. Two-stream RNN/CNN for action recognition in 3D videos. In Proceedings of the 2017 IEEE/RSJ International Conference on Intelligent Robots and Systems (IROS), Vancouver, BC, Canada, 24–28 September; IEEE: Piscataway, NJ, USA, 2017; pp. 4260–4267.

41. Li, C.; Zhong, Q.; Xie, D.; Pu, S. Skeleton-based action recognition with convolutional neural networks. In Proceedings of the 2017 IEEE International Conference on Multimedia & ExpoWorkshops (ICMEW), Hong Kong, China, 10–14 July 2017; IEEE: Piscataway, NJ, USA, 2017; pp. 597–600.

42. Li, C.; Xie, C.; Zhang, B.; Han, J.; Zhen, X.; Chen, J. Memory attention networks for skeleton-based action recognition. *IEEE Trans. Neural Netw. Learn. Syst.* **2021**, 1–15. [CrossRef]

43. Tang, Y.; Tian, Y.; Lu, J.; Li, P.; Zhou, J. Deep progressive reinforcement learning for skeleton-based action recognition. In Proceedings of the IEEE Conference on Computer Vision and Pattern Recognition, Salt Lake City, UT, USA, 18–23 June 2018; pp. 5323–5332.

44. Song, Y.F.; Zhang, Z.; Wang, L. Richly activated graph convolutional network for action recognition with incomplete skeletons. In Proceedings of the 2019 IEEE International Conference on Image Processing (ICIP), Taipei, Taiwan, 22–25 September 2019; IEEE: Piscataway, NJ, USA, 2019; pp. 1–5.

45. Wang, M.; Ni, B.; Yang, X. Learning multi-view interactional skeleton graph for action recognition. *IEEE Trans. Pattern Anal. Mach. Intell.* **2020**, 1. [CrossRef]

46. Peng, W.; Hong, X.; Chen, H.; Zhao, G. Learning graph convolutional network for skeleton-based human action recognition by neural searching. In Proceedings of the AAAI Conference on Artificial Intelligence, New York, NY, USA, 7–12 February 2020; Volume 34, pp. 2669–2676.

47. Shi, L.; Zhang, Y.; Cheng, J.; Lu, H. Skeleton-based action recognition with multi-stream adaptive graph convolutional networks. *IEEE Trans. Image Process.* **2020**, *29*, 9532–9545. [CrossRef]

 *sensors*

Article

Improved Feature-Based Gaze Estimation Using Self-Attention Module and Synthetic Eye Images

Jaekwang Oh [1], Youngkeun Lee [1], Jisang Yoo [1] and Soonchul Kwon [2,*]

[1] Department of Electronic Engineering, Kwangwoon University, Seoul 01897, Korea; dhworhkd11@kw.ac.kr (J.O.); yklee1308@kw.ac.kr (Y.L.); jsyoo@kw.ac.kr (J.Y.)
[2] Graduate School of Smart Convergence, Kwangwoon Univeristy, Seoul 01897, Korea
* Correspondence: ksc0226@kw.ac.kr; Tel.: +82-2-940-8637

Abstract: Gaze is an excellent indicator and has utility in that it can express interest or intention and the condition of an object. Recent deep-learning methods are mainly appearance-based methods that estimate gaze based on a simple regression from entire face and eye images. However, sometimes, this method does not give satisfactory results for gaze estimations in low-resolution and noisy images obtained in unconstrained real-world settings (e.g., places with severe lighting changes). In this study, we propose a method that estimates gaze by detecting eye region landmarks through a single eye image; and this approach is shown to be competitive with recent appearance-based methods. Our approach acquires rich information by extracting more landmarks and including iris and eye edges, similar to the existing feature-based methods. To acquire strong features even at low resolutions, we used the HRNet backbone network to learn representations of images at various resolutions. Furthermore, we used the self-attention module CBAM to obtain a refined feature map with better spatial information, which enhanced the robustness to noisy inputs, thereby yielding a performance of a 3.18% landmark localization error, a 4% improvement over the existing error and A large number of landmarks were acquired and used as inputs for a lightweight neural network to estimate the gaze. We conducted a within-datasets evaluation on the MPIIGaze, which was obtained in a natural environment and achieved a state-of-the-art performance of 4.32 degrees, a 6% improvement over the existing performance.

Citation: Oh, J.; Lee, Y.; Yoo, J.; Kwon, S. Improved Feature-Based Gaze Estimation Using Self-Attention Module and Synthetic Eye Images. *Sensors* **2022**, *22*, 4026. https://doi.org/10.3390/s22114026

Academic Editor: Jing Tian

Received: 21 April 2022
Accepted: 22 May 2022
Published: 26 May 2022

Publisher's Note: MDPI stays neutral with regard to jurisdictional claims in published maps and institutional affiliations.

Keywords: gaze estimation based on feature; eye landmark detection; self-attention; synthetic eye images

1. Introduction

Accurately estimating gaze direction plays a major role in applications, such as the analysis of visual attention, research on consumer behavior, augmented reality (AR), and virtual reality (VR). Because inference results are more improved by using deep-learning models than other approaches, they can be applied to advanced technologies, such as autonomous driving [1] and smart glasses [2], and can overcome challenges in the medical field [3]. Using these deep-learning models is quite helpful, but it is difficult to train them due to lighting conditions and insufficient and poor-quality datasets. Moreover, the value of gaze datasets is very expensive and complicated to process. To alleviate this problem, we propose a model that extracts eye features using UnityEyes [4], high-quality synthetic data. An exact position of a feature is obtained from the enhanced model by using a self-attention module. Subsequently, gaze estimation is performed through using high-level eye features, which is less restrictive as it does not utilize complex information, such as full-face information and head poses used for gaze estimation [5–7].

Recently, deep-learning-based eye-tracking technology has been developed mainly through appearance-based methods [5–11] that use eye images or face images. These appearance-based models currently perform particularly well in a controlled, environment in which there are no disturbances, such as noise in an input frame. However, these models have some drawbacks. First, the cost of datasets is very high, and the quality of data has a significant impact on the training of the model. Second, most models are black-box solutions, which pose the challenge of locating and understanding points for improvement. This study reduces the dependency of the feature map, which is difficult to interpret and approaches a feature-based method that can estimate a gaze vector through accurate feature points after acquiring landmarks obtained from an image. Refs. [12,13] used a stacked hourglass model [14] to extract a few eyelid and iris points.

In this study, we reinforced and used an advanced model, called HRNet [15], which shows state-of-the-art performance in the pose estimation task to extract high-quality landmarks. In pixel-wise fields, such as pose estimation and landmark detection, the resolutions and sizes of images have huge impacts on performance. Therefore, we extracted landmarks with a high accuracy by remodeling the model using a self-attention module [16–18]. CBAM [18], a self-attention technology, helps to generate a refined feature map that better encodes positional information using channel and spatial attention.

Because we aimed to estimate a gaze vector centered on a landmark extraction, a labeled gaze vector and eye-landmark dataset were essential. However, because gaze data are very expensive and difficult to generate, it is more difficult to obtain a dataset that provides both high-resolution images and landmarks simultaneously. Therefore, UnityEyes, a synthetic eye-image dataset with eye landmarks, was adopted as a training dataset through high-resolution images and an automatic labeling system. The model was trained by processing 32 iris and 16 eyelid points from the eye image obtained by fixing the head pose. Figure 1 shows the predicted heatmaps during the training We evaluated landmark and gaze performance by composing a test set for UnityEyes and performed a gaze performance evaluation using MPIIGaze [11], which has real environment settings.

Our paper is organized as follows. We first summarize related work in Section 2. In Section 3, the proposed gaze estimation method is explained. Section 4 describes the datasets used in the experiments. The experiment results are provided in Section 5. Finally, Section 6 presents our discussion on this study, and Section 7 presents the conclusion.

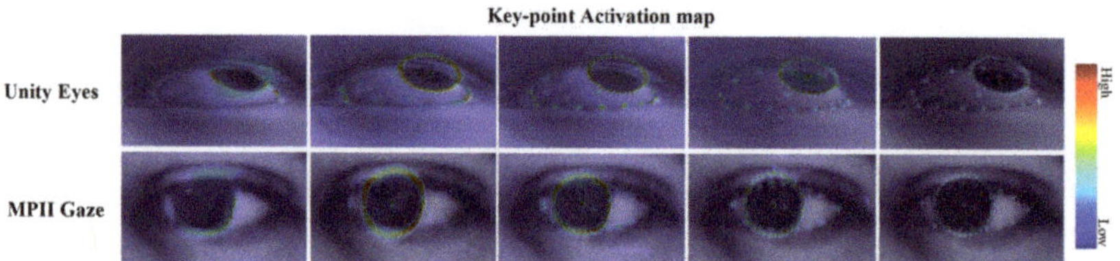

Figure 1. From left to right, the predicted heatmaps are shown as the training epoch increases. The heatmaps have high confidence scores where the landmarks are most likely to be located, and the right bar represents the color space corresponding to the confidence score.

2. Related Work

The gaze estimation method is a research topic of great interest as it has excellent applicabilities to real environments. As it can be applied to various fields, obtaining and creating accurate gaze values and gaze estimations with less constraints are challenging tasks. In this section, we provide a brief overview of the research related to our method. The studies in each subsection are summarized in Tables 1–3, respectively.

2.1. Feature-Based Method

Feature-based estimation [13,19–22], a method of gaze estimation, mainly uses unique features that have geometric relationships with the eye. Existing research studies have focused on objects that have a strong visual influences on images.

Roberto et al. [19] used the saliency method to estimate visual gaze behavior and used gaze estimation devices to compensate for errors caused by incorrect calibrations, thereby reducing restrictions caused by user movements. However, the use of these devices increases the error rate as the head moves and interferes with gaze detection.

Some researchers added multiple cameras to compute and focus head movements in multiple directions, extending the influence of eye information and head posture information [20,21]. Head pose has a significant effect on the gaze and requires many restrictions.

To avoid this problem, studies dealing with the static head-pose problem were conducted [13,22] using the convolutional neural network (CNN) model, in which images from a single camera are used to perform gaze estimations based on landmarks as they are less restrictive features. This makes it less difficult to build an experimental environment because there is no need for separate camera calibrations. As only eye images are used for gaze estimations, the dependence of the eye landmark feature vector is increased. After acquiring eye landmarks using a CNN model, a gaze is inferred using support-vector regression (SVR) [23].

Bernard et al. [24] used two gaze maps; one represented the eyeball region and the other represented the iris region. A gaze vector was regressed through the positional relationship between the two gaze maps.

Table 1. A summary table of the feature-based gaze estimation method.

Author	Methodology	Highlights	Limitations
Roberto et al. [19]	Saliency method	Fixing the shortcoming of low-quality monocular head and eye trackers	Controlled poses
Manolova et al. [20]	SDM algorithm	Estimating accurate gaze direction based on 3D head positions using a Kinect	Multiple device settings
Lai et al. [21]	CFB and GFB approaches	Integrating CFB and GFB to provide a robust and flexible system	Multiple camera settings
Wood et al. [22]	CNN and 3D head scans	Providing synthetic eye-image datasets with landmarks, a head pose and a gaze direction annotation	Weak to unmodelled occlusions
S. Park et al. [13]	Hourglass network and SVR	Estimating an accurate gaze vector based on eye landmarks in wild settings	Computational costs
Bernard et al. [24]	Capsule network	Utilizing two beat maps where one represents the eyeball and the other represents the iris	Computational costs

2.2. Landmark Detection

We used a deep-learning-based pose-estimation model as a tool to acquire eye region features. The landmark-detection task includes a key-point detection to detect a skeleton representing the body structure and a facial-landmark detection to extract landmarks in the face; this is a field that requires a large number of datasets depending on the domain. Some models have a direct regression structure based on the deep neural network and have predicted key points [25]. The predicted key-point positions are progressively improved using feedback on the error prediction.

Some researchers proposed a heatmap generation method through using soft-max in a manner that can be fully differentiated [26]. The convolutional pose machine [27] predicts a heatmap with intermediate supervision to prevent vanishing gradients, which are detrimental to deep-learning models. A new network architecture called the stacked hourglass [14] proved that repeated bottom-up and top-down processing with intermediate supervision is an important process for improving performance. A network structure that used high to low sub-networks in parallel is one of the networks that are currently showing the best performance [15]. For a spatially accurate heatmap estimation, high-resolution learning is maintained throughout the entire process; unlike the stack hourglass, it does not use intermediate heatmap supervision, which makes it efficient in terms of complexity and parameters and can generate highly information-rich feature outputs through a multi-scale feature fusion process. Yang et al. [28] introduced a transformer for key-point detections using HRNet as a backbone that extracts a feature map. It causes a performance improvement over existing performance through multi-head self-attention but is computationally demanding.

Table 2. A summary table of the landmark detection.

Author	Methodology	Highlights	Limitations
Toshev et al. [25]	DNN	Introducing a cascade of direct DNN regressors for landmark detection	Overfitting problem
Luvizon et al. [26]	CNN	Using the soft-max function to convert feature maps directly into landmark coordinates in a fully differentiable framework	Limited memory resources
Wei et al. [27]	CPMs	Providing a natural learning-objective function that enforces intermediate supervision to adder the difficulty of vanishing gradients	Enabled for only a single object
Newell et al. [14]	Stacked hourglass	Processing repeated bottom-up and top-down sampling used in conjunction with intermediate supervision	Limited input-image size
Sun et al. [15]	HRNet	Generating highly information-rich feature outputs through a multi-scale feature fusion process	Limited input image size
Yang et al. [28]	TransPose	Introducing a transformer for key-point detections to yield performance improvements through multi-head self-attention	Computationally expensive

2.3. Attention Mechanism

Attention mechanisms in computer vision aim to selectively focus on the prominent parts of an image to better capture the human visual structure. Several attempts have been made to improve the performance of CNN models in large-scale classification tasks. Residual attention networks improve feature maps through encoder–decoder style attention modules and are very robust to noisy inputs. The attention module reduces the complexity and parameters by dividing calculations into channels and spaces instead of performing calculations in the typical three-dimensional space manner in addition to achieving a significant effect.

The squeeze-and-excitation module [16] proposes an attention module to exploit the relationship between channels. Channel weights are generated through average pooling or max pooling to apply attention to each channel. BAM [17] adds a spatial attention module in addition to the above channel method and places it in the bottleneck to create richer features. CBAM [18] is not only located in each bottleneck of a network but also forms a convolution block to configure the network. In addition, the performance is increased empirically by using the sequential processing method for channel and spatial attention; this method has an empirically better performance than using only the channel unit and has achieved state-of-the-art performance in the classification field.

The self-attention module with such a flexible structure has been applied to many tasks, such as image captioning and visual question answering [29]. The self-attention module is widely used in detection and key-point detection in which spatial information is important [30,31].

Table 3. A summary table of the attention mechanism.

Author	Methodology	Highlights	Limitations
Hu et al. [16]	SENet	Providing a novel architectural unit focusing on the channel relationships on feature maps	Lack of information on pixel-wise relationships
J. Park et al. [17]	BAM	Providing a module which infers an attention map along two separate pathways (channel and spatial)	Computational complexity
S. Woo et al. [18]	CBAM	Introducing a lightweight and general module that can be integrated into any CNN architecture	Computational complexity

3. Proposed Method

3.1. Overview of Gaze Estimation Based on Landmark Features

In this section, we introduce a network structure and a process for extracting a rich and accurate landmark feature vector from an eye image and then estimating a gaze based on it. A series of procedures for estimating a proposed gaze is shown in Figure 2. Eye images can simply be acquired from a single camera. If a frame contains a full-face image, the frame must be cropped to a 160×96 sized image centered on the eye area using the face detection algorithm [32]. The image is converted into black and white image for simple processing. This can enhance the performance output of the infrared camera.

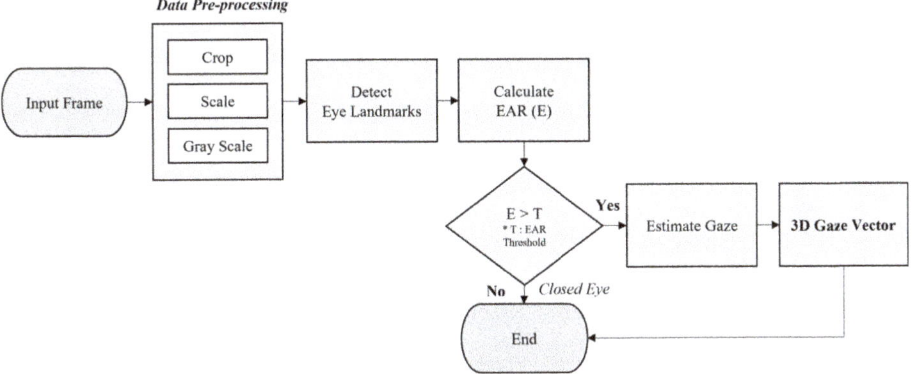

Figure 2. Overall flowchart of our feature-based gaze-estimation system.

To obtain eye-feature vectors from processed images, we selected HRNet as a baseline model that can generate feature maps containing rich information through fusions with various feature maps while maintaining a high image resolution. HRNet showed the best performance in the key-point detection task, proving its utility. We modified HRNet by additionally using the self-attention module CBAM. Channel-wise and spatial-wise weights were applied to infer the most important channels in the 3D feature map and most important spatial points in the channels. Section 5.1 shows that the proposed model achieved higher landmark accuracy than models in previous studies.

The EAR [33] threshold (T), which can be different for each individual, was set using the initial 30 input frames. The EAR ratio value (E) was calculated for each frame; if the calculated EAR ratio value was less than the threshold value, it was judged that there was no need to estimate gaze because the eyes were closed. By reducing false-positive errors, it was possible to proceed with a gaze estimation that had a computational advantage. In some cases, 3D gaze regressions use SVR, but we proceeded by constructing an optimal MLP. The architecture configurations of these models have the advantage of being able to proceed one step when learning.

The most important task of our proposed method was to acquire a high-level landmark feature that affects the EAR ratio and gaze. Before training the model, we were faced with the problem of a lack of a dataset, which is a chronic problem of deep-learning models, adversely affects their training, and can result in over-fitting. Landmark datasets are especially expensive, and only a few datasets include both a gaze and a landmark. To avoid this problem, we used a large set of UnityEyes synthetic data for training the dataset. UnityEyes synthetic data is a dataset that includes annotations, such as rich eye landmarks and gazes, by modeling a 3D eyeball based on an actual eye shape created by using the game engine Unity. The models [4] trained with this synthetic dataset showed good performances and had a lot of information and high resolutions; therefore, they are very suitable for processing and applications.

3.2. Architecture of Proposed Landmark-Detection Model

We used a feature vector for gazes with large amounts of eye landmarks obtained through the model from the input frame that contains eye information from a single camera. To increase the gaze accuracy, it was important to generate a high-level feature, and we set the advancement of the model that extracted a heatmap output most similar to the correct answer as the main goal of this study. Previous studies [12,13] that used eye landmarks as features mainly adopted [14] the production of feature outputs. However, because the feature map is restored through decoding after passing forward from high resolution to low resolution, it is weak in expression learning at a high resolution. Because our model requires extracting more eye landmarks from small-sized eye images than previous studies [13], feature learning at high resolution, which has high sensitivity to positional information in an image space, was necessary. Therefore, we adopted HRNet, which maintains multi-resolution learning (including a high-resolution), as a baseline model.

The basic structure of HRNet consists of 4 steps and 4 stages. Each step creates a feature that doubles the number of channels with half the resolution of the previous step. Each stage consists of a residual block and an exchange unit, and the feature map of each step is processed in parallel. The exchange unit is an information exchange process through fusion between feature maps of each step through fusion, and the second, third, and fourth stages have one, four, and three units, respectively. At the end of each stage, there are feature fusion and transition processes that increase a step by generating a feature map that is half the previous size. Fusion between multi-scale features includes an up-sampling process that uses 1×1 convolution, a nearest-neighbor interpolation in the bottom-up path, and a down-sampling process that uses several 3×3 convolution blocks with strides of 2 in the top-down path.

$$input = \{X_1, X_2, ...X_r\} \qquad output = \{Y_1, Y_2, ...Y_r\}$$

$$Y_k = \sum_{i=1}^{r} F(X_i, k) \quad , \quad F(X_i, k) = \begin{cases} identify\ connection, & if\ i = k \\ up\ sampling, & if\ i < k \\ down\ sampling, & if\ i > k \end{cases} \qquad (i, k \leq r) \qquad (1)$$

Equation (1) describes the feature fusion process. For input $\{X_1, X_2, ...X_r\}$ of different resolutions, output features $\{Y_1, Y_2, ...Y_r\}$ are generated through an element-wise summation of features after down-sampling and up-sampling. r represents resolution numbers; if r is the same, the widths and resolutions of the input and output are the same. At the end of the 4th stage, all step information is concatenated to create the feature block $F_b\{[Y_1^4; Y_2^4; Y_3^4; Y_4^4]\}$ and to head to the prediction head.

To solve the problem of the typically acquired eye image having a small resolution, we introduced an additional residual block layer composed of a 3×3 convolution to the model to create feature (F_o) of the origin resolution that stores the information of the largest resolution. Through the summation of F_o and up-sampled F_b, more spatially accurate features are created.

Because the heat map, which is the final result of the network, requires accurate spatial information for each channel, we applied CBAM, a self-attention technique, to the normal residual and convolution blocks of each stage. Architecture of the modified network is illustrated in Figure 3. These techniques (adding the residual CBAM layer and applying CBAM to all stages of the residual block) improved the landmark-detection performance, which is described in Section 5.1.

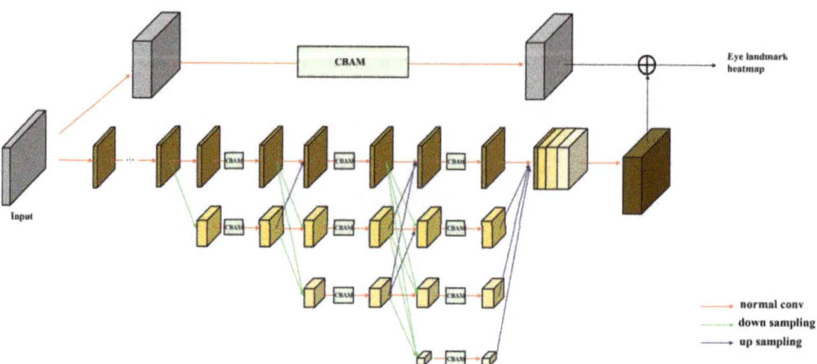

Figure 3. Our landmark-detection network architecture used to extract feature map.

3.3. Network Engineering with the Self-Attention Module CBAM

Attention mechanisms have been widely used for feature selection using multi-modal relationships. Refining the feature maps using attention module helps the network and causes it to perform well and become robust to noisy inputs. Based on empirical results, such as those in [16,17], the CBAM self-attention module has developed rapidly and showed higher accuracy than existing modules in the image classification task through various structure and processing experiments. We judged that the positional information of the refined feature would improve the performance; therefore, we applied the residual block of the network by replacing the CBAM block. The architecture of CBAM is illustrated in Figure 4.

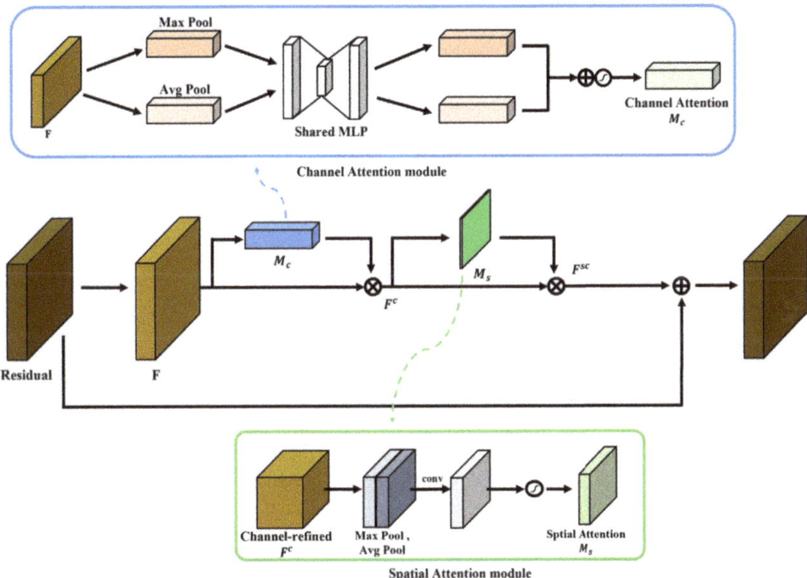

Figure 4. Convolutional block attention module (CBAM) architecture in residual block.

CBAM adds two sub-networks that consist of channel attention and spatial attention networks to the basic residual block. Feature $F \in \mathbb{R}^{C \times H \times W}$ is generated through a 3×3 convolution of the residual, which is the output of the previous block. F goes through the channel attention and spatial attention networks sequentially. First, in the case of channel attention, the two types of channel-wise pooling, that is, max pooling and average pooling, are performed to obtain weight parameters for channels. Feature vectors $F_{max} \in \mathbb{R}^{C \times 1 \times 1}$ and $F_{avg} \in \mathbb{R}^{C \times 1 \times 1}$, generated through pixel-wise pooling, share an MLP that has a bottleneck structure with the advantages of parameter reduction and generalization and are merged using element-wise summation. Finally, the product is normalized using sigmoid function to obtain the meaningful weights $M_c(F) \in \mathbb{R}^{C \times 1 \times 1}$ and generate $F^c \in \mathbb{R}^{C \times H \times W}$ by multiplying $M_c(F)$ and F. The above process is described by using Equation (2).

$$M_c(F) = F_{sigmoid}(MLP(F_{max}) + MLP(F_{avg})),$$
$$F^c = M_c(F) \bigotimes F$$

(2)

Subsequently, using the channel-refined feature (F^c) as an input, $M_s(F^c)$ is generated through the spatial attention module.

$$M_s(F^c) = F_{sigmoid}(Conv_{7\times7}([F_{max}; F_{avg}])),$$
$$F^{sc} = M_c(F^c) \otimes F^c, \tag{3}$$
$$output = Residual \oplus F$$

In Equation (3), spatial weight feature $M_s(F) \in \mathbb{R}^{1 \times H \times W}$ is made by using sequential process pooling, concatenation, 7×7 convolution and normalizing using with sigmoid function, then $F^{sc} \in \mathbb{R}^{C \times H \times W}$ is merged by multiplying $M_s(F^c) \in \mathbb{R}^{1 \times H \times W}$ and F^c. The output of blocks that are merged by using the element-summation residual and F^{sc} is refined with a focus on 'what' and 'where', respectively. Because we applied this module to the residual block of the processing stage in parallel, the output at each stage contains very rich information and encodes channel information at each pixel over all spatial locations due to attention and fusion.

We applied the CBAM module of the additional residual layer and additionally applied the CBAM module to all steps of the stage. We showed performance improvement through the normalized mean error (NME) value, which is a key-point-detection performance value. Detailed outcome indicators are described in Section 5.

3.4. Gaze-Estimation-Based Eye Landmarks with EAR

We estimated gaze vectors based on an eye feature that consists of a total of 50 eye landmarks (1 from an eye center, 1 from an iris center, 16 from an eyelid, and 32 from an iris). We extracted high-accuracy eye landmark localization while optimizing and improving the network. As the quality of the landmark extracted by the network improved, the gaze regression performance also improved empirically. Existing feature-based studies [13,34] mainly used the SVR for gaze regressions. We empirically confirmed that the difference between the SVR and multi-layer perceptron (MLP) performance is very small and that the MLP performance is relatively good. The MLP simply contains two hidden layers and uses Leaky ReLU [35] as an activation function. In addition, when the MLP is used, there is the advantage that landmark detection and gaze estimation are possible in one-stage training. The MLP contains two hidden layers and uses Leaky ReLU as an activation function. The co-ordinates used as the inputs are normalized to the distance between the eye endpoints, and all eye points are translated with respect to the eye center coordinates.

To reduce false positives and increase efficiency, we utilized an EAR value. The EAR value was calculated to decide whether an eye was closed or not using 16 eyelid points. We introduced a new EAR metric based on a method that uses 6 points because we could obtain richer, high-quality eyelid points. Figure 5 shows the measured lengths of an eye using images that include a closed eye. We measured the horizontal length through the p_1 and p_9 points out of a total of 16 points $\{p_1, p_2, ... p_{16}\}$, and the average value of the remaining seven pairs of points $\{(p_2, p_{16}), (p_3, p_{15})...(p_8, p_{10})\}$ was defined as the vertical length. The EAR was calculated using Equation (4).

$$EAR = \frac{\sum_{n=1}^{7} \| p_{n+1} - p_{17-n} \|_1}{7 \| p_1 - p_9 \|_1} \tag{4}$$

Because the EAR varies considerably from user to user, we set the EAR threshold (T) to half the median value after receiving the EARs for the initial 30 frames' inputs. Then, if the measured EAR was smaller than T, the network did not estimate gaze.

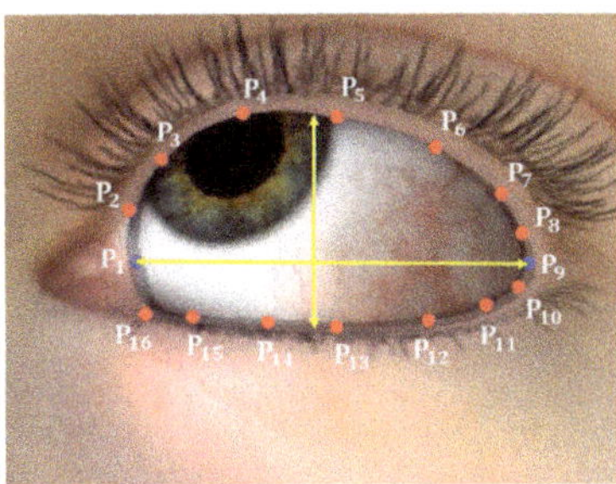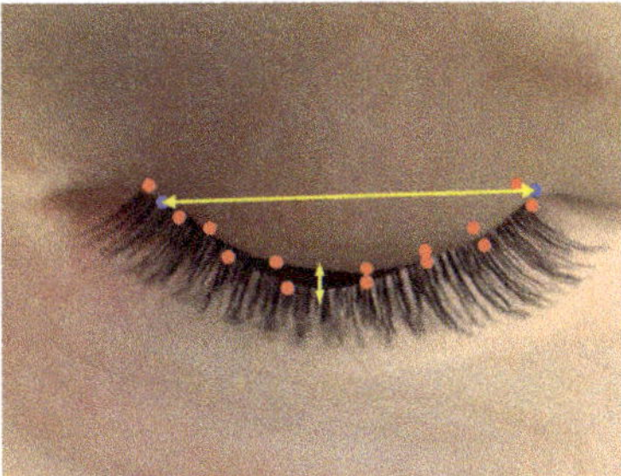

Figure 5. EAR was calculated through the displayed landmark coordinates (from P_1 to P_{16}). The blue dots are used to represent the width of the eye and the red the height. The eye on the right is in a state at which there is no need to estimate the gaze.

3.5. Learning

Two losses, a landmark loss and a gaze loss, were required for the training of our proposed network. The method of regressing the heatmap, which is the probability of the existence of each feature point using a CNN model, has fewer parameters than the method of directly regressing the feature point coordinates and can avoid the problem of over-fitting. However, it is difficult to precisely detect units below the decimal point because heatmap regression acquires integer co-ordinates through an arg-max operation in the process of converting heatmap into coordinates. We used integral regression [36] to properly compensate for the above two shortcomings. The integral regression module removes negative values by applying AB ReLU operation to the heatmap and divides the operation by the total sum to normalize it. As shown in Equation (5), all values of $\hat{H}$ are between 0 and 1 and the total sum becomes 1; therefore, it is defined as a probability distribution. Subsequently, the co-ordinates of each feature point in the heatmap can be obtained through the expected value calculation.

$$\hat{H}_c(x,y) = \frac{F_{ReLU}(H_c(x,y))}{\sum_i \sum_j (H_c(i,j)}$$

$$predicted\ coordinates = \begin{cases} x_c = \sum_i \sum_j i\hat{H}_c(i,j) \\ y_c = \sum_i \sum_j j\hat{H}_c(i,j) \end{cases}$$

$$(5)$$

Therefore, the final landmark cost function consists of the mean squared error (MSE) loss between the output and the ground-truth heatmap, the L1 loss of the ground-truth co-ordinate and the co-ordinates obtained by using the expected value operation. H' is the predicted heatmap, H is the ground-truth heatmap, (x',y') is the co-ordinate predicted through the integral module, and (x,y) is the ground-truth co-ordinate.

$$Loss_{heatmap} = \sum_i \sum_x \sum_y \parallel H'_i(x,y) - H_i(x,y) \parallel_2^2 \quad, Loss_{coordinates} = \sum_i \parallel (x',y') - (x,y) \parallel_1$$

$$(6)$$

$$Loss_{landmark} = Loss_{heatmap} + Loss_{coordinates}$$

To compare each gaze performance, experiments were conducted using several methods. There are two frequently used methods of gaze regression. The first is a method of directly regressing a 3D vector and the second is a method of encoding a 3D normal vector into 2D space pitch (θ) and yaw (φ) regression. The pitch and yaw are the angles between the pupil and the eyeball, which can explain the positional relationship. The positional relationship between an eyeball and a pupil is illustrated in Figure 6. We found that the generalization was better when a 2D angle vector was encoded empirically and cosine distance loss and MSE were used as cost functions, and the best performance was obtained in MSE. (Pitch, yaw), that is, (θ', φ'), is the predicted 2D gaze and (θ, φ) is the ground-truth 2D gaze.

$$pitch(\theta) = \arcsin(y), \quad yaw(\varphi) = \arctan(\frac{x}{z})$$

$$Loss_{gaze} = \parallel (\theta', \varphi') - (\theta, \varphi) \parallel_2^2 \tag{7}$$

We trained our model using a UnityEyes dataset that consists of 80,000 images, and each validation and test used 10,000 images. We used black and white $1 \times 160 \times 96$ images and set batch size to 16. We used the Adam optimizer. The learning schedule followed the settings in [15]. We used pre-trained data on ImageNet. The base learning rate was set as 4×10^{-4} and decreased by every 25 epochs. Specifications of the PC used in the experiment were an Intel Core i9-11900K, 3.5 GHz CPU, and NVIDIA RTX 3090 GPU with 24 GB of memory for training.

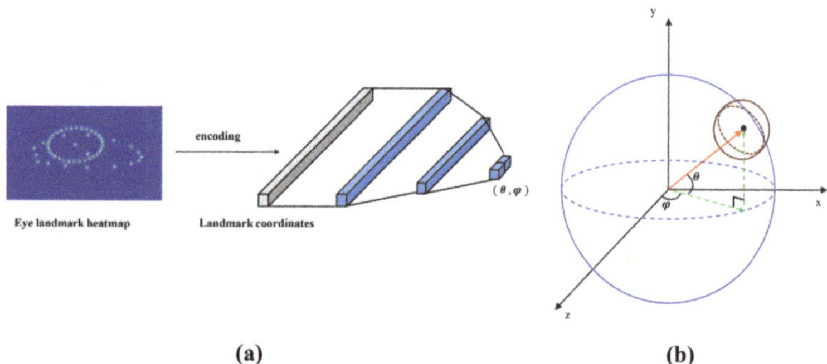

(a)   (b)

Figure 6. (**a**) illustrates the simple network architecture for gaze estimation and (**b**) shows the relationship between the pupil and the eyeball. Gray embedding vector encode the landmarks coordinates. Gaze vector (red) can be explained through a pitch (θ) and yaw (φ).

4. Description of the Dataset

This section describes the dataset used for network training and evaluation. Figure 7 shows the original forms of the utilized datasets.

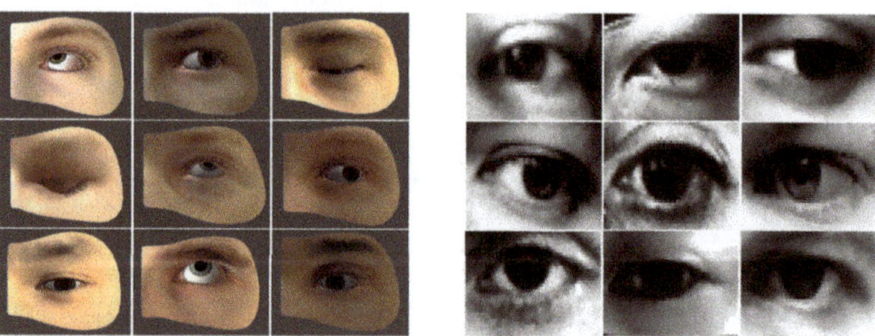

Figure 7. Samples from two datasets: left is UnityEyes and right is MPIIGaze.

4.1. UnityEyes

In a real-world setting, datasets for gaze estimation are very expensive to acquire, do not support eye landmarks, or are very poor; therefore, they are inadequate for training a network. We selected UnityEyes synthetic datasets for training to solve the above problem. UnityEyes creates an eye model by manipulating several parameters using the Unity game engine and provides high-resolution 2D images from the camera position, high-quality 3D eye coordinates, and a 3D gaze vector. We also processed rich annotations and utilized them for network learning. Previous studies [4,13] showed good performance using synthetic datasets.

An eye landmark provided in UnityEyes is presented in Figure 8. A total of 53 eye landmarks consisting of 16 eye edges, 7 caruncles, and 32 iris edges were used. We used all the labeled eye and iris edges while ignoring the caruncles because it was judged that they would have no effect on gaze. Subsequently, the eyes and iris centers, which were mean values of all the eyes and iris edges, were added to configure the ground -truth with a total of 50. It was possible to create a resolution of 640 × 480 up to 4K, and we cropped an 800 × 600 image to a 160 × 96 size.

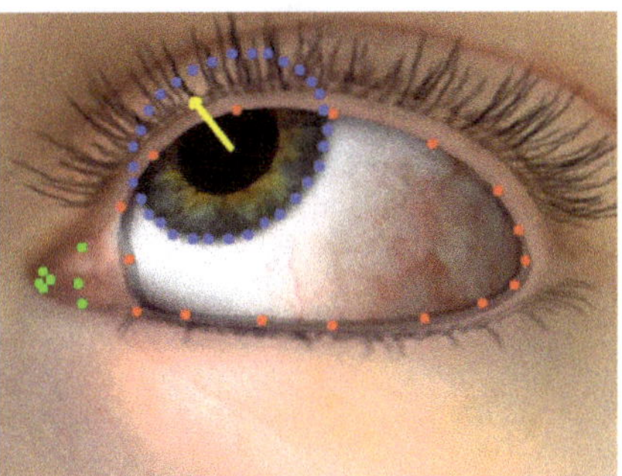

Figure 8. An annotated sample from UnityEyes. The red, green, and blue points are 16 eye edges, 7 caruncles, and 32 iris edges, respectively. The yellow arrow represents the 3D gaze direction.

4.2. MPIIGaze

The MPIIGaze datasets were recorded using a laptop for several months over the daily lives of 15 experimental participants. The datasets were representative evaluations and very suitable for judging the performance of networks in uncontrolled settings. They were proposed in 2015 and include a head pose vector, gaze vector, full-face image, and 60 × 36 normalized image required for evaluations. The datasets also provide eye landmark co-ordinates of six eye edges and one iris center but do not provide enough for within-dataset leave-one-person-out evaluations [11]. That is, the proposed network was trained using the data of 14 participants (3000 images consisting of left and right eyes) and then validated on the data of an excluded person. Thus, we made a labeling tool for rich eye landmarks as one is required by a neural network that is robust against noise.

Figure 9 illustrates an overview of this labeling tool. First, when a user draws a point on both endpoints of an eye, a line connecting the two points and straight lines dividing the line into eight equal parts are created. Subsequently, the remaining eye-edge co-ordinates are obtained by dotting the points at which the eight straight lines and eye region overlap. At this time, a correction effect is applied so that the drawn points lie on the straight lines. When a total of 16 eye edges are completed, the user specifies an elliptical area that can include the iris area. When eight dots are obtained, an ellipse is generated that best contains the iris through the RANSAC algorithm [37]. Users obtain 32 iris edges spaced at regular intervals from the ellipse. The datasets and code on 12 April 2022 are available to the public: https://github.com/OhJaeKwang/Eye_Region_Labeling.

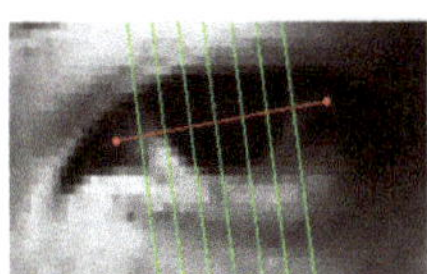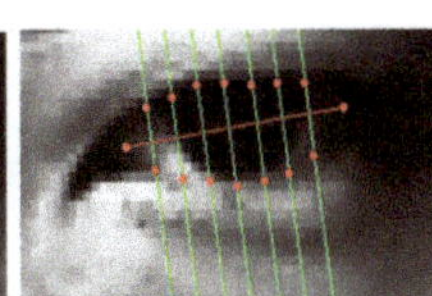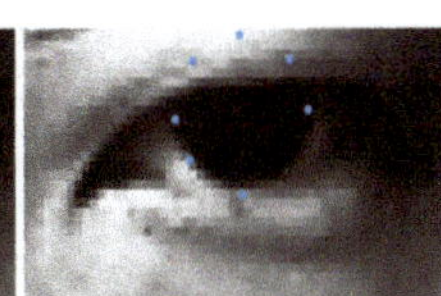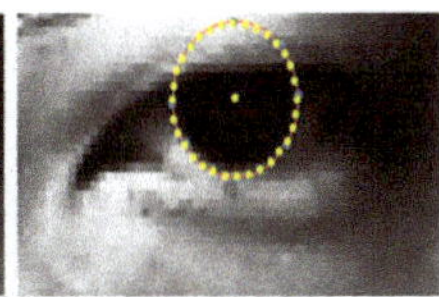

Figure 9. From left to right presents the sequential labeling process. The red and green lines and blue dots are tools for landmark coordinates, and the red and yellow dots represent annotations. Outputs are comprised of 16 eye edges, 32 iris edges, 1 eye center, and 1 iris center.

5. Experiments

In this section, we describe the experiments conducted in this study. The experiments included an evaluation considering the performance of landmark detection and gaze estimation with respect to MPIIGaze (45K) and UnityEyes (10K). The NME and mean angle error (MAE) were adopted as metrics for evaluation.

5.1. Landmark-Detection Accuracy

There are several metrics for evaluating the accuracies of landmarks, but we adopted the NME [38], which is main metric used in facial landmark detection and is the most relevant among them. The NME represents the average Euclidean distance between an estimated landmark position (P') and a corresponding ground truth (P). The NME is calculated using Equation (8), where N is the number of images, L is the number of landmarks, and d is defined as average eye width of a test set for the normalization factor.

$$NME = \frac{1}{N} \sum_{i=1}^{N} \frac{\sum_{j=1}^{L} \| P'_{i,j} - P_{i,j} \|_2^2}{L \times d} \tag{8}$$

We compared our approach to a baseline model. Two approaches were introduced as follows. The first added a CBAM residual layer and the second applied CBAM to convolution blocks of all stages. We used the model parameters trained on UnityEyes. The landmark detection results of our approaches and the baseline model (HRNet-W18) with respect to MPIIGaze and UnityEyes are shown in Table 4. HRNet-w18 and HRNet-w32 are lightweight models of HRNet, and 18 and 32 indicated the channel multiples of the last stage. In the results, each approach showed a better performance than the existing model, and the final model achieved an approximately 4% higher NME score compared to HRNet-W18 on all datasets. Graphs showing the ratios of the test sets according to the NME value are presented in Figure 10. Similarly, the AUC [39] value (the area under the curve) demonstrated that the two approaches using self-attention improved performance.

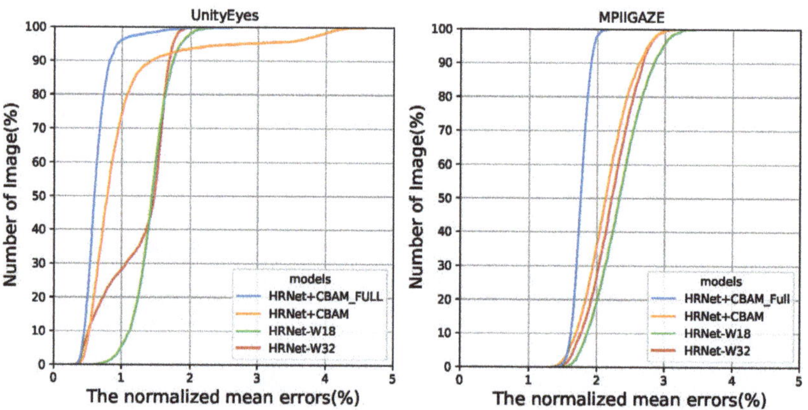

Figure 10. Comparisons of the cumulative error distribution curves of the test datasets. We compared our method with baseline approaches (HRNet). HRNet+CBAM and HRNet+CBAM_FULL denote adding a residual CBAM layer and applying CBAM to all stages of the residual blocks, respectively.

Because the MPIIGaze dataset before pre-processing consisted of a very low resolution of 60×36, we interpolated it with a 160×96 dataset and processed the result; we judged that the performance with respect to MPIIGaze was inferior to UnityEyes due to noise generated during this process, problems of poor quality, and reliability of labeling.

Table 4. The approaches that applied a CBAM module improved the quantitative metric.

Method	UnityEyes		MPIIGaze	
	NME (%) ↓	AUC ↑	NME (%) ↓	AUC ↑
HRNet-W18	7.21	75.95	11.71	60.93
HRNet-W32	6.69	78.79	11.13	62.91
HRNet + CBAM	4.95	83.49	10.72	64.20
HRNet + CBAM_FULL	3.18	89.39	8.82	70.59

5.2. Gaze Estimation Accuracy

Our method, which showed the best landmark performance, achieved an angle error of 1.7 for 10,000 UnityEyes test sets. Subsequently, we compared various systems for within-dataset evaluation (leave-one-person-out strategy) to the MPIIGaze dataset using an MAE that represents the differences in the angles of two unit vectors. The results of the models evaluated using the MPIIGaze dataset, usage techniques, and information used as inputs are included in Table 5. Our method achieved a competitive degree of error in the above experiment. Fine-tuning the model parameters pre-trained on UnityEyes using MPIIGaze improved the performance by approximately 6.80% (from 4.64° to 4.32°), and our approach surpassed the baseline method (from 4.60° to 4.32°). We improved the performance by approximately 6.04% compared to the baseline model. Additionally, unlike the appearance method, our method was less constrained by registration conditions and had better usability in that we could create high-level landmarks. This result showed that the performance improvement of landmark detection had an effect on gaze regression. We show the qualitative predictions of our gaze estimation system with respect to UnityEyes and MPIIGaze in Figure 11. It was observed that it acquired high-level features even for noisy MPIIGaze data and had good gaze accuracy.

Table 5. Comparing the MAE, representation, and registration of several methods evaluated using MPIIGaze. (*: baseline method).

Method	MAE (°)	Representation	Registration
RF [10]	7.99°	Appearance	Eyes and head pose
Mnist [11]	6.30°	Appearance	Single eye
GazeNet [8]	5.83°	Appearance	Single eye
AR-Net [9]	5.65°	Appearance	Eyes
ARE-Net [9]	5.02°	Appearance	Eyes
* S. Park et al. [13]	4.60°	Feature and gaze regression network	Single eye
S. Park et al. [12]	4.50°	Appearance	Single eye
FARE-Net [6]	4.41°	Appearance	Face, eyes
Ours	4.32°	Feature and gaze regression network	Single eye

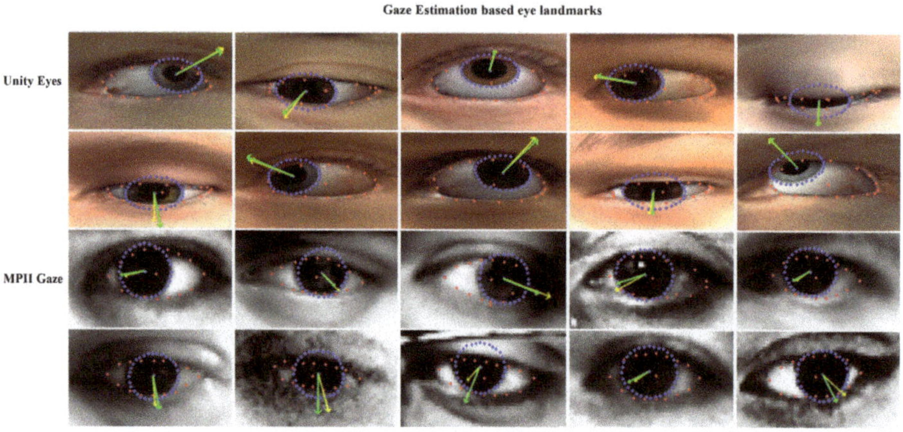

Figure 11. Results of our gaze estimation system with respect to UnityEyes and MPIIGaze test sets. The red, blue points represent the iris edge and the eye edges, respectively. Ground-truth gaze is represented by green arrows and predicted gaze is represented by yellow arrows.

6. Discussion

Applications that utilize gaze information practically and visually provide novelty and satisfaction to users, so it is essential to improve the accuracy of predicted information. To achieve a performance improvement, we conducted a study by adapting the feature-based method, which is better for generalizing than the appearance-based method. In prior work, features were usually hand-crafted for gaze using image-processing and model-fitting techniques. However, because these approaches make assumptions about geometry, such as the 3D eyeball and 3D head coordinates, they are sensitive to noise in uncontrolled real-world images.

In this study, we proposed a gaze estimation method using a more accurate and detailed eye region where eye landmarks represent the locations of the iris and eyelid. We used the UnityEyes dataset, which has high quality annotation that helped the representation learning of our network.

Since we assumed that the accuracy of gaze estimation increases as the confidence of the landmarks intended to be used as features increases, we tried to develop an advanced landmark-detection model. We also assumed that the feature map of the layer should represent meaningful location information and proposed a method combining the self-attention module with the model. The first results suggested that adding the self-attention module improves the inference accuracy. In particular, the best performance improvements were seen with negligible overheads when the module was applied to all layers. Moreover, since the inference accuracy for low-quality MPIIGaze had increased, it was shown to be robust to the noise of the input data. Then, we were able to confirm that the performance of landmark and gaze were proportional through considering the second result. We obtained a meaningful study, but there difficulties were encountered during the study.

We had to train the models on the real-world MPIIGaze dataset for the evaluation. Unavoidably, in order for our network to learn, we needed to take landmark annotations unconditionally. However, MPIIGaze didn't have as many as we needed. Consequently, we made a labeling tool and labeled MPIIGaze (45K) using it. An unsupervised domain adaptation [40] can solve this limitation. It does not require annotations on the target domain and is used for only feature training for a target. Using generative adversarial networks (GAN), the method for a fusion between datasets from different domains might help a model to perform transfer learning well [41]. To alleviate the limitation, the hope is that our work will apply these skills to our method.

7. Conclusions

In this study, we proposed a feature-based gaze system that achieved a higher accuracy than existing models trained on the same datasets by introducing a network to extract high-level landmarks. Contrary to the existing methods, we predicted a heatmap with richer representations from the transferred multi-scale features using HRNet to obtain more accurate and more spatially precise eye features. Moreover, we achieved the best performance improvement by applying a self-attention module that emphasized meaningful features in the principal dimensions, which were the channel and spatial axes of the feature map, in addition to achieving efficient computational and parameter overheads. Using UnityEyes, which supports a high-level annotation and a high resolution, we were able to extract more and greater landmarks, and these richer landmarks resulted in a competitive gaze accuracy for a within-dataset evaluation with respect to MPIIGaze. Additionally, our method had less restrictive registration conditions and great utility in providing landmarks.

During the experiment, we found that the transfer learning of the model through various real-world gaze datasets was superior to the results of the model trained with only UnityEyes. However, our model required numerous landmark annotations, and there was no dataset that satisfied this requirement. To solve this problem, we used a labeling tool in this study. However, in the next study, we plan to apply the unsupervised domain adaptation technique to optimize the model using UnityEyes and real-environment datasets without using a key-point annotation simultaneously.

Author Contributions: Conceptualization, J.Y. and S.K.; investigation, J.O.; methodology, J.O.; resources, J.O.; validation, Y.L.; visualization, Y.L.; writing—original draft preparation, Y.L.; writing—review and editing, J.O. All authors have read and agreed to the published version of the manuscript.

Funding: This research was supported by the Basic Science Research Program through the National Research Foundation of Korea (NRF), funded by the Ministry of Education (NRF-2020R1F1A1069079). This work was supported by the National Research Foundation of Korea (NRF) grant funded by the Korean government (MSIT). The present research has been conducted by the Research Grant of Kwangwoon University in 2021.

Institutional Review Board Statement: Not applicable.

Informed Consent Statement: Not applicable.

Data Availability Statement: Not applicable.

Conflicts of Interest: The authors declare no conflict of interest.

References

1. Wu, M.; Louw, T.; Lahijanian, M.; Ruan, W.; Huang, X.; Merat, N.; Kwiatkowska, M. Gaze-based intention anticipation over driving manoeuvres in semi-autonomous vehicles. In Proceedings of the 2019 IEEE/RSJ International Conference on Intelligent Robots and Systems (IROS) IEEE, Macao, China, 4–8 November 2019; pp. 6210–6216.
2. Ahn, S.; Lee, G. Gaze-assisted typing for smart glasses. In Proceedings of the 32nd Annual ACM Symposium on User Interface Software and Technology, New Orleans, LA, USA, 20–23 October 2019; pp. 857–869.
3. Kim, J.; Lee, Y.; Lee, S.; Kim, S.; Kwon, S. Implementation of Kiosk-Type System Based on Gaze Tracking for Objective Visual Function Examination. *Symmetry* **2022**, *14*, 499. [CrossRef]
4. Wood, E.; Baltrušaitis, T.; Morency, L.P.; Robinson, P.; Bulling, A. Learning an appearance-based gaze estimator from one million synthesised images. In Proceedings of the Ninth Biennial ACM Symposium on Eye Tracking Research & Applications, Charleston, SC, USA, 14–17 March 2016; pp. 131–138.
5. Fischer, T.; Chang, H.J.; Demiris, Y. Rt-gene: Real-time eye gaze estimation in natural environments. In Proceedings of the European Conference on Computer Vision (ECCV), Munich, Germany, 8–14 September 2018; pp. 334–352.
6. Cheng, Y.; Zhang, X.; Lu, F.; Sato, Y. Gaze estimation by exploring two-eye asymmetry. *IEEE Trans. Image Process.* **2020**, *29*, 5259–5272. [CrossRef] [PubMed]
7. Biswas, P. Appearance-Based gaze estimation using attention and difference mechanism. In Proceedings of the IEEE/CVF Conference on Computer Vision and Pattern Recognition, Nashville, TN, USA, 20–25 June 2021; pp. 3143–3152.
8. Zhang, X.; Sugano, Y.; Fritz, M.; Bulling, A. Mpiigaze: Real-world dataset and deep appearance-based gaze estimation. *IEEE Trans. Pattern Anal. Mach. Intell.* **2017**, *41*, 162–175. [CrossRef] [PubMed]

9. Cheng, Y.; Lu, F.; Zhang, X. Appearance-based gaze estimation via evaluation-guided asymmetric regression. In Proceedings of the European Conference on Computer Vision (ECCV), Munich, Germany, 8–14 September 2018; pp. 100–115.

10. Sugano, Y.; Matsushita, Y.; Sato, Y. Learning-by-synthesis for appearance-based 3d gaze estimation. In Proceedings of the IEEE Conference on Computer Vision and Pattern Recognition, Columbus, OH, USA, 23–28 June 2014; pp. 1821–1828.

11. Zhang, X.; Sugano, Y.; Fritz, M.; Bulling, A. Appearance-based gaze estimation in the wild. In Proceedings of the IEEE Conference on Computer Vision and Pattern Recognition, Boston, MA, USA, 7–12 June 2015; pp. 4511–4520.

12. Park, S.; Spurr, A.; Hilliges, O. Deep pictorial gaze estimation. In Proceedings of the European Conference on Computer Vision (ECCV), Munich, Germany, 8–14 September 2018; pp. 721–738.

13. Park, S.; Zhang, X.; Bulling, A.; Hilliges, O. Learning to find eye region landmarks for remote gaze estimation in unconstrained settings. In Proceedings of the 2018 ACM Symposium on Eye Tracking Research & Applications, Warsaw, Poland, 14–17 June 2018; pp. 1–10.

14. Newell, A.; Yang, K.; Deng, J. Stacked hourglass networks for human pose estimation. In *European Conference on Computer Vision*; Springer: Berlin/Heidelberg, Germany, 2016; pp. 483–499.

15. Sun, K.; Xiao, B.; Liu, D.; Wang, J. Deep high-resolution representation learning for human pose estimation. In Proceedings of the IEEE/CVF Conference on Computer Vision and Pattern Recognition, Long Beach, CA, USA, 15–20 June 2019; pp. 5693–5703.

16. Hu, J.; Shen, L.; Sun, G. Squeeze-and-excitation networks. In Proceedings of the IEEE Conference on Computer Vision and Pattern Recognition, Salt Lake City, UT, USA, 18–23 June 2018; pp. 7132–7141.

17. Park, J.; Woo, S.; Lee, J.Y.; Kweon, I.S. A simple and light-weight attention module for convolutional neural networks. *Int. J. Comput. Vis.* **2020**, *128*, 783–798. [CrossRef]

18. Woo, S.; Park, J.; Lee, J.Y.; Kweon, I.S. Cbam: Convolutional block attention module. In Proceedings of the European Conference on computer vision (ECCV), Munich, Germany, 8–14 September 2018; pp. 3–19.

19. Valenti, R.; Sebe, N.; Gevers, T. What are you looking at? *Int. J. Comput. Vis.* **2012**, *98*, 324–334. [CrossRef]

20. Manolova, A.; Panev, S.; Tonchev, K. Human gaze tracking with an active multi-camera system. In Proceedings of the International Workshop on Biometric Authentication, Sofia, Bulgaria, 23–24 June 2014; Springer: Berlin, Germany, 2014; pp. 176–188.

21. Lai, C.C.; Shih, S.W.; Hung, Y.P. Hybrid method for 3-D gaze tracking using glint and contour features. *IEEE Trans. Circuits Syst. Video Technol.* **2014**, *25*, 24–37. [CrossRef]

22. Wood, E.; Baltrusaitis, T.; Zhang, X.; Sugano, Y.; Robinson, P.; Bulling, A. Rendering of eyes for eye-shape registration and gaze estimation. In Proceedings of the IEEE International Conference on Computer Vision, Washington, DC, USA, 7–13 December 2015; pp. 3756–3764.

23. Smola, A.J.; Schölkopf, B. A tutorial on support vector regression. *Stat. Comput.* **2004**, *14*, 199–222. [CrossRef]

24. Bernard, V.; Wannous, H.; Vandeborre, J.P. Eye-Gaze Estimation using a Deep Capsule-based Regression Network. In Proceedings of the 2021 International Conference on Content-Based Multimedia Indexing (CBMI), Lille, France, 28–30 June 2021; pp. 1–6.

25. Toshev, A.; Szegedy, C. Deeppose: Human pose estimation via deep neural networks. In Proceedings of the IEEE Conference on Computer Vision and Pattern Recognition, Columbus, OH, USA, 23–28 June 2014; pp. 1653–1660.

26. Luvizon, D.C.; Tabia, H.; Picard, D. Human pose regression by combining indirect part detection and contextual information. *Comput. Graph.* **2019**, *85*, 15–22. [CrossRef]

27. Wei, S.E.; Ramakrishna, V.; Kanade, T.; Sheikh, Y. Convolutional pose machines. In Proceedings of the IEEE Conference on Computer Vision and Pattern Recognition, Las Vegas, NV, USA, 27–30 June 2016; pp. 4724–4732.

28. Yang, S.; Quan, Z.; Nie, M.; Yang, W. Transpose: Keypoint localization via transformer. In Proceedings of the IEEE/CVF International Conference on Computer Vision, Montreal, BC, Canada, 11–17 October 2021; pp. 11802–11812.

29. Guo, L.; Liu, J.; Zhu, X.; Yao, P.; Lu, S.; Lu, H. Normalized and Geometry-Aware Self-Attention Network for Image Captioning. In Proceedings of the IEEE/CVF Conference on Computer Vision and Pattern Recognition (CVPR), Seattle, WA, USA, 13–19 June 2020.

30. Perreault, H.; Bilodeau, G.A.; Saunier, N.; Héritier, M. Spotnet: Self-attention multi-task network for object detection. In Proceedings of the 2020 17th Conference on Computer and Robot Vision (CRV), Ottawa, ON, Canada, 13–15 May 2020; pp. 230–237.

31. Santavas, N.; Kansizoglou, I.; Bampis, L.; Karakasis, E.; Gasteratos, A. Attention! A lightweight 2d hand pose estimation approach. *IEEE Sensors J.* **2020**, *21*, 11488–11496. [CrossRef]

32. Viola, P.; Jones, M. Rapid object detection using a boosted cascade of simple features. In Proceedings of the 2001 IEEE Computer Society Conference on Computer Vision and Pattern Recognition, CVPR 2001, Kauai, HI, USA, 8–14 December 2001; Volume 1, p. I.

33. Cech, J.; Soukupova, T. Real-Time eye blink detection using facial landmarks. *Cent. Mach. Perception, Dep. Cybern. Fac. Electr. Eng. Czech Tech. Univ. Prague* **2016**, 1–8.

34. Yu, S. Harr Feature Cart-Tree Based Cascade Eye Detector Homepage. Available online: http://yushiqi.cn/research/eyedetection (accessed on 14 April 2022).

35. Dubey, A.K.; Jain, V. Comparative study of convolution neural network's relu and leaky-relu activation functions. In *Applications of Computing, Automation and Wireless Systems in Electrical Engineering*; Springer: Berlin/Heidelberg, Germany, 2019; pp. 873–880.

36. Sun, X.; Xiao, B.; Wei, F.; Liang, S.; Wei, Y. Integral human pose regression. In Proceedings of the European Conference on Computer Vision (ECCV), Munich, Germany, 8–14 September 2018, pp. 529–545.
37. Fischler, M.A.; Bolles, R.C. Random sample consensus: A paradigm for model fitting with applications to image analysis and automated cartography. *Commun. ACM* **1981**, *24*, 381–395. [CrossRef]
38. Xu, Z.; Li, B.; Geng, M.; Yuan, Y.; Yu, G. Anchorface: An anchor-based facial landmark detector across large poses. *AAAI* **2021**, *1*, 3.
39. Kumar, A.; Marks, T.K.; Mou, W.; Wang, Y.; Jones, M.; Cherian, A.; Koike-Akino, T.; Liu, X.; Feng, C. LUVLi Face Alignment: Estimating Landmarks' Location, Uncertainty, and Visibility Likelihood. In Proceedings of the IEEE/CVF Conference on Computer Vision and Pattern Recognition, Seattle, WA, USA, 13–19 June 2020; pp. 8236–8246.
40. Jiang, J.; Ji, Y.; Wang, X.; Liu, Y.; Wang, J.; Long, M. Regressive domain adaptation for unsupervised keypoint detection. In Proceedings of the IEEE/CVF Conference on Computer Vision and Pattern Recognition, Nashville, TN, USA, 20–25 June 2021; pp. 6780–6789.
41. Shrivastava, A.; Pfister, T.; Tuzel, O.; Susskind, J.; Wang, W.; Webb, R. Learning From Simulated and Unsupervised Images Through Adversarial Training. In Proceedings of the IEEE Conference on Computer Vision and Pattern Recognition (CVPR), Honolulu, HI, USA, 21–26 July 2017.

Article

Online Self-Calibration of 3D Measurement Sensors Using a Voxel-Based Network

Jingyu Song and Joonwoong Lee *

Department of Industrial Engineering, Chonnam National University, 77 Yongbong-ro, Buk-gu, Gwangju 61186, Korea
* Correspondence: joonlee@chonnam.ac.kr

Abstract: Multi-sensor fusion is important in the field of autonomous driving. A basic prerequisite for multi-sensor fusion is calibration between sensors. Such calibrations must be accurate and need to be performed online. Traditional calibration methods have strict rules. In contrast, the latest online calibration methods based on convolutional neural networks (CNNs) have gone beyond the limits of the conventional methods. We propose a novel algorithm for online self-calibration between sensors using voxels and three-dimensional (3D) convolution kernels. The proposed approach has the following features: (1) it is intended for calibration between sensors that measure 3D space; (2) the proposed network is capable of end-to-end learning; (3) the input 3D point cloud is converted to voxel information; (4) it uses five networks that process voxel information, and it improves calibration accuracy through iterative refinement of the output of the five networks and temporal filtering. We use the KITTI and Oxford datasets to evaluate the calibration performance of the proposed method. The proposed method achieves a rotation error of less than $0.1°$ and a translation error of less than 1 cm on both the KITTI and Oxford datasets.

Keywords: online self-calibration; convolutional neural network; voxel information

Citation: Song, J.; Lee, J. Online Self-Calibration of 3D Measurement Sensors Using a Voxel-Based Network. *Sensors* **2022**, *22*, 6447. https://doi.org/10.3390/s22176447

Academic Editor: Jing Tian

Received: 8 August 2022
Accepted: 23 August 2022
Published: 26 August 2022

Publisher's Note: MDPI stays neutral with regard to jurisdictional claims in published maps and institutional affiliations.

1. Introduction

Multi-sensor fusion is performed in many fields, such as autonomous driving and robotics. A single sensor does not guarantee reliable recognition in complex and varied scenarios [1]. Therefore, it is difficult to cope with various autonomous driving situations using only one sensor. Conversely, fusing two or more sensors supports reliable environmental perception around the vehicle. In multi-sensor fusion, one sensor compensates for the shortcomings of the other sensor [2]. In addition, multi-sensor fusion expands the detection range and improves the measurement density compared with using a single sensor [3]. Studies based on multi-sensor fusion include 3D object detection, road detection, semantic segmentation, object tracking, visual odometry, and mapping [4–9]. Moreover, most large datasets that are built for autonomous driving research [10–13] provide data measured by at least two different sensors. Importantly, multi-sensor fusion is greatly affected by the calibration accuracy of the sensors used. While the vehicle is driving, the pose or position of the sensors mounted on the vehicle may change for various reasons. Therefore, for multi-sensor fusion, it is essential to perform the online calibration of sensors to accurately recognize changes in sensor pose or changes in the positions of the sensors.

Extensive academic research on multi-sensor calibration has been performed [2,3]. Many traditional calibration methods [14–17] use artificial markers, including checkerboards, as calibration targets. The target-based calibration algorithms are not suitable for autonomous driving because they involve processes that require manual intervention. Some of the calibration methods currently used focus on fully automatic and targetless online self-calibration [3,18–24]. However, most online calibration methods perform calibration only when certain conditions are met, and their calibration accuracy is not as high as the target-based offline methods [1]. The latest online calibration methods [1,2,25–27] based on deep learning use optimization through gradient descent, large-scale datasets, and CNNs to overcome the limitations of the previous online methods. In particular, the latest research based on CNNs has shown suitable results. Compared with previous methods, CNN-based online self-calibration methods do not require strict conditions and provide excellent calibration accuracy when they are run online.

Many CNN-based LiDAR-camera calibration methods use an image for calibration. In this case, the point cloud of the LiDAR is projected onto the image. Then, 2D convolution kernels are used to extract the features of the inputs.

In this study, we propose a CNN-based multi-sensor online self-calibration method. This method estimates the values of six parameters that describe rotation and translation between sensors that are capable of measuring 3D space. The combinations of sensors that are subject to calibration in our proposed method are: a LiDAR and stereo camera and a LiDAR and LiDAR. One of the two sensors is set as the reference sensor and the other as the target sensor. In the combination of LiDAR and stereo camera, the stereo camera is set as the reference sensor.

The CNN we propose is a network that uses voxels instead of using image features. Therefore, we convert the stereo image into 3D points called pseudo-LiDAR points to feed the stereo image into this network. Pseudo-LiDAR points and actual LiDAR points are expressed in voxel spaces through voxelization. Then, 3D convolution kernels are applied to the voxels to generate features that can be used for calibration parameter regression. In particular, the attention mechanism [28] included in our proposed network confirms the correlation between the input information of the two sensors. The research fields that use voxels are diverse, including shape completion, semantic segmentation, multi-view stereoscopic vision, object detection, etc. [29–32].

The amount of data in public datasets is insufficient to perform online self-calibration. Therefore, existing studies have assigned random deviations to the values of known parameters and have evaluated the performance of online self-calibration based on how accurately the algorithm proposed in the respective study predicts this deviation. This approach is commonly referred to as miscalibration [1,2,25]. To sample the random deviation, we choose the rotation range and translation range as $\pm20°$ and ±1.5 m, respectively, as in [1]. In this study, we train five networks on a wide range of miscalibrations and apply iterative refinement to the outputs of the five networks and temporal filtering over time to increase the calibration accuracy. The KITTI dataset [10] and Oxford dataset [12] are used to conduct the research of the proposed method. The KITTI dataset is used for online LiDAR-stereo camera calibration, and the Oxford dataset is used for online LiDAR-LiDAR calibration.

The rest of this paper is organized as follows. Section 2 provides an overview of existing calibration studies. Section 3 describes the proposed method. Section 4 presents the experimental results for the proposed method, and Section 5 draws conclusions.

2. Related Work

This section provides a brief overview of traditional calibration methods. In addition, we introduce how CNN-based calibration methods have been improved. Specifically, the calibrations covered in this section are LiDAR-camera calibration and LiDAR-LiDAR calibration.

2.1. Traditional Methods

Traditional methods of calibration that use targets mainly use artificial markers. One example is the LiDAR-camera calibration method that uses a polygonal planar board [14]. This method first finds the vertices of the planar board in the image and in the LiDAR point cloud. Then, the corresponding points between the vertices of the image and the vertices of the point cloud are searched, and a linear equation is formulated. Finally, this method uses singular value decomposition to solve the linear equation and obtain calibration parameters. Another example is the LiDAR-camera calibration method that uses a planar chessboard [15]. The edge information in an image and the Perspective-n-Point algorithm are used to find the plane of the chessboard that appears in the image. Then, distance filtering and a random sample consensus (RANSAC) algorithm are used to obtain the chessboard plane information from the LiDAR point cloud. After obtaining the plane information, this method aligns the normal vectors of the planes with obtaining the rotation parameters between LiDAR and the camera. The translation parameters between LiDAR and camera are calculated by minimizing the distance between the plane searched from the LiDAR points, and the rotated plane searched from the image. Similarly, the algorithm in [16] is an automatic LiDAR-LiDAR calibration that uses planar surfaces. This method uses three planes. The planes are obtained through a RANSAC algorithm. The calibration of LiDAR-LiDAR is formulated as a nonlinear optimization problem by minimizing the distance between corresponding planes. This approach adopts the Levenberg-Marquardt algorithm for nonlinear optimization. Another LiDAR-LiDAR calibration in [17] uses two poles plastered with retro-reflective tape to easily identify them in the point cloud. This method first uses a threshold to find the reflected points on the pole. The searched point cloud is expressed as a linear equation that represents a line. The points and the linear equation are then used to solve the least squares problem. The data obtained by solving the least squares problem are the calibration parameters.

Among the targetless online LiDAR-camera calibration methods, some methods use edge information [18,19]. In these methods, when the LiDAR-camera calibration is correct, the edges of the depth map of the LiDAR points are projected onto the image, and the edges of the image are naturally aligned. As another example, there is a method that uses the correlation between sensor data, as in [20,21]. The method in [20] uses the reflectance information from the LiDAR and the intensity information from the camera for calibration. According to this method, the correlation between reflectivity and intensity is maximized when the LiDAR-camera calibration is accurate. The method in [21] uses the surface normal of the LiDAR points and the intensity of image pixels for calibration. There is also a method based on the hand-eye calibration framework [22], which estimates the motion of the LiDAR and camera, respectively, and uses this information for calibration. A targetless online LiDAR-LiDAR calibration is introduced in [23]. This method first performs a rough calibration from an arbitrary initial pose. Then, the calibration parameters are corrected through an iterative closest point algorithm and are further optimized using an octree-based method. Other LiDAR-LiDAR calibration methods are presented in [3,24]. These methods are based on the hand-eye calibration framework, and the motion of each LiDAR is estimated. This information is used for calibration.

2.2. CNN-Based Methods

RegNet [25], the first CNN-based online LiDAR-camera self-calibration method, adopted a three-step convolution consisting of feature extraction, feature matching, and global regression. RegNet uses the decalibration of a given point cloud to train the proposed CNN and also uses five networks to predict the six-degree of freedom (6-DoF) extrinsic parameters for five different decalibration ranges.

CalibNet [5], the CNN-based online LiDAR-camera self-calibration method, proposed a geometrically supervised deep network that was capable of automatically predicting the 6-DoF extrinsic parameters. The end-to-end training is performed by maximizing the photometric and geometric consistencies. Here, photometric consistency is obtained between two depth maps constructed by projecting a given point cloud onto the input image with the 6-DoF parameters predicted by the network and the ground-truth 6-DoF parameters. Similarly, geometric consistency is calculated between two kinds of 3D points obtained by transforming the point cloud into 3D space with the predicted 6-DoF parameters and the ground-truth 6-DoF parameters.

LCCNet [1], which represents a significant improvement over previous CNN-based methods, is a CNN-based online LiDAR-camera self-calibration method. This network considers the correlation between the RGB image features and the depth image projected from point clouds. Additional CNN-based online LiDAR-camera self-calibration methods are presented in [26,27]. They utilize the semantic information extracted by CNN to perform more robust calibration even under changes in lighting and noise [26].

To the best of our knowledge, no deep learning-based LiDAR-LiDAR or LiDAR-stereo camera online self-calibration method has yet been reported. In this paper, we propose, for the first time, a deep learning-based method that is capable of online self-calibration between such sensor combinations.

3. Methodology

This section describes the preprocessing of stereo images and LiDAR point clouds, the structure of our proposed network, the loss function for network training, and the postprocessing of the network output. These descriptions are commonly applied to the calibration of the LiDAR-stereo camera and LiDAR-LiDAR combinations. We chose a LiDAR as the target sensor in the LiDAR-stereo camera and LiDAR-LiDAR combinations, and the rest of the sensors as the reference sensor.

3.1. Preprocessing

In order to perform online self-calibration with the network we designed, several processes, including data preparation, were performed in advance. This section describes these processes. We assume that sensors targeted for online calibration are capable of 3D measurement. Therefore, we use point clouds that are generated by these sensors. In the LiDAR-LiDAR combination, this premise is satisfied, but in the case of the LiDAR-stereo camera combination, this premise is not satisfied, so we obtain a 3D point cloud from the stereo images. The conversion of the stereo depth map to 3D points and the removal of the 3D points, which are covered in the next two subsections, are not required for the LiDAR-LiDAR combination.

3.1.1. Conversion of Stereo Depth Map to 3D Points

A depth map is built from stereo images through stereo matching. In this paper, we obtain the depth map using the method in [33] that implements semi-global matching [34]. This depth map composed of disparities is converted to 3D points, which are called pseudo-LiDAR points, as follows:

$$P = \begin{bmatrix} X & Y & Z \end{bmatrix} = \begin{bmatrix} \frac{base \cdot (u - c_u)}{disp} & \frac{base \cdot (v - c_v)}{disp} & \frac{f_u \cdot base}{disp} \end{bmatrix} \tag{1}$$

where c_u, c_v, and f_u are the camera intrinsic parameters, u and v are pixel coordinates, $base$ is the baseline distance between cameras, and $disp$ is the disparity obtained from the stereo matching.

3.1.2. Removal of Pseudo-LiDAR Points

The pseudo-LiDAR points are too many in number compared with the points measured by a LiDAR. Therefore, we, in this paper, reduce the quantity of pseudo-LiDAR points through a spherical projection, which is implemented using the method presented in [35] as follows:

$$\begin{bmatrix} p \\ q \end{bmatrix} = \begin{bmatrix} \frac{1}{2}(1 - \text{atan}(-X, Z)/\pi) \cdot w \\ (1 - (\text{asin}(-Y \cdot r^{-1}) + f_{up}) \cdot f^{-1}) \cdot h \end{bmatrix} \tag{2}$$

where (X, Y, Z) are 3D coordinates of a pseudo-LiDAR point, (p, q) are the angular coordinates, (h, w) are the height and width of the desired projected 2D map, r is the range of each point, and $f = f_{up} + f_{down}$ is the vertical field of view (FOV) of the sensor. We set f_{up} to 3° and f_{down} to −25°. Here, the range 3° to −25° is the vertical FOV of the LiDAR used to build the KITTI benchmarks [10]. The pseudo-LiDAR points become a 2D image via this spherical projection. Multiple pseudo-LiDAR points can be projected onto a single pixel in the 2D map. In this case, only the last projected pseudo-LiDAR point is left, and the previously projected pseudo-LiDAR points are removed.

3.1.3. Setting of Region of Interest

Because the FOVs of the sensors used are usually different, we determine the region of interest (ROI) of each sensor and perform calibration only with data belonging to this ROI. However, the ROI cannot be determined theoretically but can only be determined experimentally. We determine the ROI of the sensors by looking at the distribution of data acquired with the sensors.

We provide an example of setting the ROI using data provided by the KITTI [10] and Oxford [12] datasets. For the KITTI dataset, which was built using a stereo camera and LiDAR, the ROI of the stereo camera is set to [Horizon: -10 m–10 m, Vertical: -2 m–1 m, Depth: 0 m–50 m], and the ROI of the LiDAR is set to the same values as the ROI of the stereo camera. For the Oxford dataset, which was built using two LiDARs, the ROI of the LiDAR is set to [Horizon: -30–30 m, Vertical: -2–1 m, Depth: -30–30 m].

3.1.4. Transformation of Point Cloud of Target Sensor

In this paper, the miscalibration method used in previous studies [1,2,25] is used to perform the calibration of the stereo camera-LiDAR and LiDAR-LiDAR combination. In the KITTI [10] and Oxford [12] datasets we use, the values of six extrinsic parameters between two heterogeneous sensors and the 3D point clouds generated by them are given. Therefore, we can transform the 3D point cloud created by one sensor into a new 3D point cloud using the values of these six parameters. If we assign arbitrary deviations to these parameters, we can retransform the transformed point cloud in another space. At this time, if a calibration algorithm accurately finds the deviations that we randomly assign, we can move the retransformed point cloud to the position before the retransformation.

In order to apply the aforementioned approach to our proposed online self-calibration method, a 3D point $P = [x, y, z] \in \mathbb{R}^3$ measured by the target sensor is transformed by Equation (3) as follows:

$$\widehat{P'} = RT_{mis} RT_{init} \widehat{PT} = \begin{bmatrix} & R_{mis} & & T_{mis} \\ 0 & 0 & 0 & 1 \end{bmatrix} \begin{bmatrix} & R_{init} & & T_{init} \\ 0 & 0 & 0 & 1 \end{bmatrix} \widehat{PT} \tag{3}$$

$$RT_{gt} = RT_{mis}^{-1} \tag{4}$$

$$R_{init} = \begin{bmatrix} \cos(R_z) & -\sin(R_z) & 0 \\ \sin(R_z) & \cos(R_z) & 0 \\ 0 & 0 & 1 \end{bmatrix} \begin{bmatrix} \cos(R_y) & 0 & \sin(R_y) \\ 0 & 1 & 0 \\ -\sin(R_y) & 0 & \cos(R_y) \end{bmatrix} \begin{bmatrix} 1 & 0 & 0 \\ 0 & \cos(R_x) & -\sin(R_x) \\ 0 & \sin(R_x) & \cos(R_x) \end{bmatrix} \tag{5}$$

$$T_{init} = \begin{bmatrix} T_x & T_y & T_z \end{bmatrix}^T \tag{6}$$

$$R_{mis} = \begin{bmatrix} \cos(\theta_z) & -\sin(\theta_z) & 0 \\ \sin(\theta_z) & \cos(\theta_z) & 0 \\ 0 & 0 & 1 \end{bmatrix} \begin{bmatrix} \cos(\theta_y) & 0 & \sin(\theta_y) \\ 0 & 1 & 0 \\ -\sin(\theta_y) & 0 & \cos(\theta_y) \end{bmatrix} \begin{bmatrix} 1 & 0 & 0 \\ 0 & \cos(\theta_x) & -\sin(\theta_x) \\ 0 & \sin(\theta_x) & \cos(\theta_x) \end{bmatrix} \tag{7}$$

$$T_{mis} = \begin{bmatrix} \tau_x & \tau_y & \tau_z \end{bmatrix}^T \tag{8}$$

where P' is the transformed point of P, and superscript T represents the transpose. $\widehat{P}$ and $\widehat{P'}$ are expressed with homogeneous coordinates. RT_{gt}, described in Equation (4), is the transformation matrix we want to predict with our proposed method. RT_{gt} is used as the ground truth when the loss for training is calculated. In Equation (5), each of the parameters R_x, R_y, and R_z describes the angle rotated about the x-, y-, and z-axes between the two sensors. In Equation (6), T_x, T_y, and T_z describe the relative displacement between two sensors along the x-, y-, and z-axes. In this study, we assume that the values of the six parameters R_x, R_y, R_z, T_x, T_y, and T_z are given. In Equations (7) and (8), the parameters θ_x, θ_y, θ_z, τ_x, τ_y, and τ_z represent the random deviations for R_x, R_y, R_z, T_x, T_y, and T_z, respectively. Each of these six deviations is sampled randomly with equal probability within a predefined range of deviations described next. In Equations (5)–(8), R_{init} and

R_{mis} are rotation matrices and T_{init} and T_{mis} are translation vectors. The transformation by Equation (3) is performed only on points belonging to a predetermined ROI of the target sensor.

We set the random sampling ranges for θ_x, θ_y, θ_z, τ_x, τ_y, and τ_z the same as in previous studies [1,25] as follows: (rotational deviation: $-\theta-\theta$, translation deviation: $-\tau-\tau$), Rg1 = {θ: ±20°, τ: ±1.5 m}, Rg2 = {θ: ±10°, τ: ±1.0 m}, Rg3 = {θ: ±5°, τ: ±0.5 m}, Rg4 = {θ: ±2°, τ: ±0.2 m}, and Rg5 = {θ: ±1°,τ: ±0.1 m}. Each of Rg1, Rg2, Rg3, Rg4, and Rg5 set in this way is used for training each of the five networks named Net1, Net2, Net3, Net4, and Net5. One deviation range is assigned to one network training. Training for calibration starts with Net1 assigned to Rg1, and it continues with networks assigned to progressively smaller deviation ranges. The network mentioned here is described in Section 3.2.

3.1.5. Voxelization

We first perform a voxel partition by dividing the 3D points obtained by the sensors into equally spaced 3D voxels, as was performed in [36]. This voxel partition requires a space that limits the 3D points acquired by a sensor to a certain range. We call this range a voxel space. We consider the length of a side of a voxel, which is a cube, as a hyper-parameter, and denote it as S. In this paper, the unit of S is expressed in cm. A voxel can contain multiple points, of which up to three are randomly chosen, and the rest are discarded. Here, it is an experimental decision that we leave only up to three points per voxel. Referring to the method in [37], the average coordinates along the x-, y-, and z-axes of the points in each voxel are then calculated. We build three initial voxel maps, F_x, F_y, and F_z, using the average coordinates for each axis. For each sensor, these initial voxel maps become the input to our proposed network. Section 3.2 describes the network.

In this paper, we set the voxel space to be somewhat larger than the predetermined ROI of the sensor, considering the range of deviation. For example, in the case of the KITTI dataset, the voxel space of the stereo input is set as [horizontal: $-15-15$ m, vertical: $-15-15$ m, depth: 0–55 m], and the voxel space of the LiDAR input is set to the same size as the voxel space of the stereo input. In contrast, the voxel space of the 3D points generated by the two LiDARs in the Oxford dataset is set to [width: $-40-40$ m, height: $-15-15$ m, depth: $-40-40$ m]. The points outside of the voxel space are discarded.

3.2. Network Architecture

We propose a network of three parts, which are referred to as a feature extraction network (FEN), an attention module (AM), and an inference network (IN). The overall structure of the proposed network is shown in Figure 1. The input of this network is the F_x, F_y, and F_z for each sensor built from voxelization, and the output is seven numbers, three of which are translation-related parameter values, and the other four are rotation-related quaternion values. The network is capable of end-to-end training because every step is differentiable.

3.2.1. FEN

Starting from the initial input voxel maps F_x, F_y, and F_z, FEN extracts features for use in predicting calibration parameters by performing 3D convolution on 20 layers. The number of layers, the size of the kernel used, the number of kernels used in each layer, and the stride applied in each layer are experimentally determined. The kernel size is $3 \times 3 \times 3$. There are two types of stride, 1 and 2, which are used selectively for each layer. The number of kernels used in each layer is indicated at the bottom of Figure 1. This number corresponds to the quantity of the feature volume created in the layer. In the deep learning terminologies, this quantity is called channels. Convolution is performed differently depending on the stride applied to each layer. When stride 1 is applied, submanifold convolution [38] is performed, and when stride 2 is applied, general convolution is performed. General convolution is performed on all voxels with or without a value, but submanifold convolution is performed only when a voxel with a value corresponds to the central cell of the kernel. In addition, batch normalization (BN) [39] and rectified linear unit (ReLU) activation functions are sequentially applied after convolution in the FEN.

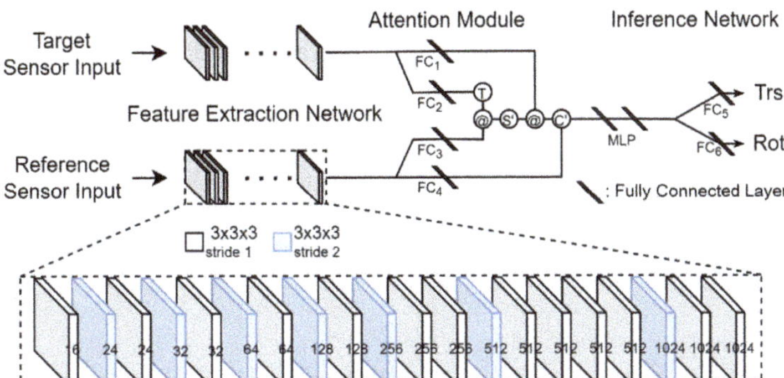

Figure 1. Overall structure of the proposed network. In the attention module, the T within a circle represents the transpose of a matrix; @ within a circle represents a matrix multiplication; S' within a circle represents the soft max function; C' within a circle represents concatenation. In the inference network, Trs and Rot represent the translation and rotation parameters predicted by the network, respectively.

We want the proposed network to perform robust calibration for large rotational and translational deviations between two sensors. To this end, a large receptive field is required. Therefore, we included seven layers with a stride of 2 in the FEN.

The final output of the FEN is 1024 feature volumes. The number of cells in the feature volume depends on the size of the voxel, but we let V be the number of cells in the feature volume. At this time, because each feature volume can be reconstructed as a V-dimensional column vector, we represent 1024 feature volumes as a matrix F of dimension $V \times 1024$. The outputs of FENs for the reference and target sensors are denoted by F_r and F_t, respectively.

3.2.2. AM

It is not easy to match the features extracted from the FEN through convolutions because the point clouds from the LiDAR-stereo camera combination are generated differently. Even in the LiDAR-LiDAR combination, if the FOVs of the two LiDARs are significantly different, it is also not easy to match the features extracted from the FEN through convolutions. Moreover, because the deviation range of rotation and translation is set large to estimate calibration parameters, it becomes difficult to check the similarity between the point cloud of the target sensor and the point cloud of the reference sensor.

Inspired by the attention mechanism proposed by Vaswani et al. [28], we solve these problems: we design an AM that implements the attention mechanism, as shown in Figure 1. The AM calculates an attention value for each voxel of the reference sensor input using the following procedure.

The AM has four fully connected layers (FCs): FC_1, FC_2, FC_3, and FC_4. A feature is input into these FCs, and a transformed feature is output. We denote the outputs of FC_1, FC_2, FC_3, and FC_4 as matrices M_1, M_2, M_3, and M_4, respectively. Each FC has 1024 input nodes. Here, the number 1024 is the number of feature volumes extracted from the FEN. The FC_1 and FC_4 have G/2 output nodes, and the FC_2 and FC_3 have G output nodes. These FCs transform 1024 features to G or G/2 features. Here, G is a hyper-parameter. If the sum of the elements in a row of matrix F, which is the output of the FEN, is 0, the row vector is not input to FC. We apply layer normalization (LN) [40] and the ReLU function to the output of these FCs so that the final output becomes nonlinear. The output M_2 of FC_2 is a matrix of dimension $V_t \times G$, and the output M_3 of FC_3 is a matrix of dimension $V_r \times G$.

Here, V_r and V_t are the number of rows in which there is at least one element with a feature value among the elements in each row of F_r and F_t, respectively. Therefore, V_r and V_t can be different for each input. However, we fix the values of V_r and V_t because the number of input nodes of the multi-layer perceptron (MLP) of the IN following the AM cannot be changed every time. In order to fix the values of V_r and V_t, we input all the data to be used in the experiments into the network and set the values when they are the largest, but we make them a multiple of 8. This is because V_r and V_t are also hyper-parameters. If the actual V_r and V_t are less than the predetermined V_r and V_t, the elements of the output matrices of FCs will be filled with zeros. The output M_1 of FC_1 is a matrix of dimension $V_t \times G/2$, and the output M_4 of FC_4 is a matrix of dimension $V_r \times G/2$.

- Computation of attention score by dot product

An attention score is obtained from the dot product of a row vector of M_3 and a column vector of M_2^T. This score is the same as the cosine similarity. The matrix A_S is obtained through the dot products of all row vectors of M_3 and all column vectors of M_2^T are called an attention score matrix. The dimension of the matrix A_S ($A_S = M_3 \cdot M_2^T$) is $V_r \times V_t$.

- Generation of attention distribution by softmax

We apply the softmax function to each row vector of A_S and obtain the attention distribution. The softmax function calculates the probability of each element of the input vector. We call this probability an attention weight, and the matrix obtained by this process is the attention weight matrix A_W of dimension $V_r \times V_t$.

- Computation of attention value by dot product

An attention value is obtained from the dot product of a row vector of A_W and a column vector of the matrix M_1. A matrix A_V obtained through the dot products of all row vectors of A_W and all column vectors of M_1 is called an attention value matrix. The dimension of the matrix A_V ($A_V = A_W \cdot M_1$) is $V_r \times G/2$.

Finally, we concatenate the attention value matrix A_V and the matrix M_4. The resulting matrix from this final process is denoted as A_C ($A_C = [A_V\ M_4]$) and has dimension $V_r \times G$; this matrix becomes the input to the IN. The reason we set the output dimension of FC_1 and FC_4 to $G/2$ instead of G is to save memory and reduce processing time.

3.2.3. IN

The IN infers rotation and translation parameters. The IN consists of an MLP and two fully connected layers, FC_5 and FC_6. The MLP is composed of an input and an output layer, as well as a single hidden layer. The input layer has $V_r \times G$ nodes, and the hidden and output layers have 1024 nodes, respectively. Therefore, when we input A_C, the output of the AM, into the MLP, we make A_C a flat vector. In addition, this MLP has no bias input, and it uses ReLU as an activation function. Moreover, LN is performed on the weighted sums that are input to nodes in the hidden layer and output layer, and ReLU is applied to the normalization result to obtain the output of these nodes. The output of the MLP becomes the input to the FC_5 and FC_6. The MLP plays the role of dimension reduction in the input vector.

We do not apply a normalization or an activation function to the FC_5 and FC_6. FC_5 produces three translation-related parameter values, which are τ_x^p, τ_y^p, and τ_z^p, and FC_6 produces four rotation-related quaternion values, which are q_0, q_1, q_3, and q_4.

3.3. Loss Function

To train the proposed network, we use a loss function as follows:

$$L = \lambda_1 L_{rot} + \lambda_2 L_{trs} \tag{9}$$

where L_{rot} is a regression loss related to rotation, L_{trs} is a regression loss related to translation, and hyper-parameters λ_1 and λ_2, respectively, are their weights. We use the quaternion distance to regress the rotation. The quaternion distance is defined as:

$$L_{rot} = \text{acos}(2(\frac{q_p}{|q_p|} \cdot \frac{q_{gt}}{|q_{gt}|})^2 - 1) \tag{10}$$

where $\cdot$ represents the dot product, $|\cdot|$ indicates the norm, and q_p and q_{gt} indicate a vector of the quaternion parameters predicted by the network and the ground-truth vector of quaternion parameters, respectively. From RT_{gt} of Equation (4), we obtain the four quaternion values. These four quaternion values are used for rotation regression as the ground truth.

For the regression of the translation vector, the smooth $L1$ loss is applied. The loss L_{trs} is defined as follows:

$$L_{trs} = \tfrac{1}{3}\left(smooth_{L1}\left(\tau_x^p - \tau_x^{gt}\right) + smooth_{L1}\left(\tau_y^p - \tau_y^{gt}\right) + smooth_{L1}\left(\tau_z^p - \tau_z^{gt}\right)\right)$$
$$smooth_{L1}(x) = \begin{cases} \frac{x^2}{2\beta} & \text{if } |x| < \beta \\ |x| - \frac{\beta}{2} & \text{otherwise} \end{cases} \tag{11}$$

where the superscripts p and gt represent prediction and ground truth, respectively, β is a hyper-parameter and is usually taken to be 1, and $|\cdot|$ represents an absolute value. The parameters τ_x^p, τ_y^p and τ_z^p are inferred by the network, and τ_x^{gt}, τ_y^{gt}, and τ_z^{gt} are obtained from RT_{gt} of Equation (4).

3.4. Postprocessing
3.4.1. Generation of a Calibration Matrix from a Network

Basically, postprocessing is performed to generate the calibration matrix RT_{pred} that is shown in Equation (12). The rotation matrix R_{pred} and translation vector T_{pred} in Equation (12) are generated by the quaternion parameters q_0, q_1, q_2, and q_3, and translation parameters τ_x^p, τ_y^p, and τ_z^p inferred from the network we built, as shown in Equations (13) and (14).

$$RT_{pred} = \begin{bmatrix} R_{pred} & T_{pred} \\ 0 \quad 0 \quad 0 & 1 \end{bmatrix} \tag{12}$$

$$R_{pred} = \begin{bmatrix} 1 - 2(q_2^2 + q_3^2) & 2(q_1 q_2 - q_0 q_3) & 2(q_0 q_2 + q_1 q_3) \\ 2(q_1 q_2 + q_0 q_3) & 1 - 2(q_1^2 + q_3^2) & 2(q_2 q_3 - q_0 q_1) \\ 2(q_1 q_3 - q_0 q_2) & 2(q_0 q_1 + q_2 q_3) & 1 - 2(q_1^2 + q_2^2) \end{bmatrix} \tag{13}$$

$$T_{pred} = \begin{bmatrix} \tau_x^p & \tau_y^p & \tau_z^p \end{bmatrix}^T \tag{14}$$

$$\theta_x^p = atan2\left(R_{pred}(3,2), R_{pred}(3,3)\right)$$
$$\theta_y^p = atan2\left(-R_{pred}(3,1), \sqrt{R_{pred}(3,2)^2 + R_{pred}(3,3)^2}\right) \tag{15}$$
$$\theta_z^p = atan2\left(R_{pred}(2,1), R_{pred}(1,1)\right)$$

Equation (15) shows how to calculate the rotation angle about each of the x-, y-, and z-axes from the rotation matrix R_{pred}. In Equation (15), (r,c) indicates the row index r and column index c of the matrix R_{pred}. The angle calculation described in Equation (15) is used to convert a given rotation matrix into Euler angles.

3.4.2. Calculation of Calibration Error

To evaluate the proposed calibration system, it is necessary to calculate the error of the predicted parameters. For this, we calculate the transformation matrix RT_{error}, which contains the errors of the predicted parameters by Equation (16). RT_{mis} and RT_{online} in

Equation (16) are calculated by Equations (3) and (17), respectively. In Equation (17), each of RT_1, RT_2, RT_3, RT_4, and RT_5 is a calibration matrix predicted by each of the five networks, Net1, Net2, Net3, Net4, and Net5. The calculation of these five matrices is described in detail in 3.4.3. From RT_{error}, we calculate the error of the rotation-related parameters using Equation (18) and the error of the translation-related parameters using Equation (19).

$$RT_{error} = RT_{online} \cdot RT_{mis} \tag{16}$$

$$RT_{online} = RT_5 \cdot RT_4 \cdot RT_3 \cdot RT_2 \cdot RT_1 \tag{17}$$

$$\begin{aligned} \theta_x^e &= atan2(RT_{error}(3,2),\ RT_{error}(3,3)) \\ \theta_y^e &= atan2\left(-RT_{error}(3,1), \sqrt{RT_{error}(3,2)^2 + RT_{error}(3,3)^2}\right) \\ \theta_z^e &= atan2(RT_{error}(2,1), RT_{error}(1,1)) \end{aligned} \tag{18}$$

$$\begin{aligned} \tau_x^e &= RT_{error}(1,4) \\ \tau_y^e &= RT_{error}(2,4) \\ \tau_z^e &= RT_{error}(3,4) \end{aligned} \tag{19}$$

In Equations (18) and (19), (r,c) indicates the row index r and column index c of the matrix RT_{error}.

In the KITTI dataset, the rotation angle about the x-axis, the rotation angle about the y-axis, and the rotation angle about the z-axis correspond to pitch, yaw, and roll, respectively. In contrast, in the Oxford dataset, they correspond to roll, pitch, and yaw, respectively.

3.4.3. Iterative Refinement for Precise Calibration

The training uses all five deviation ranges, but the evaluation of the proposed method is performed with randomly sampled deviations only in Rg1, which is the largest deviation range. Using this sampled deviation, the transformation matrix RT_{mis} is formed as shown in Equations (3), (7), and (8). Then, a point cloud prepared for evaluation is initially transformed using Equation (3). By inputting this transformed point cloud into the trained Net1, the values of parameters that describe translation and rotation are inferred. With these inferred values, we obtain the RT_{pred} of Equation (12). This RT_{pred} becomes RT_1. We multiply the initial transformed points by this RT_1 to obtain new transformed points, and we input these new transformed points into the trained Net2 to obtain RT_{pred} from Net2. This new RT_{pred} becomes RT_2. In this way, the input points to the current network are multiplied by RT_{pred}, which is the output of the current network, to obtain new transformed points for use as the input to the next network; this process of obtaining new RT_{pred} by inputting them into the next network is repeated until Net5. For each point cloud prepared for evaluation as described above, a calibration matrix (RT_i, $i = 1, \cdots, 5$) is obtained from each of the five networks, and the final calibration matrix RT_{online} is obtained by multiplying the calibration matrices as shown in Equation (17). The iterative transformation process of the point cloud for evaluation as described above is expressed as follows:

$$\widehat{P'_1} = RT_{mis} RT_{init} \widehat{P^T} \tag{20}$$

$$\widehat{P'_i} = RT_{i-1} \widehat{P'_{i-1}},\ i = 2, \ldots, 5 \tag{21}$$

3.4.4. Temporal Filtering for Precise Calibration

Calibration performed with only a single frame can be vulnerable to various forms of noise. According to [25], this problem can be improved by analyzing the results over time. For this purpose, N. Schneider et al. [25] check the distribution of the results over all evaluation frames while maintaining the value of the sampled deviation used for the first frame. They take the median over the whole sequence, which enables the best performance on the test set. They sample the deviations from Rg1. They repeat 100 runs of this experiment, keeping the sampled deviations until all test frames are passed and resampling the deviations at the start of a new run.

It is good to analyze the results obtained over multiple frames. However, applying all the test frames to temporal filtering has a drawback in the context of autonomous driving. In the case of the KITTI dataset, the calibration parameter values are inferred from the results obtained from processing about 4500 frames, which takes a long time. It is also difficult to predict what will happen during this time. Therefore, we reduce the number of frames to use for temporal filtering and randomly determine the start frame for filtering among these frames. We set the bundle size of frames to 100 and performed quantitative analysis by taking the median from 100 results obtained by applying this bundle. The value of parameters from RT_{online} for each frame is obtained using Equations (14) and (15). The basis for setting the bundle size is given in Section 4.3.3.

4. Experiments

There are several tasks, such as data preparation in training and evaluation of the proposed calibration system. The KITTI dataset provides images captured with a stereo camera and point clouds acquired using a LiDAR. The dataset consists of 21 sequences (00 to 20) from different scenarios. The Oxford dataset provides point clouds acquired using two LiDARs. In addition, both datasets provide initial calibration parameters and visual odometry information.

We used the KITTI dataset for LiDAR-stereo camera calibration. We referred to the method proposed by Lv et al. [1] in using the 00 sequence (4541 frames) for testing and using the rest (39,011 frames) of the sequences for training. We used the Oxford dataset for LiDAR-LiDAR calibration. Of the many sequences in the Oxford dataset, we used the 2019-01-10-12-32-52 sequence for training and the 2019-01-17-12-32-52 sequence for evaluation. The two LiDARs that were used to build the Oxford dataset were not synchronized. Therefore, we used visual odometry information to synchronize the frames. After the synchronization, the unsynchronized frames were deleted, and our Oxford dataset consisted of 43,130 frames for training and 35,989 frames for evaluation.

We did not apply the same hyper-parameter values to all five networks (Net1 to Net5) because of the large difference in the range of allowable deviations for rotation and translation in Rg1 and Rg5. Because Net5 is trained with Rg5, which has the smallest deviation range, and is applied last in determining the calibration matrix, we trained Net5 using different hyper-parameter values from other networks. Such hyper-parameters included S, V_r, V_t, G, λ_1, λ_2, and B, which are the length of a side of a voxel, the number of voxels with data among voxels in a voxel space of the reference sensor, the number of voxels with data among voxels in a voxel space of the target sensor, the number of output nodes of the FC$_2$ and FC$_3$ in the AM, the weight of the loss function L_{rot}, the weight of the loss function L_{trs}, and the batch size, respectively.

Through the experiments with the Oxford dataset, we observed that data screening is required to enhance the calibration accuracy. The dataset was built with two LiDARs mounted at the left and right corners in front of the roof of a platform vehicle. Figure 2 shows a point cloud for one frame in the Oxford dataset. This point cloud contains points generated by scanning the surface of the platform vehicle by LiDARs. We confirmed that the calibrations performed on point clouds containing these points degrade the calibration accuracy. Therefore, to perform calibration after excluding these points, we set a point removal area to [Horizon: -5–5 m, Vertical: -2–1 m, Depth: -5–5 m] for the target sensor and [Horizon: -1.5–1.5 m, Vertical: -2–1 m, Depth: -2.5–1.5 m] for the reference sensor. Experimental results with respect to this region cropping are provided in Section 4.3.1.

We trained the network for a total of 60 epochs. We initially set the learning rate to 0.0005 and halved it when the epochs reached 30, and we halved it again when the epochs reached 40. The batch size B was determined to be within the limits allowed by the memory of the equipment used. We used one NVIDIA GeForce RTX 2080Ti graphic card for all our experiments. Adam [41] was used for model optimization, and hyper-parameters $\beta_1 = 0.9$ and $\beta_2 = 0.999$ were used.

Figure 2. Point cloud constituting one frame in the Oxford dataset. The green dots represent points obtained by the right LiDAR, and the red dots represent the points obtained by the left LiDAR.

4.1. Evaluation Using the KITTI Dataset

Figure 3 shows a visual representation of the results for performing calibration on the KITTI dataset using the proposed five networks. In this experiment, we transform a point cloud using the calibration matrix inferred from the proposed network and using the ground-truth parameters given in the dataset. We want to show how consistent these two transformation results are. Figure 3a,b show the transformation of a point cloud by randomly sampled deviations from Rg1 and the calibrated parameters given in the KITTI dataset, respectively. The left side of Figure 3c shows the transformation of the point cloud by RT_1 predicted by the trained Net1. This result looks suitable, but as shown to the right of Figure 3c, it can be seen that the points measured on a thin column were projected to positions that deviated from the column. The effect of iterative refinement appears here. Calibration does not end at Net1 but continues to Net5. Figure 3d shows the transformation of the point cloud by RT_{online} obtained after performing calibration up to Net5. By comparing the result of Figure 3d with the result shown in Figure 3c, we can see that the calibration accuracy is improved: suitable alignment even with the thin column.

Table 1 presents the average performance of calibrations performed without temporal filtering on 4541 frames for testing on the KITTI dataset. From the results shown in Table 1, we can see the effect of iterative refinement. From Net1 to Net5, the improvements are progressive. Our method achieves an average rotation error of [Roll: 0.024°, Pitch: 0.018°, Yaw: 0.060°] and an average translation error of [X: 0.472 cm, Y: 0.272 cm, Z: 0.448 cm].

Table 1. Quantitative results of calibration performed on the KITTI dataset without temporal filtering. See footnotes [1,2] for hyper-parameter settings.

Refinement Stage	Rotation Error (°)			Translation Error (cm)		
	Roll	Pitch	Yaw	X	Y	Z
After Net1 [1]	0.182	0.110	0.386	2.393	1.205	1.781
After Net2 [1]	0.112	0.068	0.176	1.513	1.356	1.663
After Net3 [1]	0.071	0.046	0.134	1.119	0.709	1.027
After Net4 [1]	0.039	0.024	0.088	0.750	0.428	0.735
After Net5 [2]	0.024	0.018	0.060	0.472	0.272	0.448

[1] $S = 5$, $(V_r, V_t) = (96, 160)$, $G = 1024$, $(\lambda_1, \lambda_2) = (1, 2)$, $B = 8$. [2] $S = 2.5$, $(V_r, V_t) = (384, 416)$, $G = 128$, $(\lambda_1, \lambda_2) = (0.5, 5)$, $B = 4$.

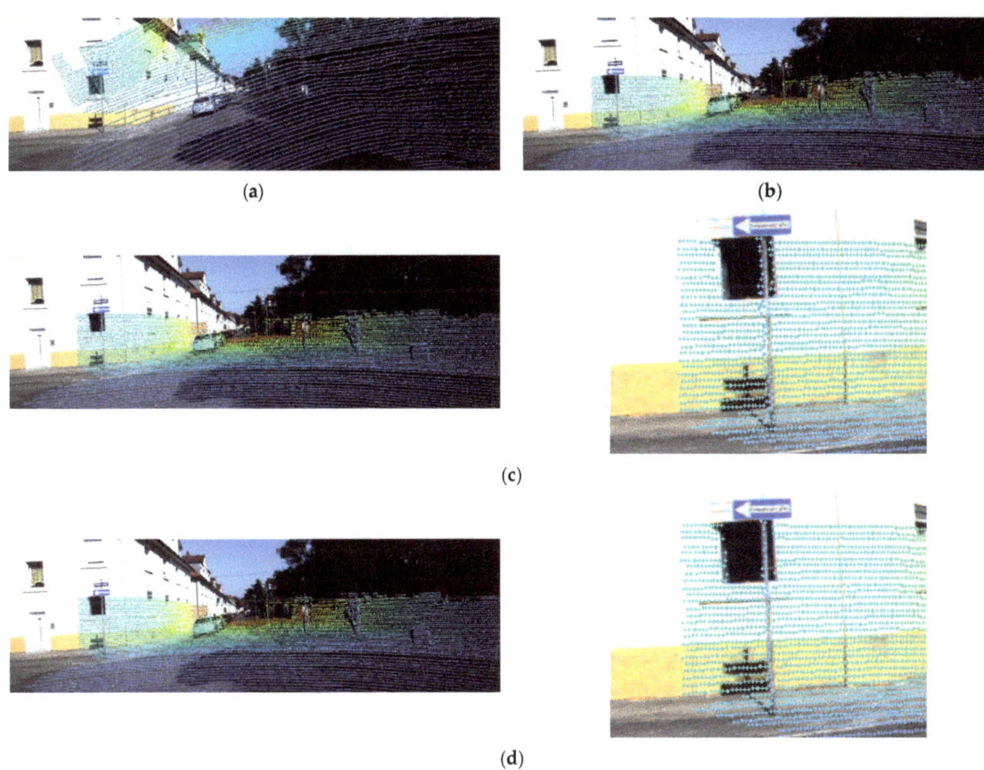

Figure 3. Results of applying the proposed method to a test frame of the KITTI dataset. (**a**) Transformation by randomly sampled deviations. (**b**) Transformation by given calibrated parameters. (**c**) Transformation by RT_1 inferred from Net1. (**d**) Transformation by RT_{online} obtained from iterative refinement by five networks.

Figure 4 shows two examples of error distribution for individual components by means of boxplots. From these experiments, we confirmed that temporal filtering provides suitable calibration results regardless of the amount of arbitrary deviation. The dots shown in Figure 4a,b are both obtained by transforming the same point cloud of the target sensor by randomly sampled deviations from Rg1, but the sampled deviations are different. As can be seen from the boxplots in Figure 4e–h, the distribution of calibration errors was similar despite the large difference in sampled deviations.

Table 2 shows the calibration results for our method and for the existing CNN-based online calibration methods. From these results, it can be seen that our method achieves the best performance. In addition, when these results are compared with the results shown in Table 1, it can be concluded that our method achieves significant performance improvement through temporal filtering. CalibNet [2] did not specify a frame bundle.

Table 2. Comparison of calibration performance between our method and other CNN-based methods.

Method	Range	Bundle Size of Frame	Rotation Error (°)			Translation Error (cm)		
			Roll	Pitch	Yaw	X	Y	Z
RegNet [25]	Rg1	4541	0.24	0.25	0.36	7	7	4
CalibNet [2]	($\pm10°$, ±0.2 m)	-	0.18	0.9	0.15	4.2	1.6	7.22
LCCNet [1]	Rg1	4541	0.020	0.012	0.019	0.262	0.271	0.357
Ours	Rg1	4541	0.002	0.011	0.004	0.183	0.068	0.183

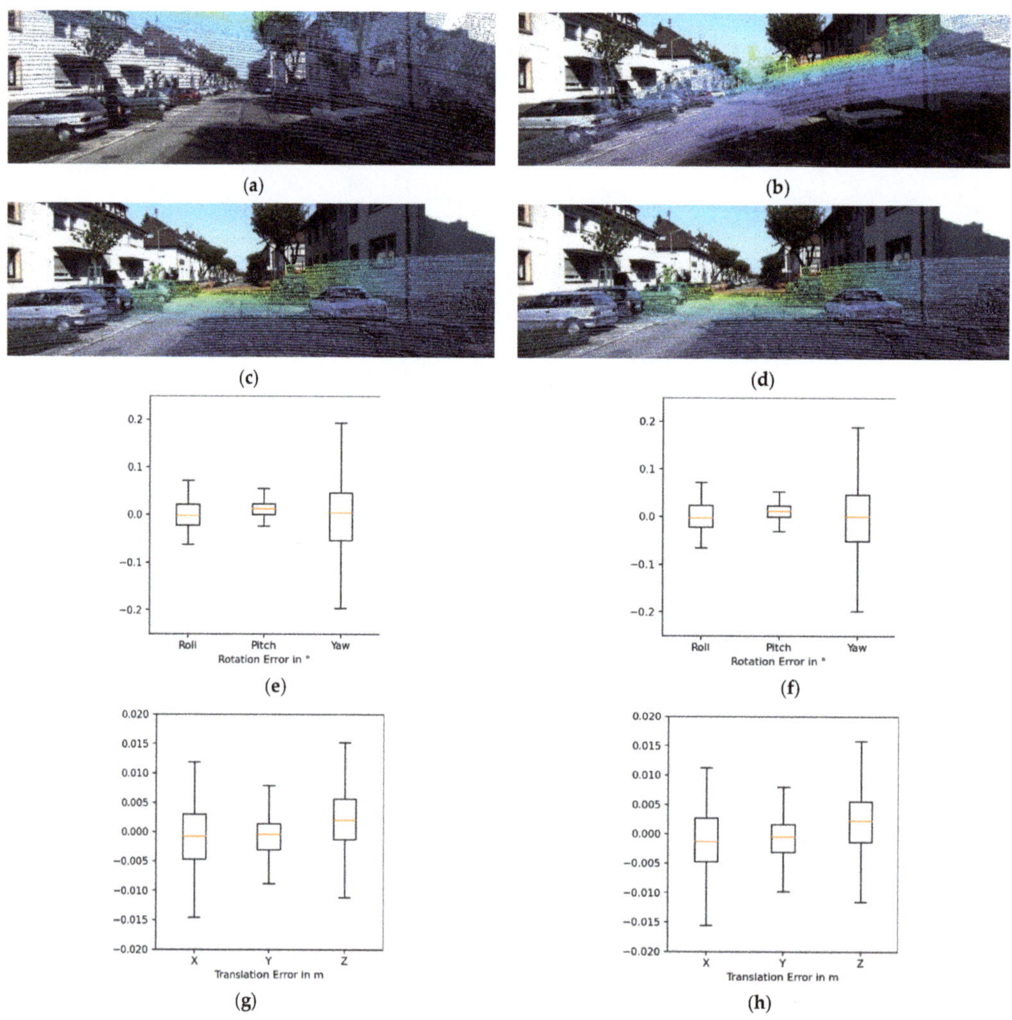

Figure 4. Calibration results and error distribution when temporal filtering was applied. (**a**) Transformation by randomly sampled deviation from Rg1. (**b**) Transformation by randomly sampled deviation from Rg1. (**c**) Calibration results from random deviations shown in (**a**). (**d**) Calibration results from random deviations shown in (**b**). (**e**) Rotation error for the results shown in (**c**). (**f**) Rotation error for the results shown in (**d**). (**g**) Translation error for the results shown in (**c**). (**h**) Translation error for the results shown in (**d**).

Figure 5 graphically shows the changes in the losses calculated by Equations (10) and (11) for training the proposed networks on the KITTI dataset. In this figure, the green graph shows the results of training with randomly sampled deviations from Rg1, and the pink graph shows the results of training with randomly sampled deviations from Rg5. The horizontal and vertical axes of these graphs represent epochs and loss, respectively. From these graphs, we can observe that the loss reduction decreases from approximately the 30th epoch. This was consistently observed, no matter what deviation range the network was trained on or what hyper-parameters were used. There were similar trends in loss reduction for rotation and translation. Given this situation, we halved the initial learning rate after the 30th epoch of training. Training was performed at a reduced learning rate for 10 epochs after the 30th epoch. After the 40th epoch, we halved the learning rate again. Training continued until the 60th epoch, and the result that produced the smallest training error among the results obtained from the 45th to the 60th epoch was selected as the training result. When Net1 was trained, the hyper-parameters were set as $S = 5$, $(V_r, V_t) = (96, 160)$, $G = 1024$, $(\lambda_1, \lambda_2) = (1, 2)$, and $B = 8$. When Net5 was trained, the hyper-parameters were set as $S = 2.5$, $(V_r, V_t) = (384, 416)$, $G = 128$, $(\lambda_1, \lambda_2) = (0.5, 5)$, and $B = 4$. In Figure 5, the training results before the 10th epoch are not shown because the loss was too large.

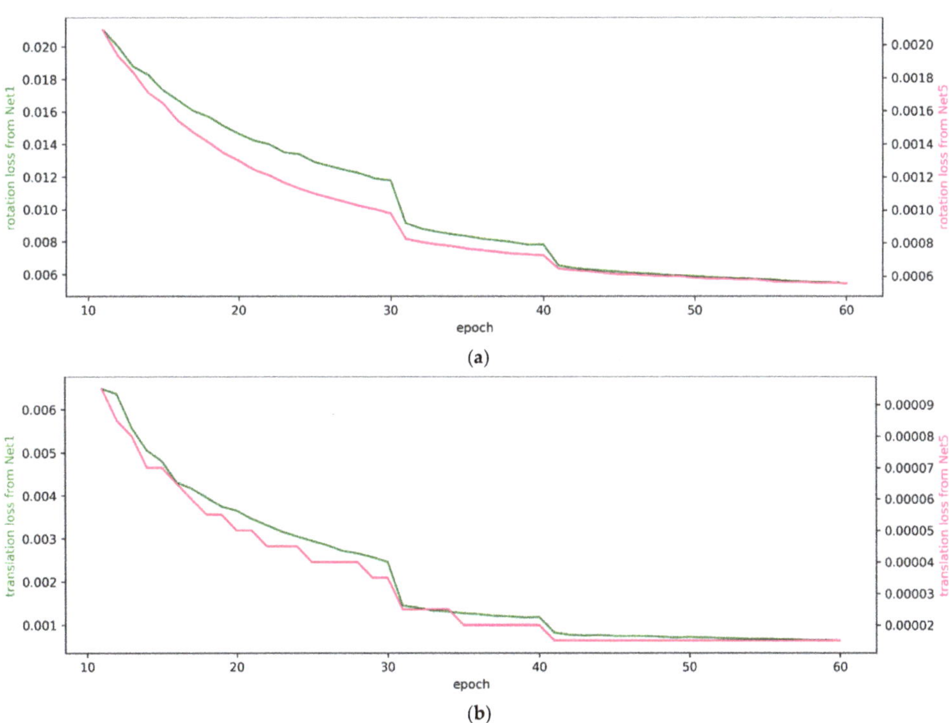

Figure 5. Changes in loss calculated during the training of Net1 and Net5 on the KITTI dataset. (**a**) L_{rot} calculated using Equation (10). (**b**) L_{trs} calculated using Equation (11).

4.2. Evaluation Using the Oxford Dataset

Figures 6 and 7 show the results of performing calibration on the Oxford dataset using the proposed five networks. In these figures, the green dots represent the points obtained by the right LiDAR, which is considered to be the target sensor, and the red dots represent the points obtained by the left LiDAR. Figure 6a,b show the results of the transformation of a point cloud from the target sensor by randomly sampled deviations from Rg1 and calibrated parameters given in the Oxford dataset, respectively. Figure 6c shows the result of the transformation of the point cloud by RT_1 inferred from the trained Net1. Figure 6d shows the result of the transformation of the point cloud by RT_{online} obtained after performing calibration up to Net5. Similar to the results of the calibration performed using the KITTI dataset, the results of Net1 look suitable, but they are not suitable when compared with the results shown in Figure 6d. The photo on the right side of Figure 6c shows that the green and red dots indicated by an arrow are misaligned. In contrast, the photo on the right side of Figure 6d shows that the green and red dots indicated by an arrow are well aligned. We show through this comparison that calibration accuracy can be improved by the iterative refinement of five networks even without temporal filtering.

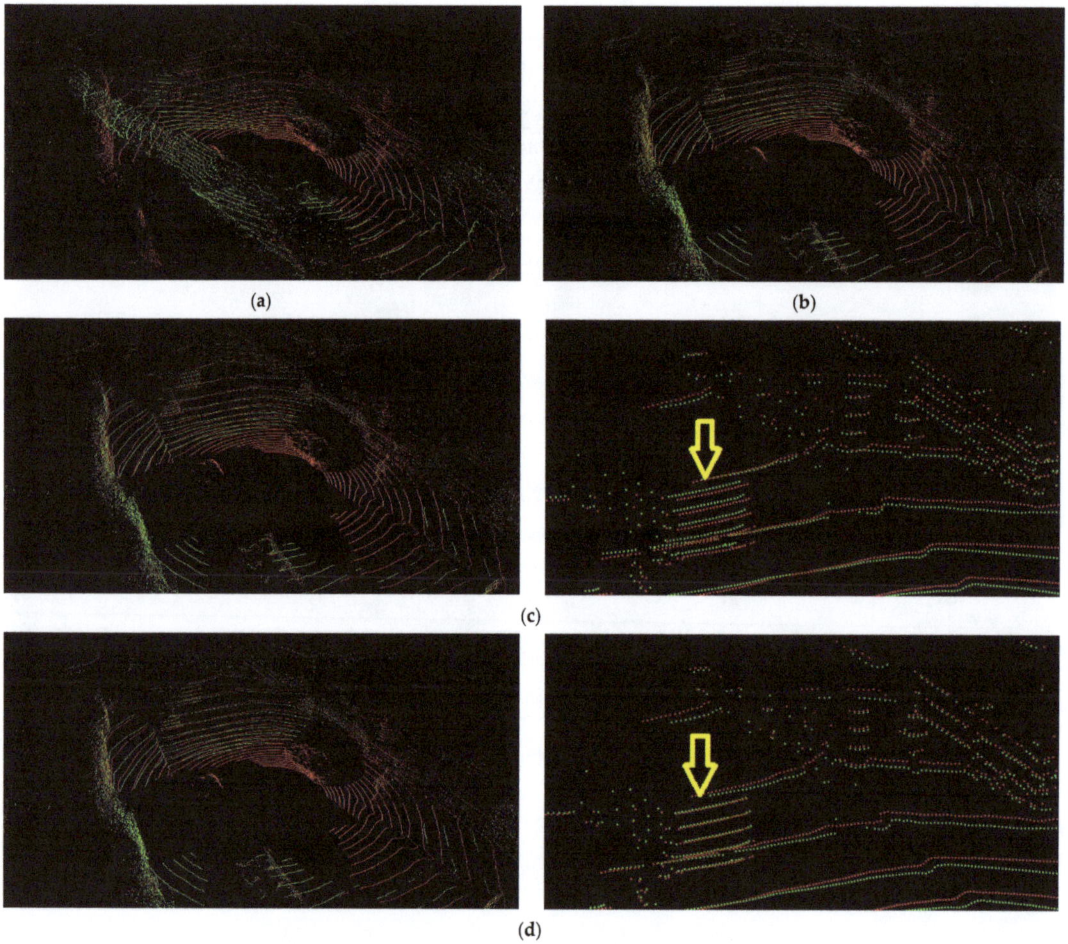

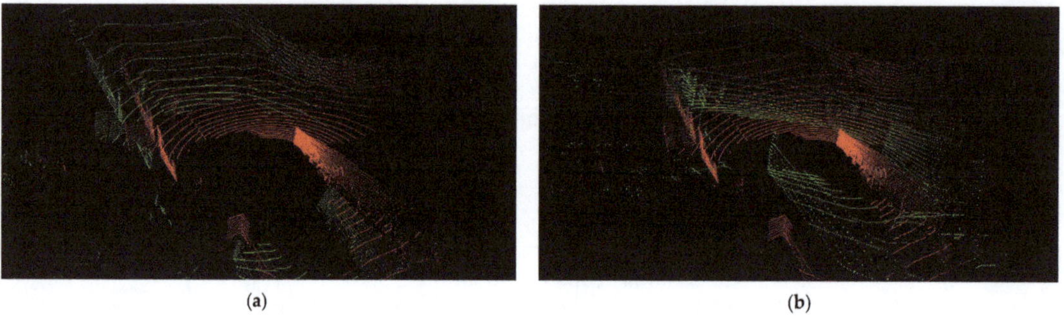

Figure 6. Results of applying the proposed method to a test frame of the Oxford dataset. (**a**) Transformation by randomly sampled deviations. (**b**) Transformation by given calibrated parameters. (**c**) Transformation by RT_1 inferred from Net1. (**d**) Transformation by RT_{online} obtained from iterative refinement by five networks.

(**a**)

(**b**)

Figure 7. *Cont.*

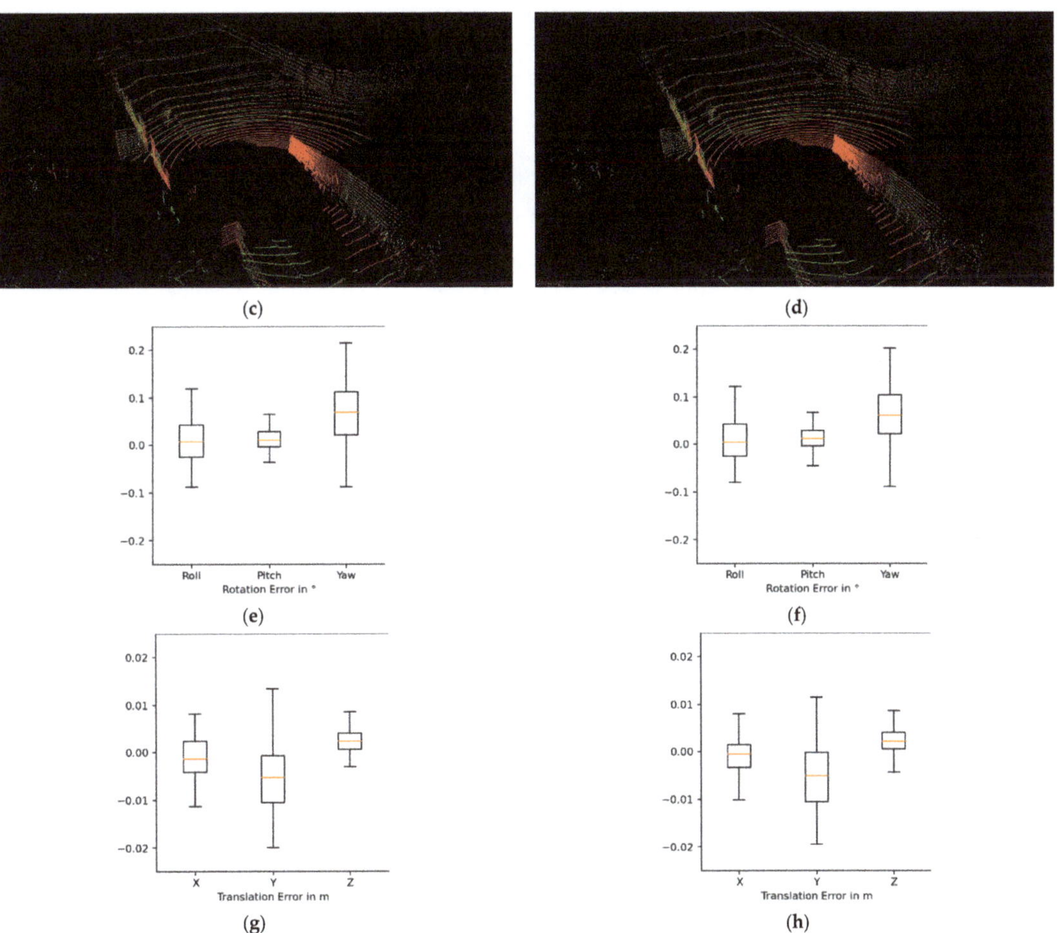

Figure 7. Calibration results and error distribution when temporal filtering was applied to the Oxford dataset. (**a**) Transformation by randomly sampled deviations from Rg1. (**b**) Transformation by randomly sampled deviations from Rg1. (**c**) Calibration results from random deviations shown in (**a**). (**d**) Calibration results from random deviations shown in (**b**). (**e**) Rotation error for the results shown in (**c**). (**f**) Rotation error for the results shown in (**d**). (**g**) Translation error for the results shown in (**c**). (**h**) Translation error for the results shown in (**d**).

Table 3 presents the average performance of calibrations performed without temporal filtering on 35,989 frames for testing in the Oxford dataset. Our method achieves an average rotation error of [Roll: 0.056°, Pitch: 0.029°, Yaw: 0.082°] and an average translation error of [X: 0.520 cm, Y: 0.628 cm, Z: 0.350 cm]. In this experiment, we applied the same hyper-parameters to all five networks. They are $S = 5$, $(V_r, V_t) = (224, 288)$, $G = 1024$, $(\lambda_1, \lambda_2) = (1, 2)$, and $B = 8$.

Table 3. Quantitative results of calibration performed on the Oxford dataset without temporal filtering.

Refinement Stage	Rotation Error (°)			Translation Error (cm)		
	Roll	Pitch	Yaw	X	Y	Z
After Net1	0.302	0.223	0.370	3.052	4.440	3.603
After Net2	0.249	0.262	0.266	1.048	2.155	2.240
After Net3	0.136	0.068	0.099	1.469	1.191	1.348
After Net4	0.072	0.036	0.073	0.632	0.809	0.985
After Net5	0.056	0.029	0.082	0.520	0.628	0.350

Figure 7 shows two examples of the error distribution of individual components by means of boxplots, as shown in Figure 4. From these experiments, we can see that temporal filtering provides suitable calibration results regardless of the amount of arbitrary deviation, even for LiDAR-LiDAR calibration. The green dots shown in Figure 7a,b are both obtained by transforming the same point cloud of the target sensor with randomly sampled deviations from Rg1, but the sampled deviations are different. As shown in Figure 7e–g, the distribution of calibration errors is similar despite the large difference in sampled deviations. In these experiments, the size of the frame bundle used in the temporal filtering was 100.

Table 4 shows the calibration performance of the proposed method with temporal filtering. Our method achieves a rotation error of less than 0.1° and a translation error of less than 1 cm. By comparing Tables 3 and 4, it can be seen that temporal filtering achieves a significant improvement in performance.

Table 4. Quantitative results of calibration on Oxford dataset with temporal filtering.

Method	Range	Bundle Size of Frame	Rotation Error (°)			Translation Error (cm)		
			Roll	Pitch	Yaw	X	Y	Z
Ours	Rg1	100	0.035	0.017	0.060	0.277	0.305	0.247

Figure 8 graphically shows the changes in the losses calculated by Equations (10) and (11) in training the proposed networks with the Oxford dataset. Compared with the results shown in Figure 5, we observed that the results from this experiment were very similar to the experimental results achieved with the KITTI dataset. Therefore, we decided to apply the same training strategy to the KITTI and Oxford datasets. However, the settings of the hyper-parameter values that were applied to the network were different. When Net1 was trained, the hyper-parameters were set as S = 5, (V_r, V_t) = (224, 288), G = 1024, (λ_1, λ_2) = (1, 2), and B = 8. When Net5 was trained, the hyper-parameters were set as S = 5, (V_r, V_t) = (224, 288), G = 1024, (λ_1, λ_2) = (0.5, 5), and B = 4.

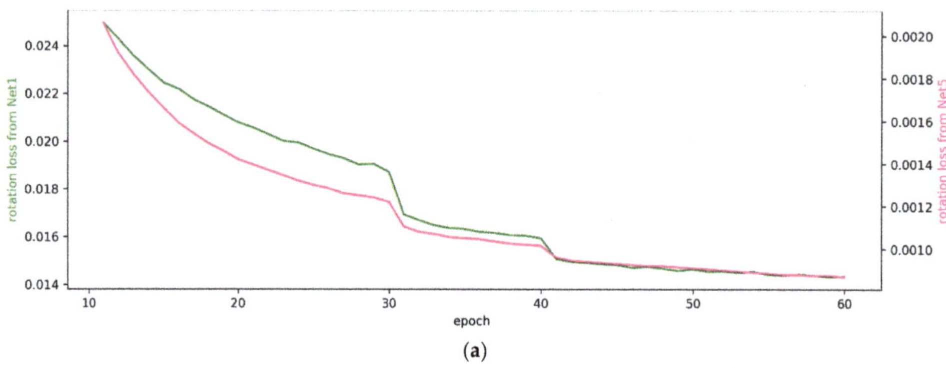

(a)

Figure 8. *Cont.*

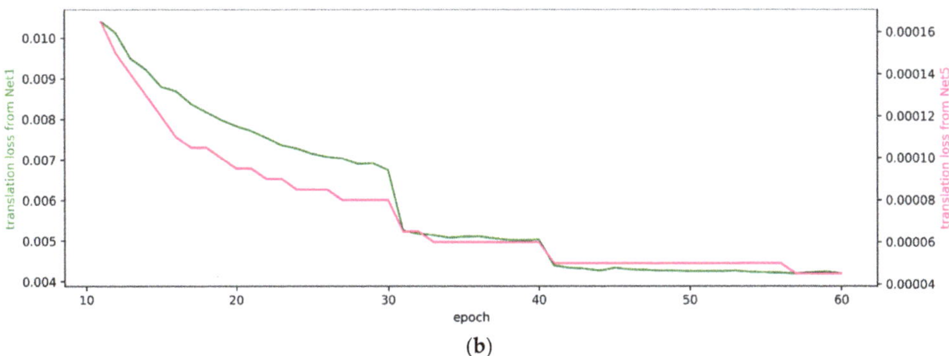

(b)

Figure 8. Changes in calculated losses during Net1 and Net5 training on the Oxford dataset. (**a**) Rotation loss L_{rot} calculated using Equation (10). (**b**) Translation loss L_{trs} calculated using Equation (11).

4.3. Ablation Studies

4.3.1. Performance According to the Cropped Area of the Oxford Dataset

At the beginning of Section 4, we mentioned the need to eliminate some points in the Oxford dataset that degraded calibration performance. To support this observation, we presented in Table 5 the results of experiments with and without the removal of those points. However, although there is a difference in the calibration performance according to the size of the removed area, it is difficult to theoretically determine the size of the area to be cropped. Table 5 shows the results of the experiments by setting the area to be cut in two ways. Through these experiments, we found that the calibration performed after removing points that caused the performance degradation generally produced better results than the calibration performed without removing those points. These experiments were performed with the trained Net5, and the hyper-parameters were as follows. S = 5, V = (224, 288), G = 1024, λ = (1, 2), and B = 8.

Table 5. Comparison of calibration performance according to the cropped area on the Oxford dataset.

Size of Area to be Cropped [Horizon, Vertical, Depth]	Rotation Error (°)			Translation Error (cm)		
	Roll	Pitch	Yaw	X	Y	Z
N/A	0.038	0.030	0.070	1.922	0.868	0.476
[−5~5 m, −2–1 m, −5~5 m]	0.033	0.027	0.062	0.538	0.668	0.564
[−10~10 m, −2–1 m, −10~10 m]	0.033	0.025	0.054	0.490	1.109	0.496

4.3.2. Performance According to the Length of a Voxel Side, S

We conducted experiments to check how the calibration performance changes according to S. Tables 6 and 7 show the results of these experiments. Table 6 shows the results for the KITTI dataset, and Table 7 shows the results for the Oxford dataset. We performed an evaluation according to S with a combination of Rg1 and Net1 and a combination of Rg5 and Net5. These experiments showed that the calibration performance improved as S became smaller. However, as S became smaller, the computational cost increased, and in some cases, the performance deteriorated. We tried to experiment with fixed values of hyper-parameters other than S, but naturally, as S decreased, the hyper-parameters V_r and V_t increased rapidly. This was a burden on the memory, and thus it was difficult to keep the batch size B at the same value. Therefore, when S was 2.5, B was 4 in the experiment performed on the KITTI dataset, and B was 2 in the experiment performed on the Oxford dataset. However, for S greater than 2.5, B was fixed at 8. In addition, there were cases where the performance deteriorated when S was very small, such as 2.5, which was considered to be the result of a small receptive field in the FEN. Even in the experiments performed on the Oxford dataset, when S was 2.5 in Net1, the training loss diverged near the 5th epoch, so the experiment could no longer be performed. For training on the KITTI dataset, S was set to 2.5 in Net5, and S was set to 5 in Net1 to Net4. However, for training on the Oxford dataset, S was set to 5 for both Net1 and Net5.

Table 6. Comparison of calibration performance according to S on the KITTI dataset.

Hyper-Parameter Setting	Rotation Error (°)			Translation Error (cm)		
	Roll	Pitch	Yaw	X	Y	Z
For the Combination of Net1 and Rg1						
$S = 10$, $V = (32, 48)$, $G = 1024$, $\lambda = (1, 2)$, $B = 8$	0.228	0.166	0.421	3.103	1.681	2.155
$S = 7.5$, $V = (48, 72)$, $G = 1024$, $\lambda = (1, 2)$, $B = 8$	0.199	0.199	0.429	2.881	1.613	2.514
$S = 5$, $V = (96, 160)$, $G = 1024$, $\lambda = (1, 2)$, $B = 8$	0.182	0.110	0.386	2.393	1.205	1.781
$S = 2.5$, $V = (384, 416)$, $G = 128$, $\lambda = (1, 2)$, $B = 4$	0.295	0.206	0.595	4.489	2.288	2.840
For the Combination of Net5 and Rg5						
$S = 10$, $V = (32, 48)$, $G = 1024$, $\lambda = (1, 2)$, $B = 8$	0.344	0.019	0.063	0.778	0.429	0.887
$S = 7.5$, $V = (48, 72)$, $G = 1024$, $\lambda = (1, 2)$, $B = 8$	0.030	0.020	0.059	0.646	0.487	0.776
$S = 5$, $V = (96, 160)$, $G = 1024$, $\lambda = (1, 2)$, $B = 8$	0.028	0.017	0.070	0.610	0.363	0.702
$S = 2.5$, $V = (384, 416)$, $G = 128$, $\lambda = (1, 2)$, $B = 4$	0.023	0.016	0.045	0.450	0.312	0.537

Table 7. Comparison of calibration performance according to S on the Oxford dataset.

Hyper-Parameter Setting	Rotation Error (°)			Translation Error (cm)		
	Roll	Pitch	Yaw	X	Y	Z
For the Combination of Net1 and Rg1						
$S = 10$, $V = (76, 96)$, $G = 1024$, $\lambda = (1, 2)$, $B = 8$	0.382	0.328	0.436	2.606	8.114	2.881
$S = 7.5$, $V = (96, 160)$, $G = 1024$, $\lambda = (1, 2)$, $B = 8$	0.415	0.263	0.433	3.542	7.574	4.151
$S = 5$, $V = (224, 288)$, $G = 1024$, $\lambda = (1, 2)$, $B = 8$	0.302	0.223	0.370	3.052	4.440	3.603
$S = 2.5$, $V = (608, 608)$, $G = 128$, $\lambda = (1, 2)$, $B = 2$	-	-	-	-	-	-
For the Combination of Net5 and Rg5						
$S = 10$, $V = (76, 96)$, $G = 1024$, $\lambda = (1, 2)$, $B = 8$	0.046	0.031	0.085	0.626	1.431	0.610
$S = 7.5$, $V = (96, 160)$, $G = 1024$, $\lambda = (1, 2)$, $B = 8$	0.031	0.228	0.535	0.448	1.357	0.457
$S = 5$, $V = (224, 288)$, $G = 1024$, $\lambda = (1, 2)$, $B = 8$	0.033	0.027	0.062	0.538	0.668	0.564
$S = 2.5$, $V = (608, 608)$, $G = 128$, $\lambda = (1, 2)$, $B = 2$	0.036	0.025	0.057	0.552	0.699	0.539

4.3.3. Performance According to the Bundle Size of Frames

We conducted experiments to observe how the calibration performance changes according to the bundle size of the frame for temporal filtering. Tables 8 and 9 show the results of these experiments. Table 8 shows the results for the KITTI dataset, and Table 9 shows the results for the Oxford dataset. We performed the experiments as presented in Section 3.4.3. Because 100 runs had to be performed, the position of the starting frame for each run was predetermined. For each run, we took the median of the values of each of the six parameters associated with rotation and translation inferred from the frames in the bundle, and we calculated the absolute difference between this median and the deviation randomly sampled from Rg1. The error of each parameter shown in Tables 8 and 9 was obtained by adding up the error of the corresponding parameters calculated for each run and dividing the sum by the number of runs. Through these experiments, we found that temporal filtering using many frames improves the overall calibration performance. However, if we look carefully at the results presented in the two tables, the effect is not shown for all parameters. Considering this observation and the processing time, the bundle size of the frame was set to 100.

Table 8. Comparison of calibration performance according to the bundle size of frames for temporal filtering on the KITTI dataset.

Bundle Size of Frames	Rotation Error (°)			Translation Error (cm)		
	Roll	Pitch	Yaw	X	Y	Z
1	0.024	0.017	0.057	0.414	0.257	0.395
10	0.009	0.013	0.018	0.210	0.102	0.245
25	0.006	0.011	0.013	0.176	0.080	0.197
50	0.004	0.011	0.008	0.170	0.070	0.190
100	0.003	0.011	0.006	0.175	0.069	0.195

Table 9. Comparison of calibration performance according to the bundle size of frame for temporal filtering on the Oxford dataset.

Bundle Size of Frames	Rotation Error (°)			Translation Error (cm)		
	Roll	Pitch	Yaw	X	Y	Z
1	0.055	0.028	0.080	0.536	0.532	0.330
10	0.049	0.024	0.066	0.363	0.335	0.305
25	0.044	0.022	0.066	0.334	0.303	0.272
50	0.039	0.019	0.065	0.290	0.286	0.269
100	0.035	0.017	0.060	0.277	0.305	0.247

5. Conclusions

In this paper, we realized a novel approach for online multi-sensor calibration implemented using a voxel-based CNN and 3D convolutional kernels. Our method aims to calibrate between sensors that can measure 3D space. In particular, the voxelization that converts the input 3D point cloud into voxel and the AM introduced to find the correlation of features between the reference and target sensors contributed greatly to the completeness of the proposed method. We demonstrated through experiments that the proposed method can perform both LiDAR-stereo camera calibration and LiDAR-LiDAR calibration. In the calibration of the LiDAR-stereo camera combination, the proposed method showed experimental results that surpassed all existing CNN-based calibration methods for the LiDAR-camera combination. We demonstrated the effects of iterative refinement on the five networks and the effects of temporal filtering through experiments. The proposed method achieved a rotation error of less than 0.1° and a translation error of less than 1 cm on both the KITTI and Oxford datasets.

Author Contributions: Conceptualization, J.S. and J.L.; methodology, J.S. and J.L.; software, J.S.; validation, J.S. and J.L.; formal analysis, J.S.; data curation, J.S.; writing—original draft preparation, J.S.; writing—review and editing, J.S. and J.L.; visualization, J.S.; supervision, J.L.; project administration, J.L.; funding acquisition, J.L. All authors have read and agreed to the published version of the manuscript.

Funding: This research was supported by Basic Science Research Program through the National Research Foundation of Korea (NRF), funded by the Ministry of Education (NRF-2016R1D1A1B02014422).

Institutional Review Board Statement: Not applicable.

Informed Consent Statement: Not applicable.

Conflicts of Interest: The authors declare no conflict of interest.

References

1. Lv, X.; Wang, B.; Dou, Z.; Ye, D.; Wang, S. LCCNet: LiDAR and Camera Self-Calibration using Cost Volume Network. In Proceedings of the 2021 IEEE/CVF Conference on Computer Vision and Pattern Recognition (CVPR), Nashville, TN, USA, 19–25 June 2021; pp. 2888–2895.
2. Iyer, G.; Ram, R.K.; Murthy, J.K.; Krishna, K.M. CalibNet: Geometrically supervised extrinsic calibration using 3D spatial transformer networks. In Proceedings of the 2018 IEEE/RSJ International Conference on Intelligent Robots and Systems (IROS), Madrid, Spain, 1–5 October 2018; pp. 1110–1117.
3. Jiao, J.; Ye, H.; Zhu, Y.; Liu, M. Robust odometry and mapping for multi-lidar systems with online extrinsic calibration. *IEEE Trans. Robot.* **2022**, *38*, 351–571. [CrossRef]
4. Zhu, H.; Deng, J.; Zhang, Y.; Ji, J.; Mao, Q.; Li, H.; Zhang, Y. VPFNet: Improving 3D Object Detection with Virtual Point based LiDAR and Stereo Data Fusion. *arXiv* **2021**, arXiv:2111.14382. [CrossRef]
5. Caltagirone, L.; Bellone, M.; Svensson, L.; Wahde, M. LIDAR–camera fusion for road detection using fully convolutional neural networks. *Robot. Auton. Syst.* **2019**, *111*, 125–131. [CrossRef]
6. Vachmanus, S.; Ravankar, A.A.; Emaru, T.; Kobayashi, Y. Multi-modal sensor fusion-based semantic segmentation for snow driving scenarios. *IEEE Sens. J.* **2021**, *21*, 16839–16851. [CrossRef]
7. Dimitrievski, M.; Veelaert, P.; Philips, W. Behavioral pedestrian tracking using a camera and lidar sensors on a moving vehicle. *Sensors* **2019**, *19*, 391. [CrossRef] [PubMed]
8. Palieri, M.; Morrell, B.; Thakur, A.; Ebadi, K.; Nash, J.; Chatterjee, A.; Kanellakis, C.; Carlone, L.; Guaragnella, C.; Agha-Mohammadi, A.A. Locus: A multi-sensor lidar-centric solution for high-precision odometry and 3d mapping in real-time. *IEEE Robot. Autom. Lett.* **2020**, *6*, 421–428. [CrossRef]
9. Sualeh, M.; Kim, G.-W. Dynamic Multi-LiDAR Based Multiple Object Detection and Tracking. *Sensors* **2019**, *19*, 1474. [CrossRef] [PubMed]
10. Geiger, A.; Lenz, P.; Urtasun, R. Are we ready for autonomous driving? The kitti vision benchmark suite. In Proceedings of the 2012 IEEE Conference on Computer Vision and Pattern Recognition (CVPR), Providence, RI, USA, 16–21 June 2012; pp. 3354–3361.
11. Sun, P.; Kretzschmar, H.; Dotiwalla, X.; Chouard, A.; Patnaik, V.; Tsui, P.; Guo, J.; Zhou, Y.; Chai, Y.; Caine, B.; et al. Scalability in perception for autonomous driving: Waymo open dataset. In Proceedings of the 2020 IEEE/CVF conference on computer vision and pattern recognition (CVPR), Seattle, WA, USA, 13–19 June 2020; pp. 2443–2451.
12. Barnes, D.; Gadd, M.; Murcutt, P.; Newman, P.; Posner, I. The oxford radar robotcar dataset: A radar extension to the oxford robotcar dataset. In Proceedings of the 2020 IEEE International Conference on Robotics and Automation (ICRA), Paris, France, 31 May–31 August 2020; pp. 6433–6438.
13. Sheeny, M.; De Pellegrin, E.; Mukherjee, S.; Ahrabian, A.; Wang, S.; Wallace, A. RADIATE: A radar dataset for automotive perception in bad weather. In Proceedings of the 2021 IEEE International Conference on Robotics and Automation (ICRA), Xi'an, China, 30 May–5 June 2021; pp. 1–7.
14. Park, Y.; Yun, S.; Won, C.S.; Cho, K.; Um, K.; Sim, S. Calibration between color camera and 3D LIDAR instruments with a polygonal planar board. *Sensors* **2014**, *14*, 5333–5353. [CrossRef] [PubMed]
15. Kim, E.S.; Park, S.Y. Extrinsic calibration between camera and LiDAR sensors by matching multiple 3D planes. *Sensors* **2019**, *20*, 52. [CrossRef] [PubMed]
16. Jiao, J.; Liao, Q.; Zhu, Y.; Liu, T.; Yu, Y.; Fan, R.; Wang, L.; Liu, M. A novel dual-lidar calibration algorithm using planar surfaces. In Proceedings of the 2019 IEEE Intelligent Vehicles Symposium (IV), Paris, France, 9–12 June 2019; pp. 1499–1504.
17. Xue, B.; Jiao, J.; Zhu, Y.; Zhen, L.; Han, D.; Liu, M.; Fan, R. Automatic calibration of dual-LiDARs using two poles stickered with retro-reflective tape. In Proceedings of the 2019 IEEE International Conference on Imaging Systems and Techniques (IST), Abu Dhabi, UAE, 8–10 December 2019; pp. 1–6.
18. Castorena, J.; Kamilov, U.S.; Boufounos, P.T. Autocalibration of lidar and optical cameras via edge alignment. In Proceedings of the 2016 IEEE International Conference on Acoustics, Speech and Signal Processing (ICASSP), Shanghai, China, 20–25 March 2016; pp. 2862–2866.
19. Kang, J.; Doh, N.L. Automatic targetless camera–lidar calibration by aligning edge with gaussian mixture model. *J. Field Robot.* **2020**, *37*, 158–179. [CrossRef]
20. Pandey, G.; McBride, J.R.; Savarese, S.; Eustice, R.M. Automatic extrinsic calibration of vision and lidar by maximizing mutual information. *J. Field Robot.* **2015**, *32*, 696–722. [CrossRef]
21. Taylor, Z.; Nieto, J. A mutual information approach to automatic calibration of camera and lidar in natural environments. In Proceedings of the 2012 Australian Conference on Robotics and Automation, Wellington, New Zealand, 3–5 December 2012; pp. 3–5.
22. Ishikawa, R.; Oishi, T.; Ikeuchi, K. Lidar and camera calibration using motions estimated by sensor fusion odometry. In Proceedings of the 2018 IEEE/RSJ International Conference on Intelligent Robots and Systems (IROS), Madrid, Spain, 1–5 October 2018; pp. 7342–7349.
23. Wei, P.; Yan, G.; Li, Y.; Fang, K.; Liu, W.; Cai, X.; Yang, J. CROON: Automatic Multi-LiDAR Calibration and Refinement Method in Road Scene. *arXiv* **2022**, arXiv:2203.03182.
24. Kümmerle, R.; Grisetti, G.; Burgard, W. Simultaneous calibration, localization, and mapping. In Proceedings of the 2011 IEEE/RSJ International Conference on Intelligent Robots and Systems, San Francisco, CA, USA, 25–30 September 2011; pp. 3716–3721.

25. Schneider, N.; Piewak, F.; Stiller, C.; Franke, U. RegNet: Multimodal sensor registration using deep neural networks. In Proceedings of the 2017 IEEE Intelligent Vehicles Symposium (IV), Los Angeles, CA, USA, 11–14 June 2017; pp. 1803–1810.
26. Zhu, Y.; Li, C.; Zhang, Y. Online camera-lidar calibration with sensor semantic information. In Proceedings of the 2020 IEEE International Conference on Robotics and Automation (ICRA), Paris, France, 31 May–31 August 2020; pp. 4970–4976.
27. Wang, W.; Nobuhara, S.; Nakamura, R.; Sakurada, K. Soic: Semantic online initialization and calibration for lidar and camera. *arXiv* **2020**, arXiv:2003.04260.
28. Vaswani, A.; Shazeer, N.; Parmar, N.; Uszkoreit, J.; Jones, L.; Gomez, A.N.; Kaiser, Ł.; Polosukhin, I. Attention is all you need. In Proceedings of the 31st Conference on Neural Information Processing Systems (NIPS 2017), Long Beach, CA, USA, 4–9 December 2017; p. 30.
29. Yan, Y.; Mao, Y.; Li, B. SECOND: Sparsely Embedded Convolutional Detection. *Sensors* **2018**, *18*, 3337. [CrossRef] [PubMed]
30. Dai, A.; Ruizhongtai Qi, C.; Nießner, M. Shape completion using 3d-encoder-predictor cnns and shape synthesis. In Proceedings of the 2017 Proceedings of the IEEE conference on computer vision and pattern recognition, Honolulu, HI, USA, 21–26 July 2017; pp. 5868–5877.
31. Ji, M.; Gall, J.; Zheng, H.; Liu, Y.; Fang, L. Surfacenet: An end-to-end 3d neural network for multiview stereopsis. In Proceedings of the 2017 Proceedings of the IEEE International Conference on Computer Vision, Venice, Italy, 22–29 October 2017; pp. 2307–2315.
32. Xu, Y.; Tuttas, S.; Hoegner, L.; Stilla, U. Voxel-based segmentation of 3D point clouds from construction sites using a probabilistic connectivity model. *Pattern Recognit. Lett.* **2018**, *102*, 67–74. [CrossRef]
33. Hernandez-Juarez, D.; Chacón, A.; Espinosa, A.; Vázquez, D.; Moure, J.C.; López, A.M. Embedded real-time stereo estimation via semi-global matching on the GPU. *Procedia Comput. Sci.* **2016**, *80*, 143–153. [CrossRef]
34. Hirschmuller, H. Accurate and efficient stereo processing by semi-global matching and mutual information. In Proceedings of the 2005 IEEE Computer Society Conference on Computer Vision and Pattern Recognition (CVPR), Boston, MA, USA, 7–12 June 2015; Volume 2, pp. 807–814.
35. Xu, C.; Wu, B.; Wang, Z.; Zhan, W.; Vajda, P.; Keutzer, K.; Tomizuka, M. Squeezesegv3: Spatially-adaptive convolution for efficient point-cloud segmentation. In Proceedings of the European Conference on Computer Vision, Glasgow, UK, 23–28 August 2020; pp. 1–19.
36. Zhou, Y.; Tuzel, O. VoxelNet: End-to-End Learning for Point Cloud Based 3D Object Detection. *arXiv* **2017**, arXiv:1711.06396.
37. Zheng, W.; Tang, W.; Chen, S.; Jiang, L.; Fu, C.W. Cia-ssd: Confident iou-aware single-stage object detector from point cloud. In Proceedings of the AAAI Conference on Artificial Intelligence, Virtual Event, 2–9 February 2021; Volume 35, pp. 3555–3562.
38. Graham, B.; van der Maaten, L. Submanifold Sparse Convolutional Networks. *arXiv* **2017**, arXiv:1706.01307.
39. Ioffe, S.; Szegedy, C. Batch normalization: Accelerating deep network training by reducing internal covariate shift. In Proceedings of the International Conference on Machine Learning, Lille, France, 6–11 July 2015; pp. 448–456.
40. Ba, J.L.; Kiros, J.R.; Hinton, G.E. Layer normalization. *arXiv* **2016**, arXiv:1607.06450.
41. Kingma, D.P.; Ba, J. Adam: A method for stochastic optimization. *arXiv* **2014**, arXiv:1412.6980.

sensors

Article

Deep-Learning-Based Adaptive Advertising with Augmented Reality

Marco A. Moreno-Armendáriz [1], Hiram Calvo [1,*], Carlos A. Duchanoy [2], Arturo Lara-Cázares [3], Enrique Ramos-Diaz [3] and Víctor L. Morales-Flores [3]

[1] Centro de Investigación en Computación, Instituto Politécnico Nacional, Ciudad de Mexico 07738, Mexico; mam_armendariz@cic.ipn.mx

[2] Gus Chat, Ciudad de Mexico 06600, Mexico; carlos.duchanoy@gus.chat

[3] Escuela Superior de Cómputo, Instituto Politécnico Nacional, Ciudad de Mexico 07738, Mexico; jlarac1300@alumno.ipn.mx (A.L.-C.); eramosd1300@alumno.ipn.mx (E.R.-D.); vmoralesf1300@alumno.ipn.mx (V.L.M.-F.)

* Correspondence: hcalvo@cic.ipn.mx

Abstract: In this work we describe a system composed of deep neural networks that analyzes characteristics of customers based on their face (age, gender, and personality), as well as the ambient temperature, with the purpose of generating a personalized signal to potential buyers who pass in front of a beverage establishment; faces are automatically detected, displaying a recommendation using deep learning methods. In order to present suitable digital posters for each person, several technologies were used: Augmented reality, estimation of age, gender, and estimation of personality through the Big Five test applied to an image. The accuracy of each one of these deep neural networks is measured separately to ensure an appropriate precision over 80%. The system has been implemented into a portable solution, and is able to generate a recommendation to one or more people at the same time.

Keywords: targeted advertising; emotion-based recommendation; augmented reality; computer vision; deep learning

Citation: Moreno-Armendáriz, M.A.; Calvo, H.; Duchanoy, C.A.; Lara-Cázares, A.; Ramos-Diaz, E.; Morales-Flores, V.L. Deep-Learning-Based Adaptive Advertising with Augmented Reality. *Sensors* **2022**, *22*, 63. https://doi.org/10.3390/s22010063

Academic Editor: Jing Tian

Received: 10 November 2021
Accepted: 21 December 2021
Published: 23 December 2021

Publisher's Note: MDPI stays neutral with regard to jurisdictional claims in published maps and institutional affiliations.

1. Introduction

Today, competition in the market for products and services is intense, so companies have been forced to adopt different strategies to differentiate themselves from the crowd and thereby attract and retain customers [1] because, although the quality of these products or services is an important point, at present the experience that is provided to the user during the acquisition of any product becomes a crucial point. Customizing products or services is a differentiation strategy that allows to satisfy better customer needs [1] to the point that it is associated with a 26% increase in profitability and a 12% increase in capitalization of the market [2].

Given the importance of differentiating companies, the objective of the system proposed in this work is to display a personalized advertisement for each potential client that passes outside the BubbleTown® establishment, using a screen to display advertising, which will use technologies such as augmented reality to show the user the recommendation and in this way draw their attention. BubbleTown® is a Mexican company with a branch in Mexico City specialized in the sale of customizable tea or yogurt-based drinks.

The objective of the system will be to analyze the client by means of an image of their face and to recommend one of the BubbleTown® products that they might like the most. To achieve the objective, artificial vision techniques will be used from cameras strategically installed in the premises, together with neural networks that will allow estimating the age, gender, and personality of the client.

A recommendation system filters personalized information, seeking to understand the user's tastes to suggest appropriate things considering the exclusive patterns of them [3]. A content-based recommendation system examines the characteristics of the products in order to identify those that might be of interest to the user. It is common to have product information stored in a database and with the description, together with the user's profile to generate the recommendation, it is possible to generate a preference profiles for the user's feedback [4]. For its part, collaborative filtering is the process in which different articles are evaluated or filtered using the opinion generated by users. For its correct operation, the system must have scores or ratings of the article to be recommended, so it requires users to assign ratings to the articles they consume [4].

Through various studies, it has been questioned whether a person's taste preference is determined by some factor, of which it has been found that age, gender, and even personality can influence these preferences. In [5,6], analysis was carried out considering age, where it was found that young people prefer sweet flavors, while with aging the preference for this flavor reduces, giving way to the preference for salty, sour, and bitter flavors; and regarding gender, in studies such as [7], it has been shown that women tend to prefer sweet flavors 10% more than men, while in [6], it was concluded that men will have greater acceptance towards acidic or bitter flavors. Last but not least, it has also been shown that there is a relationship between personality and the tendency towards some flavor, as is the case of [8] which results in certain personality traits that influence the preference of any kind of flavors.

Until recently, progress in computer vision was based on the features of manual engineering however, feature engineering is difficult, time consuming, and requires expert knowledge of the problem domain. The other problem with hand-designed features, such as background subtraction and edge detection, is that they are too scarce in terms of the information they can capture from an image [9]. Fortunately in recent years, deep learning advances have gained significant attention in fields such as image processing, so the task to obtain data regarding age, gender, and personality will not be handled through traditional techniques, but rather through deep neural networks, algorithms that today have gained importance in the area of computer systems due to their ability to learn.

This work is divided into four parts: The state of the art, methodology, results, and conclusions. It begins by giving a tour of the relevant works that are related to the areas that this work addresses in the section on the state of the art. Afterwards, the methodology section will explain the steps that were carried out to achieve the objective along with a brief explanation of each of them. Finally, in the results section, a short explanation will be given about the most relevant parts at the end of the project.

2. State of the Art

Within recommendations, there are many works that propose and achieve the task of recommending a product to a client, but there are few systems whose main focus is the generation of dynamic advertising from the detection of an individual in front of this. The Intel suite® [10,11], distributed in 2011 in the USA, is a targeted advertising device that makes use of automated systems to detect potential consumers through computer vision. Among its most striking features are the use of anonymous sensors that temporarily search and capture patterns of faces or bodies within a predetermined range of vision, in other words, the ability to detect faces; the analysis of anthropometric features so as to provide advertisements through screens, depending on the viewer, is also generated based on attributes such as the age, height, race, and gender of the viewer.

Wang et al., (2020) in [12] use their users' information, such as age, gender, location, education level, and more to create a personalized recommendation for online courses. On the other hand, some recommenders use deep learning, such as Liu et al., (2019) [13] that presents a recommender which learns from the interaction between user and product through Deep Learning, highlighting the use of convolutional neural networks.

As mentioned in the introduction, this system analyzes the faces of clients to obtain information regarding age, gender, and personality using deep learning. Since 2011, the use of a CNN for estimating age through a face was proposed for the first time [14]. A more recent work, presented by Orozco et al., (2017) [15], uses a neural network with the purpose of obtaining the gender of a person through the image of their face; for this they implemented 2 stages: Generation of candidate regions (ROI) and classification of the candidate regions in the male or female person. Another relevant work is the multipurpose convolutional network of Ranjan et al., (2017) [16]. This CNN is able to detect faces, extract key points, pose angles, determine smile expression, gender (binary classification), and estimate age, simultaneously. Another work by Xing et al., (2017) [17] carried out a diagnosis on the three types of formulations (classification, regression, or ordinal regression) to estimate age using five cost functions as well as three different multitasking architectures that include estimation age, gender, and race classification. Vasileiadis et al., (2019) [18] proposed a convolutional MobileNet network with TensorFlow Lite, which is suitable for low computational power devices that simultaneously estimates characteristics such as age, gender, race, and eye status, as well as whether the subject is smiling or has a beard, mustache, or glasses. As well as previous works, there are many more that aim to classify, through an image of a face, the age and gender of a person. In addition, other works of value are Zhang et al., (2017) [19], where the faces that appear in a video stream are detected and in [20], Liu et al., (2018), a face detection using LFDNet is presented. In [21], a probability Boltzmann machine network is used for face detection. Zhou et al., (2019) [22] presents a system using the YTF dataset, obtaining a 99.83% correct face detection, and Greco et al., (2020) [23] presented a gender recognition algorithm with a 92.70% accuracy.

Regarding personality, this is not an accessible piece of information that can be found in documents, but rather a characteristic that requires professionals and personalized research in human behavior [24], but it has been discovered that personality traits can be predicted with precision depending on the characteristics of an image, such as the saturation mean, the variation of the value, the temperature, the number of faces, or the color level (Instagram filters) [24].

In 2016, the ChaLearn dataset [25] was created for a contest whose objective was to identify the Big Five in a person through videos, composed of 10,000 videos of people speaking in front of the camera during 15 s obtained from YouTube in 720-p resolution, each tagged by Big Five using Amazon Mechanical Turk. Using deep regressions and convolutional neural networks, the ChaLearn winner combines the results of image analysis (face detection in frames) and analysis of audio characteristics (divided into N pieces) extracted from the dataset videos, to obtain a final mean precision slightly above 91% [24,25].

In [26,27], audio, images (using OpenFace), and spoken text are extracted from the videos in the ChaLearn dataset. In both, there are 3 separate components or channels for processing and extracting characteristics, one for each modality taken, and at the end the results of each component are combined to obtain a personality prediction.

Similarly, the compilation in [28] shows that the precision of jobs where only images are used versus those where they are combined with audio and even text (natural language) varies very little, at no more than 1%, and that the implemented model does not mean a great impact or increase in it.

In [24,29], a new dataset (PortraitPersonality.v2 dataset) was built from ChaLearn's, which consists of selfie-type images where only one person appears and their face is visible, labeled with the Big Five of the person in the photo. They were tested with the PortraitPersonality.v2 dataset, giving the FaceNet-1 model the best result. FaceNet is a face verification, recognition, and grouping network trained with millions of face images. Applying Transfer Learning reaches an average precision of 65.86%.

2.1. Augmented Reality

The use of augmented reality (AR) for advertising and commercial applications lies in completely replacing the need to try anything in stores, thus saving a considerable amount of time for customers, which would probably be used to observe, decide, and select a product (not always concluding in the sale of the same) and thus increase the sales possibilities of the stores [30].

AR also complements web applications by supporting the "live" observation of the objects displayed on screens, as a supplement to what is being produced. Thus, not only is the user informed about when they are "live", but they can also use it as a learning tool for future activities. In contrast to virtual reality (VR), which creates an artificial environment, AR simply makes use of the existing environment by overlaying new information on top of it. In AR, the information about the surrounding real world is made available to the user for information and/or interaction through the use of screens.

When selecting a beverage from a set of possible options, for example, it is possibly to see it first in your eyes, through a suitable AR application, a virtual glass, which has your preferred beverage with the best tasting quality and other associated characteristics such as the origin of the product, the way the product is processed, the number of calories in a unit of volume, etc.

A study of the market by Grand View Research, the market research firm, points out that this kind of application would generate a considerable increase in sales for stores and restaurants. The total worldwide market for AR is estimated to be more than US$13.4 billion by 2019 and is expected to reach US$340.16 billion in 2028, growing at a CAGR of 43.8% from 2021 to 2028 (www.grandviewresearch.com, accessed on 18 December 2021).

An early start in the realization of the commercial potential of AR was made by the launch of Hololens, a headset capable of creating a virtual vision. This device, with a screen of about one inch by two inches and a thickness of two centimeters, is a product of Microsoft, since its development is carried out in the context of HoloLens, a new project of a company dedicated to the research and development of products focused on augmented reality applications. Perhaps, the best known example is Magic Mirror [31,32], devices that are basically a long-dimensional screen where the customer can interact with various simulated objects, provided by another specific device (markers). The marketing approach used in work [30] is that the users can see their reflection in the Magic Mirror with a virtual model of clothing or a product that they would like to try on. The advantage of this system over going to the store is that once the user selects the garment for testing, they have the ability to change some details, such as color, size, and even stitching.

Another application where augmented reality interacts with the person is Snapchat Lenses [33], which is a popular mobile application that applies filters to the face, such as changing eye color, the shape of the face, adding accessories, having animations started when the mouth is opened or the eyebrows raised, as well as exchanging faces with someone else. Other functionalities are the detection of frontal faces by means of the camera of the mobile device, as well as the application of filters on a three-dimensional mask superimposed on them in real time. A Snapchat and Kohl collaboration [34] resulted in an AR feature that allows customers to visualize Kohl's products at home within the Snapchat app.

Recently, Berman et al., (2021) [35] published a self-explanatory guide on the following steps to successfully develop an AR app. One of the most important things to consider is how AR will help meet a business's marketing goals. Regarding selecting channels, wAR can be for online or in-store sales. However, one option to consider is to follow an omnichannel strategy that allows covering all types of customers. Millennials are a good target market for their affinity to new technologies. One last point to highlight is the importance of measuring the return on investment of the AR app, where one crucial aspect is to evaluate AR's success in increasing profits due to reductions in costs and increased sales.

2.2. Related Works

A brief comparison of our work with some published works and industry applications [36] is shown in Table 1. Academic papers focus on facial and gender recognition using various algorithms but do not incorporate other aspects such as Big Five personality analysis, generation of a personalized recommendation, and AR. On the other hand, the AR company apps focus on AR technology, but other elements are missing.

With the review of previous works, it can be said that, although the task of recommending a product to a client has been approached several times, few works do not require having the data of the client's preferences or history in their database to achieve the recommendation. On the other hand, combining the recommendations with the use of augmented reality is also scarce since it has focused more on other areas such as video games or applications for social networks. In most of the researched works, the similarity is that all the systems are made for previously registered users and that they interact on the commerce website— this limits the use of the recommender to the online user and forgets those who prefer interaction in physical stores, this being the motivation for this work.

Table 1. Related works in the literature and industry.

Authors	Face/Gender Recognition	Big5 Personality	Ambient Temperature	Product Recommendation	Augmented Reality	Advertising Totem
Kim et al., (2021) [37]	□	□	□	□	✓	□
Kohls (2020) [34]	□	□	□	□	✓	□
Xueyi et al., (2020) [21]	✓	□	□	□	□	□
Wayfair (2020) [38]	□	□	□	□	✓	□
Ikea (2019) [39]	□	□	□	□	✓	□
Adidas (2019) [40]	□	□	□	□	✓	□
L'Oréal (2019) [41]	□	□	□	□	✓	□
Zhou et al., (2019) [22]	✓✓	□	□	□	□	□
Asos (2019) [42]	□	□	□	□	✓	□
Hamid et al., (2018) [43]	□	□	□	□	✓	□
Liu et al., (2018) [20]	✓	□	□	□	□	□
Zara (2018) [44]	□	□	□	✓	✓	□
Hamid et al., (2016) [43]	□	□	□	✓	✓	□
Sephora (2016) [45]	□	□	□	□	✓	□
Zhang et al., (2017) [19]	✓	□	□	□	□	□
Lacoste (2014) [46]	□	□	□	✓	✓	□
Hermès (2015) [47]	□	□	□	□	✓	□
Converse (2012) [48]	□	□	□	□	✓	□
Ours	✓✓	✓	✓	✓	✓	✓

The objective of this work is to analyze the face of a client on a particular pose, that is, to show the importance of knowing the client's personality and their age. For the face recognition of a client, the approach has been made using face detection and classification methods. After this, the work presents the recommendation of products in a commerce display (totem); using the detected client's age, gender, and personality from the customer's face, the recommendation is sent by the system allowing to make it possible to use it as feedback to improve the final recommendation.

The main contributions of this work are:

- A novel deep neural network can predict the age and gender of more than one person in a selfie (see details in Section 3.7).
- A new content-based recommender can select a different beverage for each user.
- A methodology to provide a complete solution to implement an AR system that can be of help to stores that seek to boost sales using an innovative display system (cf. Figure 1).
- A publicity totem that works in soft real-time [49] can present an AR recommendation to the user.

Finally, to our knowledge, a system with all these features has not yet been developed.

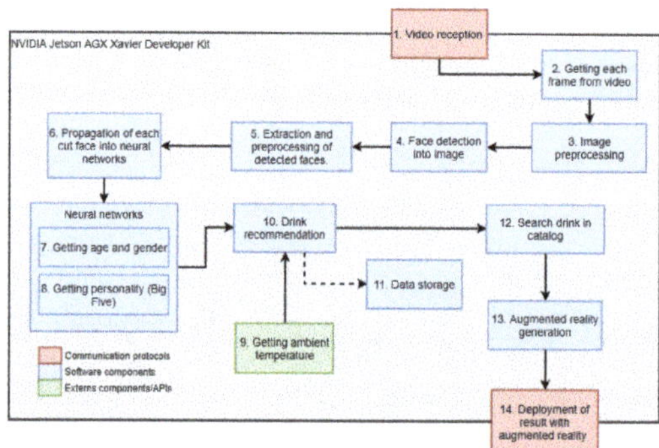

Figure 1. Global system diagram.

3. Methodology

The proposed system is composed of deep neural networks that analyze independent characteristics of the user by means of their face (age, gender, and personality), in addition to obtaining the ambient temperature, to be able to be entered into a drinks recommender system and display the recommendation obtained through a module that adds augmented reality to a screen.

Figure 1 shows three types of blocks: The green ones represent external components used for our purposes. Those in red represent actions that are carried out using communication protocols between hardware components, and the blue ones are the modules that were implemented in this work.

Next, a description is made of each of the modules that appear in the diagram in Figure 1, in addition, in the final block used the hardware is explained in a general way.

3.1. Video Reception

Video from outside the store is continuously sent from an IP camera to an NVIDIA Jetson device for processing. Previously, the network configuration must be made to communicate all the devices, as indicated in [50].

3.2. Video-to-Image Capture

Once the video has been obtained in the NVIDIA Jetson from the camera, it is necessary to extract frame by frame from it, since these are the basis of all the processing, using the OpenCV library.

3.3. Image Preprocessing

Four main preprocessing operations were performed on the frame obtained by the previous design:

1. Resize the image: The frames coming from the camera are resized to a size of 300×300.
2. Normalize the image Convert the frame to a matrix and then normalize it (divide the maximum value of a pixel in RGB color format by 255).
3. Increase the brightness: Each value of the matrix is increased.
4. Apply the transpose to the image: In order to improve the precision of face detection, the transpose operation is applied to the matrix.

3.4. Face Detection in the Image

The architecture of the neural network Single Shot Detection (SSD) [51] has several advantages, such as the ability to detect objects at different scales and resolutions, in addition to performing it at high speed. This is a perfect fit for the needs of the project, as it requires a fast response time and the ability to detect faces at various distances from the camera. Therefore, a Single Shot Detection is used based on a neural network MobileNet [52] to have an even lower processing time.

It is necessary to carry out a new custom training to adapt the detection only of faces, so using the API Tensorflow Object Detection [53] that already contains the pre-trained neural network to classify 80 types of classes, together with the transfer of learning technique, this goal is achieved. The datasets used for this process are Face Detection in Images [54] and Google Facial Expression Comparison Dataset [55]. Figure 2 shows an example [55] in which the face is painted in a red box.

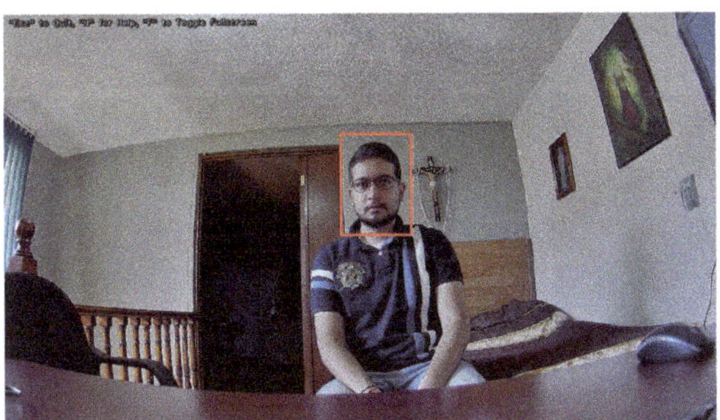

Figure 2. Detected face.

3.5. Face Extraction and Preprocessing

Since the location of the faces in the image is known, they are cut out and pre-processed in order to be independently analyze and propagate in a convolutional neural network. This process is illustrated in Figure 3.

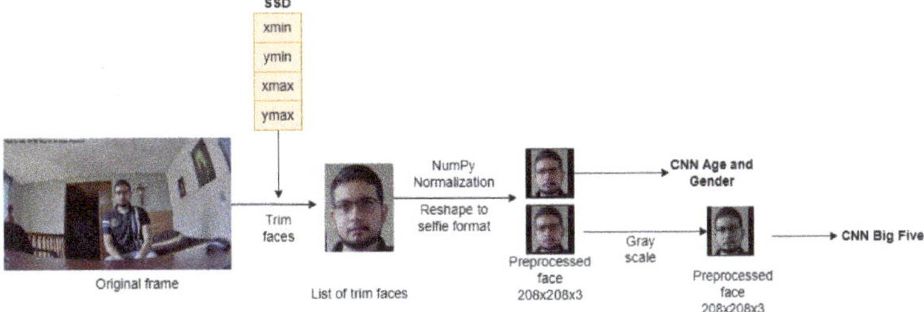

Figure 3. Extraction of the faces detected by the SSD from the frames captured by the IP camera, and their processing to be entered into the neural networks.

3.6. Propagation of Each Cropped Face in Neural Networks

Each of the faces obtained in the previous block will be propagated through two convolutional neural networks to estimate age, gender, and personality characteristics. Figures 4 and 5 show these propagations with their respective inputs and outputs.

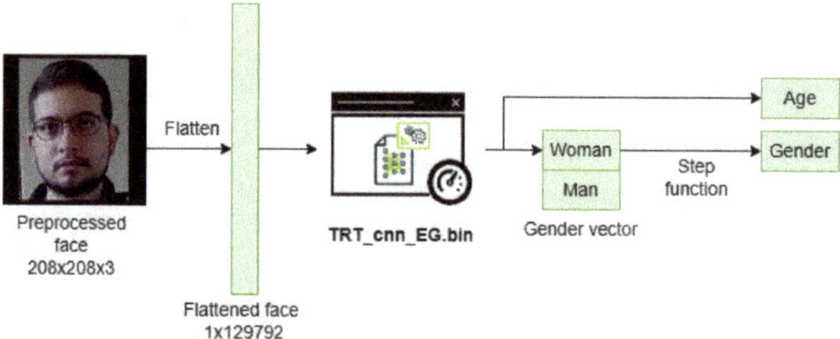

Figure 4. Diagram showing the flattened input face, the age, and gender network model in TensorRT, and its two outputs: The age obtained from the regressor, and the gender vector obtained from the classifier.

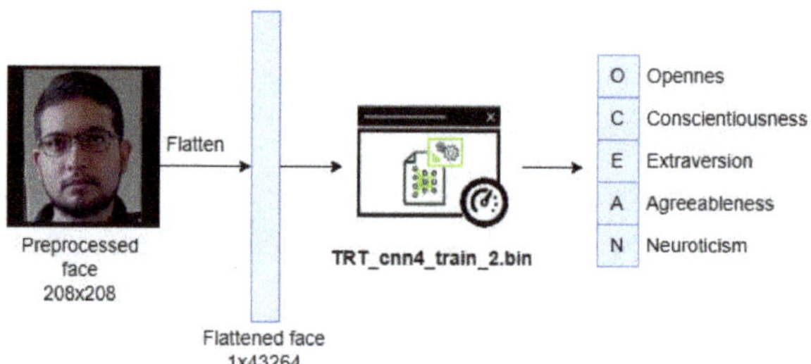

Figure 5. Diagram showing the flattened input face, the classification model: CNN-4 [24] of the Big Five network in TensorRT, and the output vector of size 5 where each position corresponds to a personality dimension as indicated.

3.7. Getting Age and Gender

To obtain age and gender from an image of a person's face in selfie form, a multitasking convolutional neural network is designed with the intention of reducing the amount of computational resources to be used. We start with the layers estimating age, which is the first part of the design. The architecture of this network is shown in Figure 6 and its hyper-parameters in Table 2.

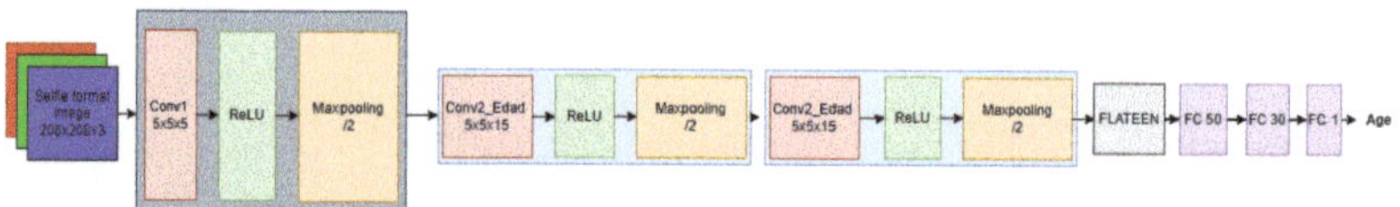

Figure 6. Proposed CNN used to estimate age.

Table 2. Hyper-parameters used to train age in a CNN.

Hyperparameter	Value
Epochs	2600
Learning Factor	1×10^{-4}
Batch Size	4

Afterwards, we carry out a knowledge transfer that will allow us to reduce the training times required to obtain the second CNN that will be responsible for determining the gender of the person. The transfer is carried out in two ways. First, the starting layer enclosed in the gray rectangle (Conv1) freezes during this workout. Second, the final weights of the individual layers are used as initial values for this training which is indicated by the red arrows, see Figure 7.

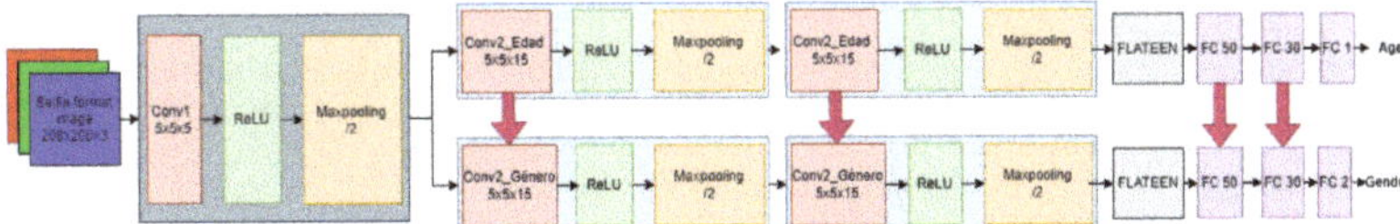

Figure 7. Designed transfer learning.

Now the training of the second part of CNN is carried out using the hyper-parameters of Table 3. For this training, the regularization of the network was necessary; for this, dropout layers were added. Thus, the final design of the CC is shown in Figure 8.

Table 3. Hyper-parameters used to train the genre rating portion of CNN multitasking.

Hyperparameter	Value
Epochs	25
Learning Factor	1×10^{-4}
Batch Size	4

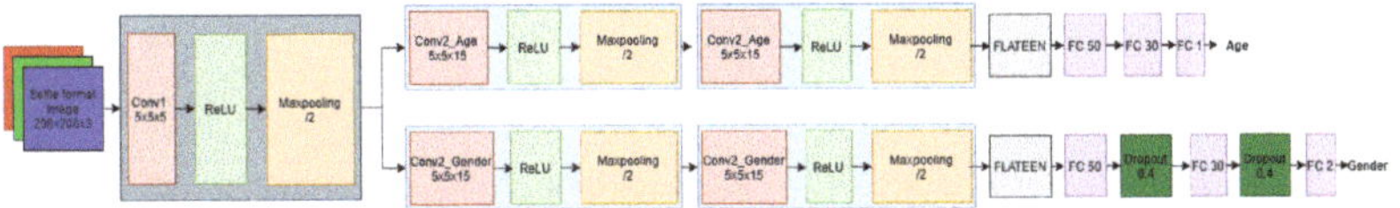

Figure 8. Final architecture of the multitasking convolutional neural network to obtain the age and gender of the person.

3.8. Obtaining the Personality (Big Five)

The estimation of personality based on his face will be measured using the Big Five model [24], which measures personality through 5 dimensions on scales from 0 to 1:

1. Openness to Experience (O);
2. Conscientiousness (C);
3. Extraversion (E);
4. Agreeableness (A);
5. Neuroticism (N).

For this step, a model previously built and developed from facial analysis in images will be taken, since these provide better results than all types of existing multimedia files (audio, text, video, and images) [27], being the model of classification: CNN-4 of the work in [24,29] the one selected to be integrated into the system, because despite the fact that there is a better model obtained in the same work called FaceNet-1 [56], its weight, computational requirement, and convergence time is very high, causing the selected hardware devices not to be able to support it, taking then the second best model of the work, whose precision obtained does not differ significantly from this, and in return offers better performance and speed. In Tables 4 and 5, a comparison of the precision in the detection of personality by Big Five is made between the various models of [56] from the images of faces, which have the following characteristics: They are in a selfie format, they are in grayscale, normalized, and their resolution is 208 × 208.

Table 4. Comparison of the best results obtained from the different Big Five models developed in [56].

Model	Average Precision
Regressor: ACoder	50.05% (with tolerance)
Regressor: CNN-2 (B)	52.06% (with tolerance)
Clasiffier: CNN-4	**65.77%**
Classiffier: FaceNet-1	**65.86%**

Table 5. Comparison between the results (each one is the percentage of precision of the detection of each personality) by dimension of personality among the best Big Five models in [56].

Model	Precision	O	C	E	A	N
Human	56.6	58.3	50.0	33.3	**83.3**	58.3
CNN-4	65.7	62.0	67.9	72.1	61.4	65.2
FaceNet-1	**65.8**	**61.4**	**69.5**	**73.2**	60.6	**64.3**

The convolutional neural network architecture of the CNN-4 classification model [56] for the detection of Big Five in the system is shown in Figure 9.

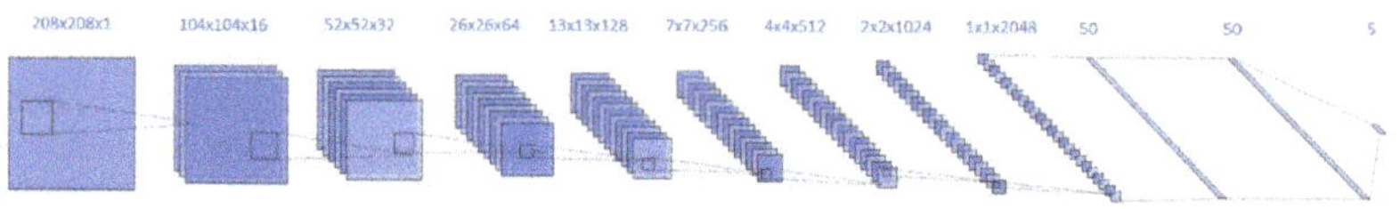

Figure 9. CNN-4 classification model architecture for Big Five [24].

3.9. Obtaining the Ambient Temperature

Although the ambient temperature is not a factor that will influence the recommendation of the drink, knowing this information, the system will have the ability to recommend the most suitable drink modality, since within the BubbleTown® catalog there are three options: Zen (hot drink), Iced (cold drink), or Frozen (Frappé). To give the system the ability to obtain the ambient temperature, the API provided by OpenWeatherMap [57] was used.

3.10. Drink Recommendation

In very simple terms, a recommendation system is an application that filters information in order to suggest appropriate things to the user [3], which for this job, the recommendation will be a drink from the BubbleTown® catalog.

To achieve the proposed task, this work uses a content-based recommender, that is, a recommendation system that examines the characteristics of the products [4] of BubbleTown® that could be of interest for the user.

The recommender works by using characteristic flavor vectors for each drink on the menu, generated with the support of the *Coffee Taster's Flavor Wheel* [58] because although there are flavor wheels specific for tea, these wheels have been built considering aroma, texture, and flavor characteristics; while the wheel generated in [58] only considers the flavor.

For the vectorization of the beverages, vectors are first generated for each element of the flavor wheel. This vectorization process is achieved by dividing the wheel into flavor classes (sweet, umami, bitter, sour, and spicy), then the ingredients are listed and a value between 1 (one) and 0 (zero) is assigned depending on the location they have within the flavor class.

With the vectorized basic flavors, the vectors that characterize the ingredients of each drink are added and with the "softmax" function of the resulting vector, what will be the characteristic flavor vector for a said drink is obtained.

Figure 10 shows in a general way how the system recommender works. It is important to note that for this recommendation process the salty taste has been eliminated because none of the BubbleTown® drinks have this flavor.

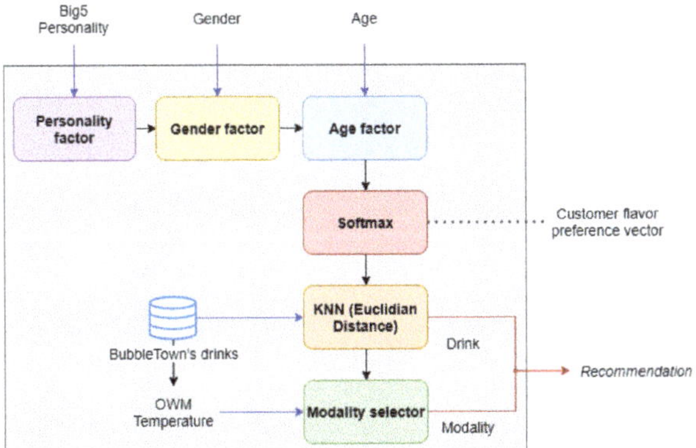

Figure 10. Diagram of the recommender operation.

- Personality factor: Once the customer's personality vector is known together with the Big Five network, the vector will be multiplied by a matrix with the values from Table 6 to obtain a matrix with the customer's taste preference based on their personality. The next step will be to add all the values for each flavor and thus a vector of size five will be obtained, where each value of the vector will correspond to a flavor class (sweet, umami, bitter, sour, and spicy).

Table 6. Average taste preference as a function of personality in values from 0 to 1. Table obtained with data from [8].

Taste Preference	O	C	E	A	N
Sweet taste	0.7243	0.4357	0.6138	0.5614	0.4036
Umami taste	0.2986	0.4900	0.5852	0.1886	0.1979
Bitter taste	0.2786	0.1307	0.1600	0.2424	0.1350
Acid taste	0.0271	0.1264	0.2381	0.0252	0.1543
Spicy taste	0.4857	0.2421	0.5971	0.6424	0.5607

- Gender factor: For this objective, Table 7 is used, so that the vector obtained after applying the personality factor is multiplied by this factor too.

Table 7. Average taste preference based on gender. Table obtained with data from [8].

Taste Preference	Female	Male
Sweet taste	0.7789	0.7292
Umami taste	0.4434	0.3752
Bitter taste	0.2530	0.3596
Acid taste	0.3783	0.4472
Spicy taste	0.5431	0.8288

- Age factor: After obtaining the estimated age, the customer will be placed in one of the four age classes and, together with Table 8, the preference vector will be adjusted of the customer's flavor when multiplied by said table. To continue with the vector format, the flavor *umami* and *spicy* are added however, in the Table both have values of 0 (zero), which means that age will not influence these flavor classes.

Table 8. Average taste preference based on age group. Table obtained with data from [6].

Taste Preference	1–17 Years	18–36 Years	37–50 Years	>50 Years
Sweet taste	0.9520	0.8500	0.8100	0.7700
Umami taste	0.0000	0.0000	0.0000	0.0000
Bitter taste	0.2652	0.2400	0.1867	0.1700
Acid taste	0.1972	0.2033	0.2067	0.2233
Spicy taste	0.0000	0.0000	0.0000	0.0000

- Softmax: After applying all the factors that influence the taste preference, the "Softmax" function will be applied to the vector in order to obtain a customer flavor preference vector.

- KNN (Euclidean distance): Since we have the client flavor preference vector, we will use the KNN or "nearest neighbor" algorithm using as a metric, the Euclidean distance to obtain a drink from the database whose flavor is the one that is most similar to that vector.
- Mode selector: From the previous step, you already have the drink that will be recommended by the system. In this step, the drink mode (zen, iced, or frozen) will be selected depending on the ambient temperature.

3.11. Data Storage

The system uses a database to store information such as the beverage catalog, the list of static advertising images or the recommendations that the system makes for each person. For this, it was decided to use a non-relational database.

The database will store the drinks that are available for purchase, the static advertising that will be projected when there is no customer in front of the system, and, finally, each of the recommendations generated by the system during its operation. The recommendations are stored together with the recognized face's features, the estimated age, its gender classification, and the personality vector. The recommended drink will also be stored with said parameters, the ambient temperature at the time of recommendation, and the most suitable drink modality.

Since the face could be considered as a sensitive data, it receives a special treatment before being stored, since it is encrypted using the AES-256 algorithm, in such a way that only system administrators can access this image and only in order to maintain or improve the system proposed through this work.

3.12. Search for the Drink in the Catalog

The database contains the characteristic vectors for each drink from the BubbleTown® catalog and the valid modalities for each drink (as explained in Section 3.10) however, the path of the images that will be used to generate the augmented reality is also stored. These images will be loaded into memory in order to create the augmented reality that shows the recommendation to the client.

3.13. Generation of Augmented Reality

One of the parts that consumes the most resources is the generation of augmented reality, since to create that feeling of "interaction" with the user it must be constantly updating itself and, considering that at the code level, manipulating the images consists of modifying an array the same size as the image resolution, too many operations are performed to produce a single image. These operations should be carried out approximately 60 times per second if it is to be made an imperceptible process for the user.

Figure 11 shows in a general way how the module in charge of generating augmented reality works. To operate the system, it is necessary to feed the module with the frame obtained from the camera, the faces that have been previously recognized and processed, as well as advertising (images of the drinks to be recommended). The first component of the augmented reality module will be in charge of adjusting the size of the thought balloon based on how close or far it is from the camera to create an effect of depth. The next step in this process will be to add the advertising to the original frame, where finally the company logo will be added in the upper left corner, a banner with the name of the drink to recommend at the bottom, and finally, the balloon of thought with the image of the drink in the upper left side of each detected face. In order to exemplify the idea for the reader, the augmented reality proposal is placed in Figure 12.

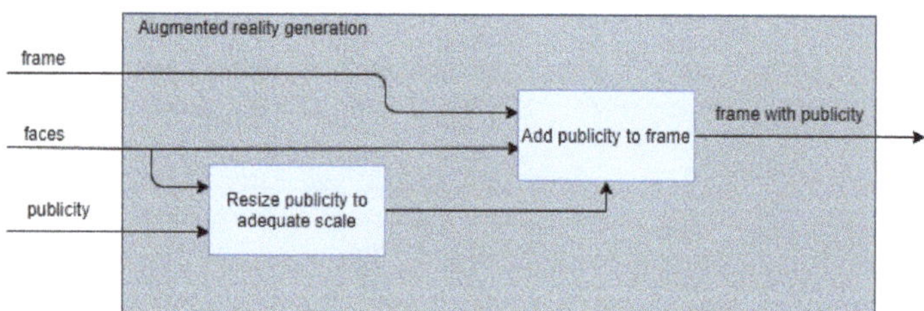

Figure 11. Augmented reality design.

Figure 12. Example of the augmented reality to be carried out.

3.14. Display of the Result with Augmented Reality

The image with the recommendation of the drink added in the form of augmented reality is displayed on the LED screen. It is important to mention that periodically, the output on the screen will be refreshed with a new frame (obtained from the camera video).

At this point, each frame is completely ready to be shown on the screen. The communication flow of the system, as well as the type of information that travels between them, the protocol of communication they use and the way to connect them are shown in Figure 13.

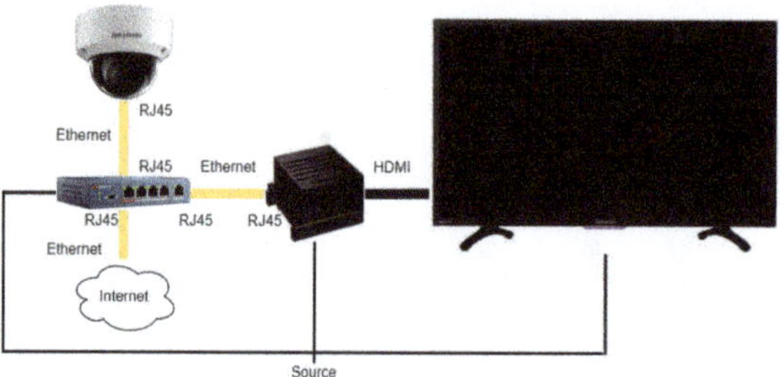

Figure 13. System hardware connection diagram.

All the functionality will be carried out within the NVIDIA Jetson Xavier, a small but powerful computer for artificial intelligence tasks [59,60] that has an ARM64 architecture and Linux operating system called Jetpack [61].

4. Validation

4.1. Face Detection

The training of the neural network Single Shot Detection was started using only the dataset Face Detection in Images [54] with 100,000 epochs, later, with the aim of reducing the error substantially, 30,000 epochs were trained with both datasets. Figure 14 represents the training of the neural network from epoch 1 to 130,000.

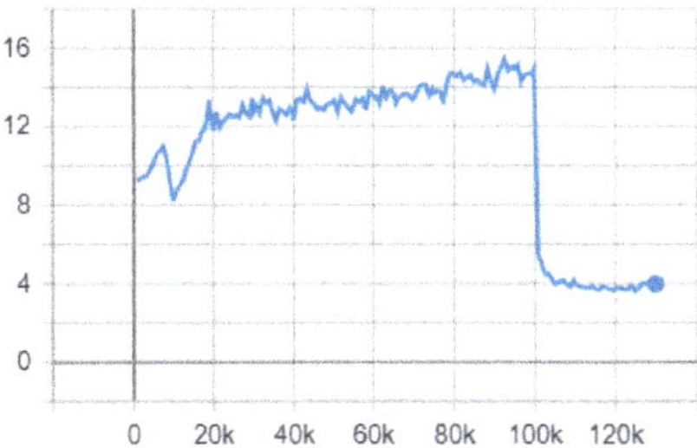

Figure 14. Neural network error Single Shot Detection, note that from epoch 100,000 the error decreases significantly.

While the error metric is a valid indicative. A confusion matrix allows to know more clearly how the network is performing, since with it, the following scenarios can be considered:

1. True positive: Is the case where the network detects a face and it is truly a face.
2. False negative: Occurs when the network detects a face but it is not actually a face.
3. False positive: Is the opposite case opposite to the false negative, it occurs when the network does not detect a face that should.

To evaluate the three previous cases, the *Intersection over the Union* (IoU) [62] is used as a metric. Table 9 shows the results of the aforementioned confusion matrix.

Table 9. Confusion matrix for face detection.

	Positive	**Negative**
True	6818	-
False	830	2617

4.2. Obtaining Age and Gender

The part of the age estimate is measured with the error obtained in the training and testing phases, since it is configured as a regressor. In the training, an average error of 0.11 years is achieved at the end of the 2600 epochs, while in the test set, the average error is 10.44 years. Figures 15 and 16 show the evolutions of the errors during the 2600 training and test periods, respectively.

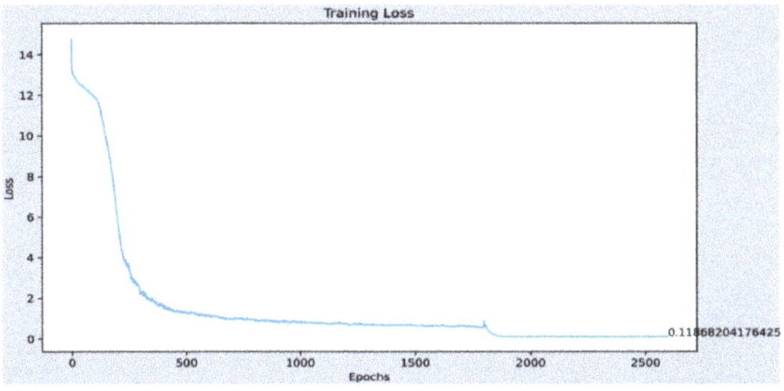

Figure 15. Evolution of the error in the training phase of the age estimation.

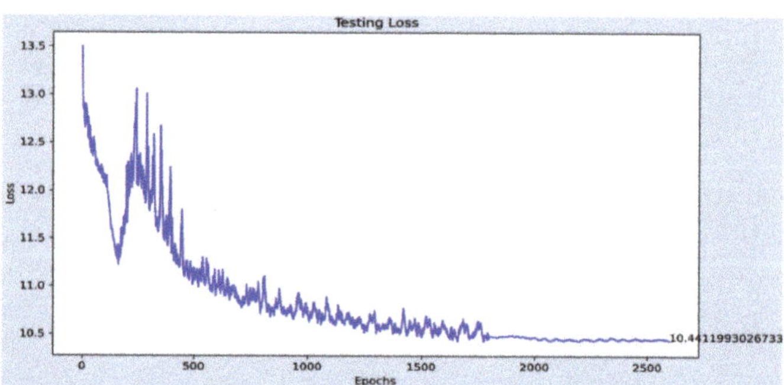

Figure 16. Evolution of the error in the test phase of the age estimate.

The classification part of the gender is measured by its accuracy, giving a result of 83.73% in the training phase and 80.09% in the testing phase. The evolution of these are shown in Figures 17 and 18, respectively.

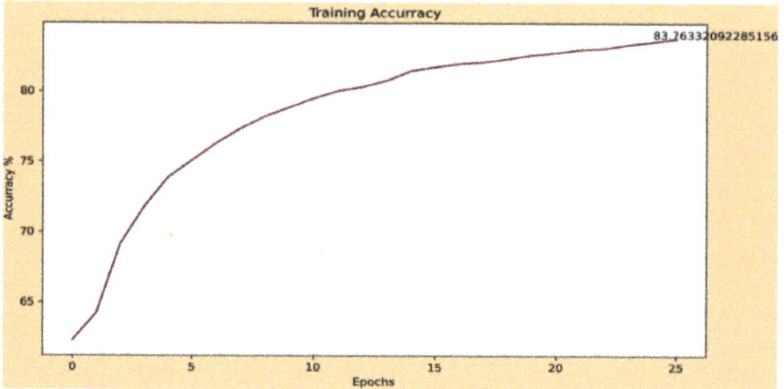

Figure 17. Evolution of precision in the training phase of gender classification.

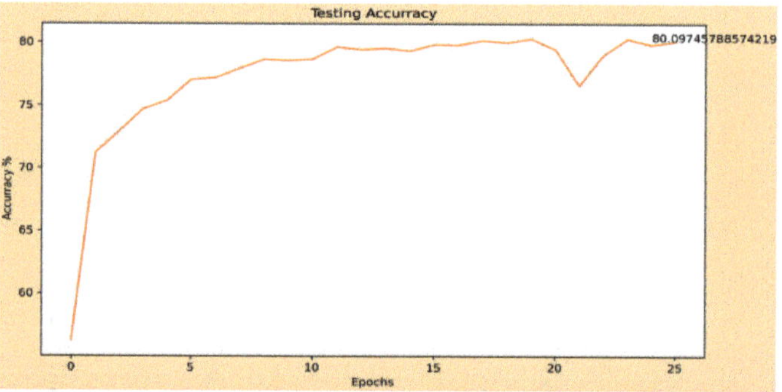

Figure 18. Evolution of precision in the test phase of gender classification.

Finally, the confusion matrix of the gender classification is presented in Table 10.

Table 10. Gender classification confusion matrix.

	Female (Target)	Male (Target)
Classified as Female	5094	1455
Classified as Male	1088	5295

4.3. Augmented Reality

In Figure 19, the image shown on the screen in a real case is shown. The company logo (Bubble Town® [63]) is displayed at the upper left part of the face, there is the image of the drink and name of the drink and modality are shown in the lower central part. However, even though the NVIDIA Jetson Xavier is a powerful device, it was observed that it is not enough for the project to flow properly, since there is a delay between the video captured by the camera and image displayed on the screen (with augmented reality).

Another important aspect to highlight in the work is the capacity of the system to be able to generate a recommendation to more than one person at the same time, see Figure 20, given that if there are several faces in the camera image, the system will recognize them and generate recommendations independently. So that each user knows the recommended drink, the thought balloons for each one, and the banner will be changing from time to time; to identify who the banner belongs to, it will have the same color as the thought balloon. The advertising totem was in operation for several days in the Laboratory of Computational Cognitive Sciences of the Center for Computing Research [64], showing that it properly worked at all times. To replicate these results, the reader can visit https://github.com/vicleo14/PublicidadBT (accessed on 20 December 2021). A short video demo can be found at https://tinyurl.com/2p8bf68s (accessed on 20 December 2021). Regarding working time, once the system detected a face, it started generating the augmented reality scene in 2.03 s on average; lastly, on a small poll with 30 users, 86.66% of them liked our beverage recommendation; see all data in Table A1 and logfile in https://tinyurl.com/59ev279p (accessed on 18 December 2021).

Figure 19. Final result of Augmented Reality.

Figure 20. Final result of augmented reality with two people using our advertising totem.

5. Conclusions

Generating targeted, different, and personalized advertising to recommend a product to the customer in an unconventional way is not a trivial task. The main reason that customer tastes can become so complex is that they become unique and unrepeatable preferences; for this reason, achieving generalization of the entire process within the system is highly complex in terms of development and effectiveness.

The raw material of this work are the images in which the clients appear. The preprocessing of the images that will be entered into the deep learning models is one of the main aspects that will influence obtaining good results, since having an adequate processing in the data will allow the model to obtain a result with high precision using fewer resources.

The personalized retraining of the Single Shot Detection with the help of transfer of learning (in order to achieve good results without the need for extensive training) and its results, as well as the execution tests of the convolutional neural network to estimate age and gender and the neural network for Big Five confirm something that is very clear in the world of artificial intelligence: No model will be completely accurate in scenarios of the real world. There is no model that is perfect, that is why the field of deep learning has had periods of oblivion throughout history, and although recently, thanks to more powerful computers, as well as large data banks, it has been a relevant of study, since precisely obtaining an ideal model is one of the objectives that these fields seek to achieve.

With all this, face detection has an acceptable performance for the purposes of the system, although it could be improved, since it must be observed that the almost null existence of data sets available for commercial use or with free licenses is the main cause of not being able to refine or perfect the Single Shot Detection to detect faces in very difficult conditions, such as, people with glasses (whether dark or transparent), with hats, scarves, with tied hair in the case of women, and recently with a mouth mask. The recommendation generated by the system, in the end, is a suggestion based on certain parameters identified in a person, and clearly the ultimate decision to accept or reject it will be with the clients. The importance of this works lies in the "aggressiveness" in which it is suggested, and since it is simply a graphic that does not compromise the decision or intentions of the buyer, in addition to how attractive augmented reality can be for a public unfamiliar with this technology, it is considered to be more likely to arouse interest in Bubble Town®beverages rather than having some rejection or negative impact due to a breach of their personal data.

A complementary part of this system is to take into account the need to safeguard people's sensitive data (faces), stored in its database, to comply with business rules, and not incur any violations to Federal Law on the Protection of Personal Data. For this reason, a privacy notice is provided along with the system, and information that could be considered sensitive is encrypted to prevent its misuse.

Author Contributions: Conceptualization, methodology, M.A.M.-A. and C.A.D.; investigation and resources, M.A.M.-A., H.C. and C.A.D.; software, visualization and data curation, A.L.-C., E.R.-D. and V.L.M.-F.; validation H.C.; formal analysis, M.A.M.-A., C.A.D. and H.C.; writing–original draft preparation, M.A.M.-A. and A.L.-C.; writing–review and editing, H.C., E.R.-D. and V.L.M.-F.; supervision, M.A.M.-A., C.A.D. and H.C.; project administration and funding acquisition, M.A.M.A. All authors have read and agreed to the published version of the manuscript.

Funding: This work has been possible thanks to the support of the Mexican government through the FORDECYT-PRONACES program Consejo Nacional de Ciencia y Tecnología (CONACYT) under grant APN2017–5241; the SIP-IPN research grants, SIP 2083, SIP 20210169, and SIP 20210189; IPN-COFAA and IPN-EDI.

Informed Consent Statement: Informed consent was obtained from all subjects involved in the study.

Data Availability Statement: The authors are committed to providing access to all the necessary information so that readers can fully reproduce the results presented in this work. For this, all the necessary information is available in the following repository at https://github.com/vicleo14/PublicidadBT (accessed on 20 December 2021). A demo is also available at https://tinyurl.com/2p8 bf68s (accessed on 20 December 2021).

Conflicts of Interest: The authors declare no conflict of interest.

Appendix A

Table A1. Results of the elapsed time to generate the AR and the satisfaction survey.

Poll Number	Elapsed Time (s)	Did You Like the Recommendation?
1	1.479222298	like
2	1.285918808	dislike
3	1.27488842	like
4	2.368936348	like
5	2.293576145	like
6	2.358893967	like
7	2.369532681	like
8	2.393731785	like
9	2.377891922	like
10	2.281593037	dislike
11	2.267432785	dislike
12	2.641290951	like
13	2.449003696	like
14	2.427659225	like
15	2.363572884	like
16	2.310142899	like
17	2.294511223	like
18	1.219994068	like
19	2.308717346	like
20	2.259387589	like
21	2.253286648	like
22	2.304634762	like
23	1.212442875	like
24	1.298417091	like
25	2.355474758	like
26	2.696813393	like
27	1.217700481	like
28	1.231303215	dislike
29	1.212442875	like
30	2.29670248	like
Average:	**2.036837222**	**86.66%**

References

1. Kwon, K.; Kim, C. How to Design Personalization in a Context of Customer Retention: Who Personalizes What and to What Extent? *Electron. Commer. Rec. Appl.* **2012**, *11*, 101–116. [CrossRef]
2. Bleier, A.; De Keyser, A.; Verleye, K. *Customer Engagement through Personalization and Customization*; Palgrave Macmillan: Cham, Switzerland, 2018; pp. 75–94.
3. Das, D.; Sahoo, L.; Datta, S. A Survey on Recommendation System. *Int. J. Comput. Appl.* **2017**, *160*, 7. [CrossRef]
4. Brusilovski, P.; Kobsa, A.; Nejdl, W. *The Adaptive Web: Methods and Strategies of Web Personalization*; Springer Science & Business Media: Berlin/Heidelberg, Germany, 2007; Volume 4321.
5. AINIA. Más del 55 por Ciento de los Millennials Prefiere el Sabor Dulce. Available online: https://www.ainia.es/noticias/prensa/mas-del-55-por-ciento-de-los-millennials-prefiere-el-sabor-dulce/?fbclid=IwAR3Wm2DDvxBTjzFg2P9FBGush3EcyNdMfkMq3JwypAyLQSyezrzqXSZV9cg (accessed on 19 December 2020).
6. Barragán, R.; Coltell, O.; Portolés, O.; Asensio, E.M.; Sorlí, J.V.; Ortega-Azorín, C.; González, J.I.; Sáiz, C.; Fernández-Carrión, R.; Ordovas, J.M.; et al. Bitter, Sweet, Salty, Sour and Umami Taste Perception Decreases with Age: Sex-specific Analysis, Modulation by Genetic Variants and Taste-preference Associations in 18 to 80 year-old subjects. *Nutrients* **2018**, *10*, 1539. [CrossRef] [PubMed]
7. Claudia, A.; Anna, S.; Raffaella, C.; Fabio, V.; Aida, T. Gender Differences In Food Choice And Dietary Intake In Modern Western Societies. *Public Health-Soc. Behav. Health* **2012**, *4*, 83–102.
8. Day, C.J. An Exploration of the Relationships Between Personality, Eating Behaviour and Taste Preference. Ph.D. Thesis, Sheffield Hallam University, Sheffield, UK, 2009.

9. Khan, S.; Rahmani, H.; Shah, S.A.A.; Bennamoun, M. A Guide To Convolutional Neural Networks For Computer Vision. *Synth. Lect. Comput. Vis.* **2018**, *8*, 1–207. [CrossRef]

10. Intel, C. Digital Signage Solutions Including Intel AIM Suite. 2011. Available online: https://aimsuite.intel.com/sites/default/files/resources/Presentation%20-%20Intel%20OSC%20Digital%20Signage%20Webinar%20-%20Oct%2013%2C%202011.pdf (accessed on 1 September 2019).

11. Intel, C. Inside AIM Suite: Pricing/Licensing. Available online: https://aimsuite.intel.com/inside-aim-suite/pricinglicensing (accessed on 1 September 2019).

12. Wang, J.; Xie, H.; Au, O.T.S.; Zou, D.; Wang, F.L. Attention-based CNN for personalized course recommendations for MOOC learners. In Proceedings of the 2020 International Symposium on Educational Technology (ISET), Bangkok, Thailand, 24–27 August 2020; pp. 180–184.

13. Liu, D.; Li, J.; Du, B.; Chang, J.; Gao, R. DAML: Dual Attention Mutual Learning between Ratings and Reviews for Item Recommendation. In *KDD '19, Proceedings of the 25th ACM SIGKDD International Conference on Knowledge Discovery & Data Mining*; Association for Computing Machinery: New York, NY, USA, 2019; pp. 344–352. [CrossRef]

14. Yang, M.; Zhu, S.; Lv, F.; Yu, K. Correspondence Driven Adaptation for Human Profile Recognitional. In Proceedings of the CVPR 2011, Colorado Springs, CO, USA, 20–25 June 2011; pp. 505–512.

15. Orozco, C.I.; Iglesias, F.; Buemi, M.E.; Berlles, J.J. Real-time Gender Recognition From Face Images Using Deep Convolutional Neural Network. In Proceedings of the 7th Latin American Conference on Networked and Electronic Media (LACNEM 2017), Valparaiso, Chile, 6–7 November 2017.

16. Ranjan, R.; Sankaranarayanan, S.; Castillo, C.D.; Chellappa, R. An All-in-one Convolutional Neural Network For Face Analysis. In Proceedings of the 2017 12th IEEE International Conference on Automatic Face & Gesture Recognition (FG 2017), Washington, DC, USA, 30 May–3 June 2017; pp. 17–24.

17. Xing, J.; Li, K.; Hu, W.; Yuan, C.; Ling, H. Diagnosing Deep Learning Models for High Accuracy Age Estimation from a Single Image. *Pattern Recognit.* **2017**, *66*, 106–116. [CrossRef]

18. Vasileiadis, M.; Stavropoulos, G.; Tzovaras, D. Facial soft biometrics detection on low power devices. In Proceedings of the IEEE/CVF Conference on Computer Vision and Pattern Recognition Workshops, Long Beach, CA, USA, 16–17 June 2019.

19. Zhang, X.; Liu, C.; Su, Z. Face detection system based on video stream. In Proceedings of the 2017 International Conference on Computer Systems, Electronics and Control (ICCSEC), Dalian, China, 25–27 December 2017; pp. 1130–1133.

20. Liu, W. Video face detection based on deep learning. *Wirel. Pers. Commun.* **2018**, *102*, 2853–2868. [CrossRef]

21. Ye, X.; Ji, B.; Chen, X.; Qian, D.; Zhao, Z. Probability Boltzmann Machine Network for Face Detection on Video. In Proceedings of the 2020 13th International Congress on Image and Signal Processing, BioMedical Engineering and Informatics (CISP-BMEI), Chengdu, China, 17–19 October 2020; pp. 138–147.

22. Zhou, Y.; Ni, H.; Ren, F.; Kang, X. Face and gender recognition system based on convolutional neural networks. In Proceedings of the 2019 IEEE International Conference on Mechatronics and Automation (ICMA), Tianjin, China, 4–7 August 2019; pp. 1091–1095.

23. Greco, A.; Saggese, A.; Vento, M. Digital signage by real-time gender recognition from face images. In Proceedings of the 2020 IEEE International Workshop on Metrology for Industry 4.0 & IoT, Roma, Italy, 3–5 June 2020; pp. 309–313.

24. Moreno Sotelo, M. Smart Prediction of Apparent Personality Traits through Big Five from Selfies. Master's Thesis, CIC-IPN, Mexico City, Mexico, 2019.

25. Ponce-López, V.; Chen, B.; Oliu, M.; Corneanu, C.; Clapés, A.; Guyon, I.; Baró, X.; Escalante, H.J.; Escalera, S. ChaLearn LAP 2016: First Round Challenge on First Impressions-Dataset and Results. In Proceedings of the European Conference on Computer Vision, Amsterdam, The Netherlands, 8–16 October 2016.

26. Gorbova, J.; Avots, E.; Lüsi, I.; Fishel, M.; Escalera, S.; Anbarjafari, G. Integrating Vision and Language for First-Impression Personality Analysis. *IEEE Multimed.* **2018**, *25*, 24–33. [CrossRef]

27. Kampman, O.; J. Barezi, E.; Bertero, D.; Fung, P. Investigating Audio, Video, and Text Fusion Methods for End-to-End Automatic Personality Prediction. In *Proceedings of the 56th Annual Meeting of the Association for Computational Linguistics (Volume 2: Short Papers)*; Association for Computational Linguistics: Melbourne, Australia, 2018; pp. 606–611.

28. Jacques Júnior, J.C.S.; Güçlütürk, Y.; Pérez, M.; Güçlü, U.; Andújar, C.; Baró, X.; Escalante, H.J.; Guyon, I.; van Gerven, M.A.J.; van Lier, R.; et al. First Impressions: A Survey on Computer Vision-Based Apparent Personality Trait Analysis. *arXiv* **2018**, arXiv:1804.08046.

29. Moreno Sotelo, M. Personality. Available online: https://github.com/miguelmore/personality (accessed on 22 December 2020).

30. Carmigniani, J.; Furht, B.; Anisetti, M.; Ceravolo, P.; Damiani, E.; Ivkovic, M. Augmented Reality Technologies, Systems and Applications. *Multimed. Tools Appl.* **2011**, *51*, 341–377. [CrossRef]

31. Mirror, M. Magic Mirror Technologies and Latest Development. Available online: https://www.magicmirror.me/Products/Photobooth-Technologies (accessed on 1 September 2019).

32. Mirrors, T.W. Smart Mirror Product Information (For 2019). Available online: https://www.twowaymirrors.com/smart-mirror/ (accessed on 1 September 2019).

33. Arcangel, A. Snapchat Filters: How Do They Work? Available online: https://dev.to/aubreyarcangel/snapchat-filters-how-do-theywork-4c83 (accessed on 1 September 2019).

34. Kohl's. Kohl's—Virtual Closet with Snapchat. Available online: https://wwd.com/fashion-news/fashion-features/snapchat-kohls-virtual-closet-coronavirus-1203629594/ (accessed on 7 December 2021).

35. Berman, B.; Pollack, D. Strategies for the Successful Implementation of Augmented Reality. *Bus. Horizons* **2021**, *64*, 621–630. [CrossRef]

36. Plotkina, D.; Dinsmore, J.; Racat, M. Improving service brand personality with augmented reality marketing. *J. Serv. Mark.* **2021**. [CrossRef]

37. Kim, M.; Choi, S.H.; Park, K.B.; Lee, J.Y. A Hybrid Approach to Industrial Augmented Reality Using Deep Learning-Based Facility Segmentation and Depth Prediction. *Sensors* **2021**, *21*, 307. [CrossRef] [PubMed]

38. Wayfair. Wayfair—View in Room. Available online: https://www.aboutwayfair.com/2020/09/augmented-reality-with-a-purpose/ (accessed on 7 December 2021).

39. Ikea. Ikea Augmented Reality App. Available online: https://www.ikea.com/au/en/customer-service/mobile-apps/say-hej-to-ikea-place-pub1f8af050 (accessed on 7 December 2021).

40. Adidas. Adidas AR Sneakers Try on App. Available online: https://www.virtualrealitymarketing.com/case-studies/adidas-ar-sneakers-try-on-app-2/ (accessed on 7 December 2021).

41. Gilliland, N. 14 Examples of Augmented Reality Brand Experiences. Available online: https://econsultancy.com/14-examples-augmentedreality-brand-marketing-experiences/ (accessed on 7 December 2021).

42. Asos. Asos Virtual Catwalk. Available online: https://www.youtube.com/watch?v=Nr6OEU-9_Vs (accessed on 7 December 2021).

43. Hamid, N.F.I.A.; Din, F.A.M.; Izham, S.; Isa, S.N.M. An Interactive Mobile Augmented Reality for Advertising Industry. In *Proceedings of the SAI Intelligent Systems Conference*; Springer: Berlin, Germany, 2016; pp. 763–770.

44. Zara. Zara Virtual Models. Available online: https://www.youtube.com/watch?v=PTiT-Y4y7AI (accessed on 7 December 2021).

45. Sephora. Sephora—Virtual Artist. Available online: https://www.sephora.sg/pages/virtual-artist (accessed on 7 December 2021).

46. Lacoste. Lacoste LCST App. https://vimeo.com/89596935 (accessed on 7 December 2021).

47. Parfumerie, H. Augmented Reality App. Available online: https://vimeo.com/160245751 (accessed on 7 December 2021).

48. Converse, C. Converse Sampler App. https://www.youtube.com/watch?v=4NzB5Cb6HNk (accessed on 7 December 2021).

49. Buttazzo, G.; Lipari, G.; Abeni, L.; Caccamo, M. *Soft Real-Time Systems*; Springer: Berlin, Germany, 2005.

50. Hikvision. User Manual UD14456B Network Camera. Available online: https://www.hikvision.com/mtsc/uploads/product/accessory/UD14456B_Baseline_User_Manual_of_Network_Camera_V5.6.0_20190430.pdf (accessed on 12 April 2020).

51. Liu, W.; Anguelov, D.; Erhan, D.; Szegedy, C.; Reed, S.; Fu, C.Y.; Berg, A.C. SSD: Single Shot Multibox Detector. In *European Conference On Computer Vision*; Springer: Berlin, Germany, 2016; pp. 21–37.

52. Tsang, S.H. Review: MobileNetV1—Depthwise Separable Convolution (Light Weight Model). 2018. Available online: https://towardsdatascience.com/review-mobilenetv1-depthwise-separable-convolution-light-weight-model-a382df364b69 (accessed on 17 April 2020).

53. Huang, J.; Rathod, V.; Chow, D.; Sun, C.; Zhu, M.; Fathi, A.; Lu, Z. Tensorflow Object Detection Api. Available online: https://github.com/tensorflow/models/tree/master/research/object_detection (accessed on 7 December 2020).

54. Mishra, D.; DataTurks. Face Detection in Images Dataset. 2018. Available online: https://www.kaggle.com/dataturks/face-detection-in-images (accessed on 7 December 2020).

55. Vemulapalli, R.; Agarwala, A. A Compact Embedding for Facial Expression Similarity. In Proceedings of the IEEE Conference on Computer Vision and Pattern Recognition (CVPR), Long Beach, CA, USA, 15–20 June 2019.

56. Moreno-Armendáriz, M.A.; Duchanoy Martínez, C.A.; Calvo, H.; Moreno-Sotelo, M. Estimation of Personality Traits From Portrait Pictures Using the Five-Factor Model. *IEEE Access* **2020**, *8*, 201649–201665. [CrossRef]

57. OpenWeather. OpenWeather Mobile App. Available online: https://openweathermap.org/ (accessed on 10 December 2020).

58. Spencer, M.; Sage, E.; Velez, M.; Guinard, J.X. Using Single Free Sorting and Multivariate Exploratory Methods to Design a New Coffee Taster's Flavor Wheel. *J. Food Sci.* **2016**, *81*, S2997–S3005. [CrossRef] [PubMed]

59. NVIDIA. Jetson AGX Xavier Developer Kit. Available online: https://developer.nvidia.com/embedded/jetson-agx-xavier-developer-kit (accessed on 8 December 2020).

60. NVIDIA. Jetson AGX Xavier and the New Era Of Autonomous Machines. Available online: http://info.nvidia.com/rs/156-OFN-742/images/Jetson_AGX_Xavier_New_Era_Autonomous_Machines.pdf (accessed on 28 March 2020).

61. NVIDIA. Jetpack. Available online: https://developer.nvidia.com/embedded/jetpack (accessed on 29 March 2020).

62. Rezatofighi, H.; Tsoi, N.; Gwak, J.; Sadeghian, A.; Reid, I.; Savarese, S. Generalized intersection over union: A metric and a loss for bounding box regression. In Proceedings of the IEEE/CVF Conference on Computer Vision and Pattern Recognition, Long Beach, CA, USA, 15–20 June 2019; pp. 658–666.

63. Town, B. Bubble Town Website. Available online: https://3498523.wixsite.com/bubbletown (accessed on 8 December 2021).

64. CIC. CIC Website. Available online: https://www.cic.ipn.mx (accessed on 8 December 2021).

MDPI
St. Alban-Anlage 66
4052 Basel
Switzerland
Tel. +41 61 683 77 34
Fax +41 61 302 89 18
www.mdpi.com

Sensors Editorial Office
E-mail: sensors@mdpi.com
www.mdpi.com/journal/sensors